Cost Analysis of ELECTRONIC SYSTEMS

Second Edition

WSPC Series in Advanced Integration and Packaging

Series Editors: Avram Bar-Cohen *(University of Maryland, USA)*
Shi-Wei Ricky Lee *(Hong Kong University of Science and Technology, ROC)*

Published

Vol. 1: Cost Analysis of Electronic Systems
by Peter Sandborn

Vol. 2: Design and Modeling for 3D ICs and Interposers
by Madhavan Swaminathan and Ki Jin Han

Vol. 3: Cooling of Microelectronic and Nanoelectronic Equipment: Advances and Emerging Research
edited by Madhusudan Iyengar, Karl J. L. Geisler and Bahgat Sammakia

Vol. 4: Cost Analysis of Electronic Systems (Second Edition)
by Peter Sandborn

WSPC Series in Advanced Integration and Packaging — Vol. 4

Cost Analysis of Electronic Systems

Second Edition

Peter Sandborn
CALCE - Center for Advanced Life Cycle Engineering
Mechanical Engineering Department
University of Maryland, USA

World Scientific
NEW JERSEY • LONDON • SINGAPORE • BEIJING • SHANGHAI • HONG KONG • TAIPEI • CHENNAI • TOKYO

Published by

World Scientific Publishing Co. Pte. Ltd.
5 Toh Tuck Link, Singapore 596224
USA office: 27 Warren Street, Suite 401-402, Hackensack, NJ 07601
UK office: 57 Shelton Street, Covent Garden, London WC2H 9HE

British Library Cataloguing-in-Publication Data
A catalogue record for this book is available from the British Library.

WSPC Series in Advanced Integration and Packaging — Vol. 4
COST ANALYSIS OF ELECTRONIC SYSTEMS
Second Edition

ISBN 978-981-3148-25-3

Printed in Singapore

Preface to the Second Edition

I received helpful criticism from numerous sources since the first edition of this book was published in 2013. In addition to the first edition's use as a graduate course text, we are now using selected chapters in an undergraduate course on engineering economics and cost modeling. Along with the inputs I have received on how to make the original topics more complete, I have also had numerous requests for new material addressing new areas.

Of course no book like this can ever be truly complete, but attempting to make it so keeps me out of trouble and gives me something to do on the weekends and evenings.

I have added two new chapters and two new appendices to this edition. The new chapter on real option analysis treats modeling of management flexibility and provides a case study on maintenance optimization. A chapter on cost-benefit analysis has also been added. This chapter comes as the direct result of many inquiries about how to model consequences (benefits, risks, etc.) concurrent with costs. The new appendices cover weighted average cost of capital and discrete-event simulation, both of these topics don't warrant a chapter, but nonetheless are useful topics for this type of book.

In addition to the new chapters and appendices, several new sections have been added to the 1st edition chapters and new problems have been added to all the chapters (and a few problems that students convinced me didn't quite make sense have been deleted).

Peter Sandborn
2016

Preface to the First Edition

Twenty years ago many engineers involved in the design of electronic systems took, at most, a secondary interest in the cost effectiveness of their design decisions; they considered that someone else's job or an issue to be addressed after the initial release of the product.[1] Today the world has changed. Every engineer in the design process for an electronic product is also tasked with understanding, or contributing to the understanding of, the economic tradeoffs associated with their decisions. Yet aside from general engineering economics that focuses on capital allocation problems, system designers have virtually no resources and obtain little or no training in cost analysis, let alone analysis that is specific to electronic systems.

Unfortunately, when engineering students were asked what they thought the cost of a product was (and assigned to determine cost estimates of products in an undergraduate capstone design course at the University of Maryland) they all too often added up the costs of procuring the bill of materials and declared that to be the cost of the product. Few students are surprised when shown a breakdown of the life-cycle costs or the cost of ownership of systems, but virtually none, even those who had taken courses in engineering economics, were equipped to competently estimate the manufacturing or life-cycle cost of a real product.

This book is an outgrowth of a course on Electronic Product and System Cost Analysis developed at the University of Maryland. Since 1999, the course has been taught as a one-semester graduate course (populated with a mix of senior-level undergraduates and graduate students) and many times in the form of an industry short course.

---

[1] Many types of electronic systems have been primarily driven by time to market rather than cost; this situation is not necessarily shared by non-electronic systems.

This book is intended to be a resource for electronic system designers who want to be able to assess the economic impact of their design decisions on the manufacturing of a system and its life cycle.

The book is oriented toward those interested in the entire electronic systems hierarchy from the bare die (integrated circuits) through the single chip packages, modules, boards, and enclosures.

This book provides an in-depth understanding of the process of predicting the cost of systems. Elements of traditional engineering economics are melded with manufacturing process modeling and life-cycle cost management concepts to form a practical foundation for predicting the real cost of electronic products.

Various manufacturing cost analysis methods are included in the book: process-flow cost modeling and parametric, cost-of-ownership, and activity-based costing. The effects of learning curves, data uncertainty, test and rework processes, and defects are considered in conjunction with these methodologies. In addition to manufacturing processes, the product life-cycle costs associated with the sustainment of systems are also addressed through a treatment of the cost impacts of reliability (sparing, availability, warranty) and obsolescence. The chapters use real-life scenarios from integrated circuit fabrication, electronic systems assembly, substrate fabrication, and electronic systems testing and support at various levels.

The chapters contain problems of varying levels of difficulty, ranging from alternative numerical values that can be used in the examples included in the chapter text to derivations of relations presented in the text and extensions of the models described. Even for the simple problems, students may have to reproduce (via spreadsheet or other methods) the examples from the text before attempting the problems. The notation (symbols) used in each chapter are summarized in the Appendix. Every attempt has been made to make the notation consistent from chapter to chapter; however, some common symbols have different meanings in different chapters.

The author is grateful to many people who have made this a much better book with their input. First, I want to thank the several hundred students who have taken courses at the University of Maryland and seem to somehow always find new and unique questions to ask every time it is taught. My graduate students, present and past, deserve appreciation for their contributions to many portions of the book. In particular I would like to acknowledge Andre Kleyner (Delphi) and Linda Newnes (University of Bath) for their contributions reading and commenting on several of the chapters. I would also like to thank my numerous colleagues at the University of Maryland and in CALCE, including Michael Pecht and Avi Bar-Cohen for encouraging the writing of this book.

Peter Sandborn
2013

Contents

Chapter 1

Introduction

Why analyze costs? Cost is an integral part of planning and managing systems. Unlike other system properties, such as performance, functionality, size, and environmental footprint, cost is always important, always must be understood, and never becomes dated in the eyes of management. As pressure increases to bring products to market faster and to lower overall costs, the earlier an organization can understand the cost of manufacturing and support, the better. All too often, managers lack critical cost information with which to make informed decisions about whether to proceed with a product, how to support a product, or even how much to charge for a product.

Cost often represents the "golden metric" or benchmark for analyzing and comparing products and systems. Cost, if computed comprehensively enough, can combine multiple manufacturability, quality, availability, and timing attributes together into a single measure that everyone comprehends.

1.1 Cost Modeling

Cost modeling is one of the most common business activities performed in an organization. But what is cost modeling, or maybe more importantly, what isn't it? The goal of cost modeling is to enable the estimation of product or system life-cycle costs. Cost analyses generally take one of two forms:

- *Ex post facto* (after the event) – Cost is often computed after expenditures have been made. Accounting represents the use of cost as an objective measure for recording and assessing the

financial performance of an organization and deals with what either has been done or what is currently being done within an organization, not what will be done in the future. The accountant's cost is a financial snapshot of the organization at one particular moment in time.
- *A priori* (prior to) – These cost estimations are made before manufacturing, operation and support activities take place.

Cost modeling is an *a priori* analysis. It is the imposition of structure, incorporation of knowledge, and inclusion of technology in order to map the description of a product (geometry, materials, design rules, and architecture), conditions for its manufacture (processes, resources, etc.), and conditions for its use (usage environment, lifetime expectation, training and support requirements) into a forecast of the required monetary expenditures. Note, this definition does not specify from whom the monetary resources will be required — that is, they may be required from the manufacturer, the customer, or a combination of both.

Engineering economics treats the analysis of the economic effects of engineering decisions and is often identified with capital allocation problems. Engineering economics provides a rigorous methodology for comparing investment or disinvestment alternatives that include the time value of money, equivalence, present and future value, rate of return, depreciation, break-even analysis, cash flow, inflation, taxes, and so forth. While it would be wrong to say that this book is not an engineering economics book (it is), its focus is on the detailed cost modeling necessary to support engineering economic analyses with the inputs required for making investment decisions. However, while traditional engineering economics is focused on the financial aspects of cost, cost modeling deals with modeling the processes and activities associated with the manufacturing and support of products and systems, i.e., determining the actual costs that engineering economics uses within its cash flow oriented decision making processes.

Unfortunately, it is news to many engineers that the cost of products is not simply the sum of the costs of the bill of materials. An undergraduate mechanical engineering student at the University of Maryland, in his final report from a design class, stated: "The sum total cost to produce each accessory is 0.34+0.29+0.56+0.65+0.10+0.17 = $2.11 [the bill of materials cost]. Since some estimations had to be made, \$2.00 will arbitrarily be added to the cost of [the] product to help cover costs not accounted for. This number is arbitrary only in the sense that it was chosen at random." Unfortunately, analyses like this are only too prevalent in the engineering community and traditional engineering economics texts don't necessarily provide the tools to remedy this problem.

Cost modeling is needed because the decisions made early in the design process for a product or system often effectively commit a significant portion of the future cost of a product. Figure 1.1 shows a representation of the product manufacturing cost commitment associated with various product development processes. Even though it is not represented in Figure 1.1, the majority of the product's life-cycle cost is also committed via decisions made early in the design process.

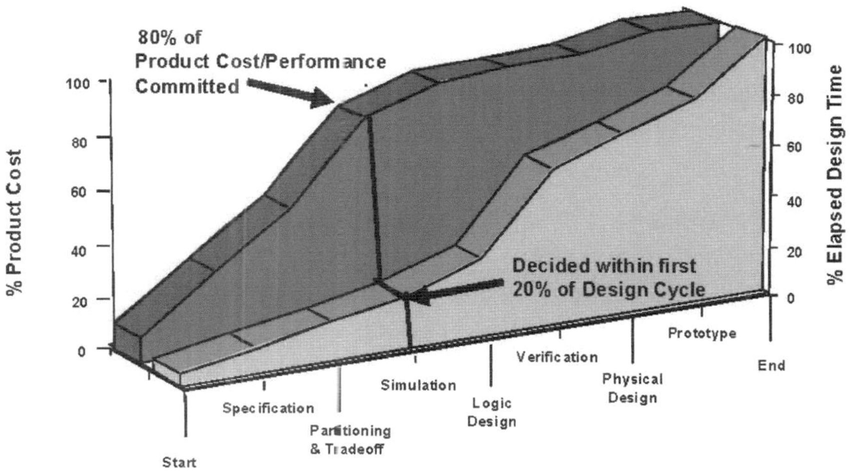

Fig. 1.1. 80% of the manufacturing cost and performance of a product is committed in the first 20% of the design cycle, [Ref. 1.1].

Cost modeling, like any other modeling activity, is fraught with weaknesses. A well-known quote from George Box, "Essentially, all models are wrong, but some are useful," [Ref. 1.2] is appropriate for describing cost modeling. First, cost modeling is a "garbage in, garbage out" activity — if the input data is inaccurate, the values predicted by the model will be inaccurate. That said, cost modeling is generally combined with various uncertainty analysis techniques that allow inputs to be expressed as ranges and distributions rather than point values (see Chapter 9). Obtaining absolute accuracy from cost models depends on having some sort of real-world data to use for calibration. To this end, the essence of cost modeling is summed up by the following observation from Norm Augustine [Ref. 1.3]:

> "Much cost estimation seems to use an approach descended from the technique widely used to weigh hogs in Texas. It is alleged that in this process, after catching the hog and tying it to one end of a teeter-totter arrangement, everyone searches for a stone which, when placed on the other end of the apparatus, exactly balances the weight of the hog. When such a stone is eventually found, everyone gathers around and tries to guess the weight of the stone. Such is the science of cost estimating."

Nonetheless, when absolute accuracy is impossible, relatively accurate costs models can often be very useful.[1]

1.2 The Product Life Cycle

Figure 1.2 provides a high-level summary of a product's life cycle. Note that not all the steps that appear in Figure 1.2 will be relevant for every type of electronic product and that more detail can certainly be added. Product life cycles for electronic systems vary widely and the treatment in this section is intended to be only an example.

---

[1] Relatively accurate cost models produce cost predictions that have limited (or unknown) absolute accuracy, but the differences between model predictions can be extremely accurate if the cost of the effects omitted from the model are a "wash" between the cases considered — that is, when errors are systematic and identical in magnitude between the cases considered. While an absolute prediction of cost is necessary to support the quoting or bidding process, an accurate relative cost can be successfully used to support making a business case for selecting one alternative over another.

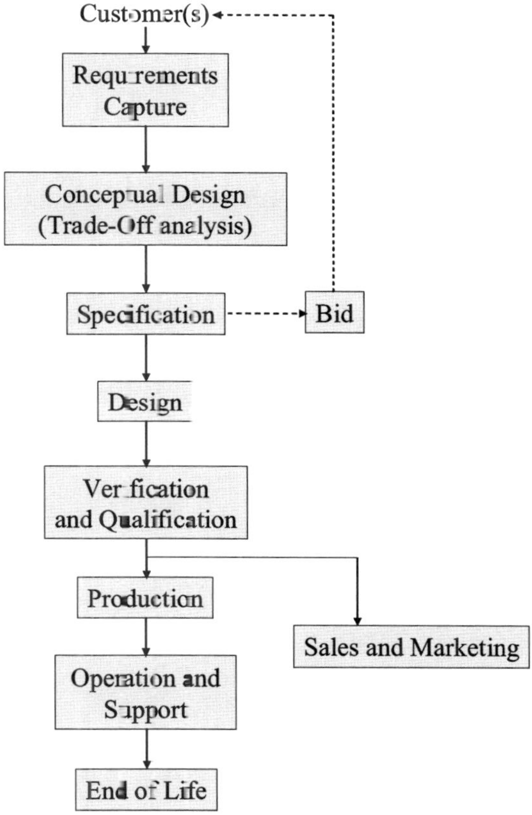

Fig. 1.2. Example product/system life cycle.

In the process shown, a specific customer provides the requirements or a marketing organization determines the requirements through interactions in the marketplace with customers and competitors. Conceptual design encompasses selection of system architecture, possibly technologies, and potentially key parts.

Specifications are engineering's response to requirements and results in a bid that goes to the customer or to the marketing organization. The bid is a cost estimation against the specifications. Design represents all the activities necessary to perform the detailed design and prototyping of the product. Verification and qualification activities determine if the design fulfills the specifications and requirements. Qualification occurs at the functional and environmental (reliability) levels, and may also include certification activities that are necessary to sell or deliver the product to the customer. Production is the manufacturing process and includes sourcing the parts, assembly, and recurring functional testing. Operation and support (O&S) represents the use and sustainment of the product or system. O&S represents recurring use — for example, power, water, or fuel — as well as maintenance, servicing the warranty, training and support for users, and liability. Sales and marketing occur concurrent with production and operation and support. Finally, end of life represents activities needed to terminate the use of the product or system, including possible disassembly and/or disposal.

A common thread through the activities in the life cycle of a product or system is that they all cost money. The product requirements are of particular interest since they ultimately determine the majority of the cost of a product or system and also represent the primary and initial inputs for cost modeling. The requirements will, of course, be refined throughout the design process, but they are the inputs for the initial cost estimation. Figure 1.3 shows the elements that go into the product requirements.

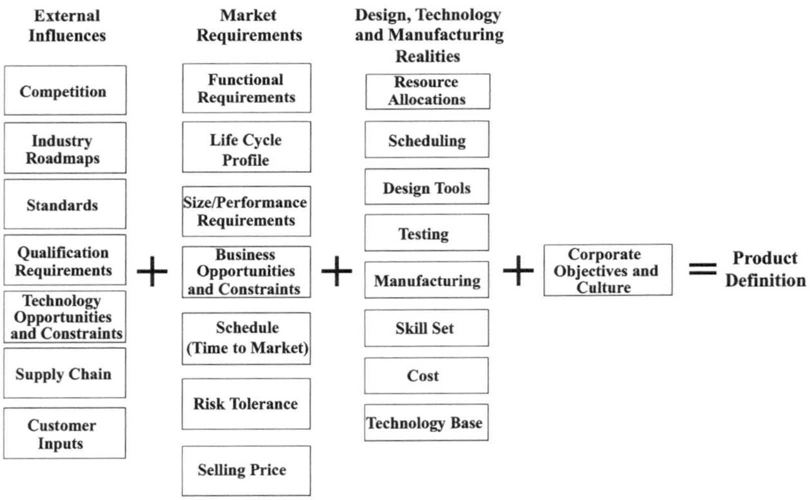

Fig. 1.3. Product/system requirements, [Ref. 1.4].

1.3 Life-Cycle Cost Scope

The factors that influence cost analysis are shown in Figure 1.4. For low-cost, high-volume products, the manufacturer of the product seeks to maximize the profit by minimizing its cost. For a high-volume consumer electronics product (e.g., a cell phone), the cost may be dominated by the bill of materials cost. However, for some products, a more important customer requirement for the product may be minimizing the total cost of ownership of the product. The total cost of ownership includes not only the cost of purchasing the product, but the cost of maintaining and using it, which for some products can be significant. Consider an inkjet printer that sells for as little as $20. A replacement ink cartridge may cost $40 or more. Although the cost of the printer is a factor in deciding what printer to purchase, the cost and number of pages printed by each ink cartridge contributes much more to the total cost of ownership of the printer. For products such as aircraft, the operation and support costs can represent as much as 80% of the total cost of ownership.

Since manufacturing cost and the cost of ownership are both important, Part I of this book focuses on manufacturing cost modeling and Part II expands the treatment to include life-cycle costs and takes a broader view of the cost of ownership.

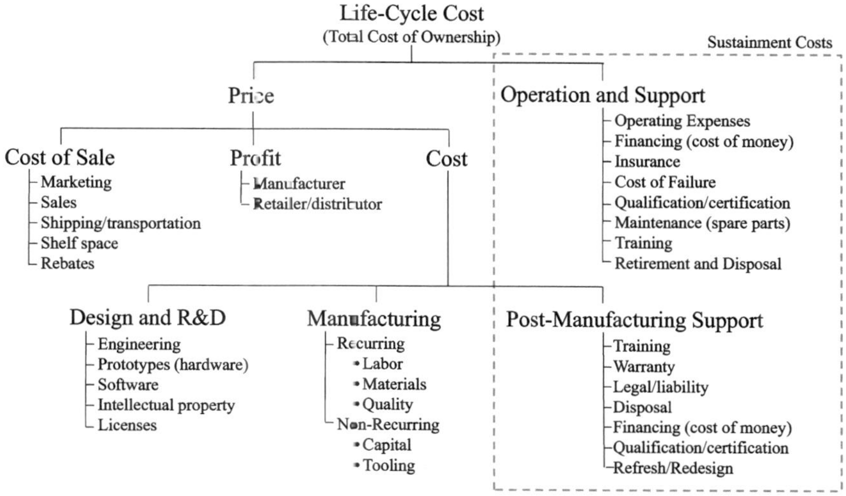

Fig. 1.4. The scope of cost analysis (after [Ref. 1.5]).

1.4 Cost Modeling Definitions

It is important to understand several basic cost modeling concepts in order to follow the technical development in this book. Many of these ideas will be expanded upon in the chapters that follow.

Price is the amount of money that a customer pays to purchase or procure a product or service.

Cost is the amount of money that the manufacturer/supporter of a product or system or the supplier of a service requires to produce and/or provide the product or service. Cost includes money, time and labor.

Profit is the difference between price and cost,

$$\text{Price} = \text{Cost} + \text{Profit} \tag{1.1}$$

Technically, profit is the excess revenue beyond cost. Profit is an accounting approximation of the earnings of a company after taxes, cash, and expenses. Note that profit may be collected by different entities throughout the supply chain of the product or system.

Recurring costs, also referred to as "variable" costs, are costs that are incurred for each unit or instance of the product or system produced. The concept of recurring cost is generally applicable to manufacturing processes. For example, the cost of purchasing a part that is assembled into each individual product is a recurring cost.

Non-recurring costs, also referred to as "fixed" costs, are charged once, independent of the quantity of products manufactured and/or supported. For example, design costs are non-recurring costs.

Labor costs are the costs of employing the people required to perform specific activities.

Tooling cost is a non-recurring cost that is dependent on the quantity of products manufactured and/or supported. Examples of tooling costs are programming and calibration costs for manufacturing equipment, training people, and the purchase or manufacture of product-specific tools, jigs, stencils, fixtures, masks, and so on.

Material costs are the cost of the materials associated with an activity. Material costs may include the purchase of more material than is used in the final product due to the waste generated during the manufacturing process, and it may include the purchase of consumable materials that are completely wasted during manufacturing, such as water.

Capital costs, also called equipment or facilities costs, are the costs of purchasing and maintaining the equipment and facilities necessary to perform manufacturing and/or support of a product or system. In some cases, the capital costs associated with standard activities or processes are incorporated in the overhead rate. Even if the capital costs are included in the overhead, specific capital costs may be included that are associated with buying unique equipment or facilities that must be created or purchased for a specific product.

Depreciation is the decrease in the value of an asset (in the context of this book, the asset is capital equipment or facilities) over time. Depreciation is used to spread the cost of an asset over time.

Direct costs can be traced directly to (or identified with) a specific cost center or object, such as a department, process, or product. Direct costs (such as labor and material) vary with the rate of output but are uniform for each unit item manufactured.

Overhead costs, also called indirect costs, are the portion of the costs that cannot be clearly associated with particular operations, products, or projects and must be prorated among all the product units [Ref. 1.6]. Overhead costs include labor costs for persons who are not directly involved with a specific manufacturing process, such as managers and secretaries; various facilities costs such as utilities and mortgage payments on the buildings; non-cash benefits provided to employees such as health insurance, retirement contributions, and unemployment insurance; and other costs of running the business such as accounting, taxes, furnishings, insurance, sick leave, and paid vacations.

In traditional cost accounting, overhead costs are allocated to a designated base. The base is often determined by direct labor hours or the sum of all the direct costs, but it can also be determined by machine time, floor space, employee count, material consumption, or some combination of these. When overhead is allocated based on direct labor hours, it is often called a **burden rate** and is used to determine either the overhead cost, C_{OH}, or a burdened labor rate, L_{RB}, as follows:

$$C_{OH} = N_{pm} b C_L \tag{1.2}$$

or

$$L_{RB} = L_R(1+b) \tag{1.3}$$

where

N_{pm} = the total number of units produced during the lifetime of the product

b = the labor burden rate (typical range: $0.3 \leq b \leq 2$)

C_L = the labor cost of manufacturing or assembly (per unit)

L_R = the labor rate (often expressed in dollars per hour), which, when converted to an annual basis, is an employee's gross annual wage.

Hidden costs are those costs that are difficult to quantify and may even be impossible to connect with any particular product. Examples of hidden costs include:

- the gain or loss of market share
- the stock price changes of a company
- the company's position in the market for future products
- impacts on competitors and their response
- cost associated with product failure and lawsuits brought against the company
- long-term health, safety, and environmental impacts that may have to be resolved in the future.

The impacts listed above are difficult to quantify in terms of cost because they require a view of the enterprise (i.e., the entire organization or company) that includes more than just one product and an analysis horizon that is longer than the manufacturing and support life of any one product. However, these costs are real and may contribute significantly to product cost.

1.5 Cost Modeling for Electronic Systems

Fundamentally, all of the topics treated in this book are applicable to non-electronic products and systems, however, taken in total, the modeling techniques discussed are those required to assess the manufacturing and life-cycle sustainment of electronic products. The following paragraphs describe attributes of electronic systems that differentiate their costs from non-electronic systems.

For electronics products such as integrated circuits, relatively few organizations have manufacturing capability because of the extreme cost of the required facilities. The cost of recurring functional testing for electronics alone can represent a very large portion of the cost of products (even high-volume products), making the modeling and analysis of recurring functional testing an important contributor to cost modeling (see Chapters 7 and 8).

For all but the highest volume products, manufacturers and supporters of electronic products have virtually no control over the supply chains for their parts. As a result, products that are manufactured and/or supported for longer than a few years experience a high frequency of technology obsolescence, which can be very expensive to resolve (see Chapter 16).

The majority of electronic products are not repaired if they fail during field use; they are thrown away (exceptions are low-volume, long-life, expensive systems). Moreover, most electronic systems are not pro-actively maintained and are traditionally subject to unscheduled ("break-fix") maintenance policies.

1.6 The Organization of this Book

This book is divided into two parts. The first part (Chapters 2-8) focuses on cost modeling for manufacturing electronic systems. Several different approaches are discussed, in addition to manufacturing yield, recurring functional testing (test economics) and rework. Demonstrations of the cost models in the first part of the book focus on the fabrication and assembly of electronic products, ranging from fabricating integrated circuits and printed circuit boards to assembling parts on interconnects. The second part of the book (Chapters 11-19) focuses on life-cycle cost analysis. Life-cycle costing addresses non-manufacturing product and system costs, including maintenance, warranty, reliability, and obsolescence. Chapters 20-22 include the broader topics of total cost of ownership of electronic products, cost-benefit analysis, and real options analysis. Additional chapters (Chapters 9 and 10) address modifications to cost modeling to account for uncertainties and learning curves. These topics are applicable to both manufacturing and life-cycle cost analyses. Appendices that treat discount rate determination and discrete-event simulation are also provided.

A rich set of references (and in some cases bibliographies) have been provided within the chapters to support the methods discussed and to provide sources of information beyond the scope of this book. In addition, problems are provided with the chapters to supplement the examples and demonstrations within the text.

References

1.1 Sandborn, P. A. and Vertal, M. (1998). Packaging tradeoff analysis: Predicting cost and performance during system design, *IEEE Design & Test of Computers*, 15(3), pp. 10-19.

1.2 Box, G. E. P. and Draper, N. R. (1987). *Empirical Model-Building and Response Surfaces* (Wiley, Hoboken, NJ).

1.3 Augustine, N. R. (1997). *Augustine's Laws*, 6th Edition (AIAA, Reston, VA).

1.4 Sandborn, P. and Wilkinson, C. (2004). Chapter 3 - Product requirements, constraints, and specifications, *Parts Selection and Management*, Ed. M. G. Pecht, (John Wiley & Sons, Inc., Hoboken, NJ).

1.5 Magrab, E. B., Gupta, S. K., McCluskey, F. P. and Sandborn, P. A. (2010). *Integrated Product and Process Design and Development - The Product Realization Process*, 2nd Edition (CRC Press, Boca Raton, FL).

1.6 Ostwald, P. F. and McLaren, T. S. (2004). *Cost Analysis and Estimating for Engineering and Management* (Pearson Prentice Hall, Upper Saddle River, NJ).

Part I

Manufacturing Cost Modeling

The objective of manufacturing cost models is to predict the manufacturing cost of a product or system. The manufacturing cost estimate is a function of application-specific details (e.g., the size of the product), technology/material details (e.g., the technology and materials required to meet thermal and electrical performance requirements), processing details (e.g., the manufacturing facilities available to fabricate and assemble the product), and accounting realities (e.g., applicable labor and overhead rates). Manufacturing costs include three basic activities: fabrication and/or assembly, recurring functional test/inspection, and diagnosis and rework (when relevant).

A hierarchy of cost modeling approaches is shown in Figure I.1. The approaches shown are defined as:

Heuristic	simple, commonsense rules or sets of rules derived from experience
Parametric	parametric (e.g., curve) fits to historical data
Analytical	closed-form models derived from first principles
Simulation	non-closed-form models derived from first principles
Emulation	a simulation that mimics the actual manufacturing process
Quote	a statement from the entity that will perform the manufacturing that constitutes a bid for performing the manufacturing and/or assembly work

As the accuracy of the modeling approach increases, so does the amount of data necessary to populate the model. Heuristic and parametric modeling approaches may not have any connection to the "physics" of the manufacturing process; they are purely based on experience and/or historical data. Analytical, simulation and emulation are derived from first principles. The basis for quotes is not necessarily known to the customer, but is also not necessarily relevant.[1]

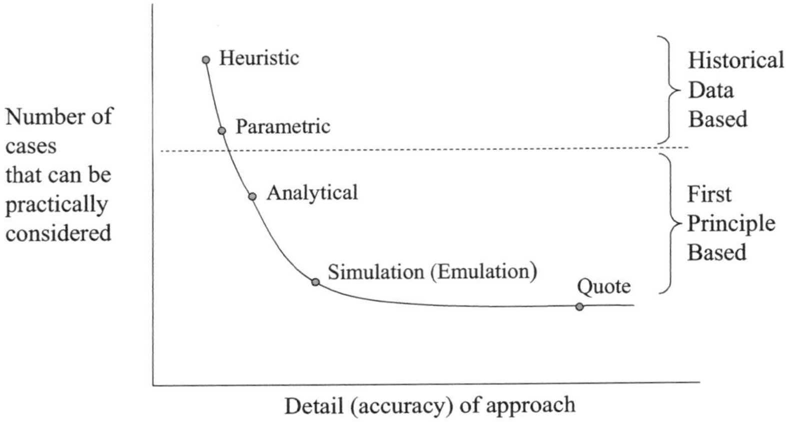

Fig. I.1. Hierarchy of manufacturing cost modeling.

One of the things that Figure I.1 is intended to indicate is that successful manufacturing cost modeling is not necessarily performed using a single methodology or approach, but potentially several approaches. During the conceptual design (tradeoff analysis) (see Figure 1.2 in Chapter 1), heuristic models may be the most practical modeling approach. Heuristics may be the least accurate approach, but it has the advantage of allowing a designer to assess a nearly unlimited number of different cases. As the product development process progresses and more detailed data about the product or system becomes available, more accurate approaches become practical. Finally, the most accurate cost estimation is the actual quote from the organization that will manufacture and/or assemble the item. However, while a quote is absolutely accurate (at least from the customer's perspective, because it represents a contract), the manufacturer is unlikely to be willing to provide quotes for more than a couple of alternatives.

---

[1] An alternative view of the hierarchy shown in Figure I.1 is presented in [Ref. I.1]. The alternative view describes the accuracy of the cost and revenue estimates versus the cost of preparing those estimates.

I.1 Classification of Products Based on Manufacturing Cost

The different types of electronic products, from a manufacturing cost perspective, include the following:

Materials-dominated products: For these, the manufacturing cost of the product is driven by the recurring cost of procuring the materials to fabricate it. For example, printed circuit board manufacturing tends to be dominated by material costs. High-volume consumer electronics usually also fit in this category when the materials are the constituent parts that are purchased from a supply chain. As an example, consider the manufacturing cost breakdown for an Apple 128GB iPhone 6+, shown in Figure I.2. The total cost (parts and assembly) is estimated to be $281, but the actual manufacturing activities (assembly) only cost $9.

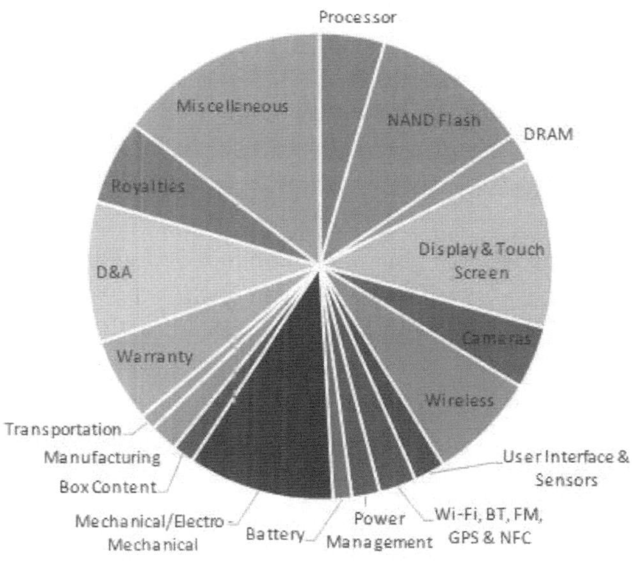

Fig. I.2. Cost breakdown for an Apple 128GB iPhone 6+ [Ref. I.2].

Labor-dominated products: For this type of product, the manufacturing cost is dominated by labor costs. This is rarely the case for the production of electronic products and would only be relevant for custom product manufacturing. However, labor-dominated non-recurring costs are relevant for software development and custom chip development (e.g., ASICs).

Equipment/facilities-dominated products: The manufacturing cost of these products is driven by the purchase and support of expensive equipment and facilities to perform the manufacturing. This is usually the case for integrated circuit manufacturing, where the cost of the chip is dominated by the cost of the fabrication facility.

Test-dominated products: In this case, the manufacturing cost of the product has a very large recurring functional testing or test/diagnosis/rework component. Examples include high density electronic packaging that uses bare die that may have significantly less than 100% yield when they enter the assembly process, and high gate-count complex integrated circuits that have very large functional test overheads.

The chapters in Part I of this book describe various approaches for manufacturing cost modeling. Examples are provided for electronic systems ranging from integrated circuit and printed circuit board fabrication to surface mount assembly. Part I also discusses manufacturing yield and its modeling within various manufacturing processes. A treatment of recurring functional testing (during manufacturing), coupled with possible diagnosis and rework, is included.

Part I closes with a discussion of uncertainty analysis and learning curves, both of which are applicable to manufacturing cost modeling.

References

I.1 Sullivan, P. H., Wicks, E. M. and Luxhoj, J. T. (2006). *Engineering Economy*, 13th Edition (Prentice Hall, Upper Saddle River, NJ).

I.2 Elmer-DeWitt, P. (2014). Morgan Stanley: China's favorite iPhone is Apple's most profitable, Sept. 23. http://fortune.com/2014/09/23/morgan-stanley-chinas-favorite-iphone-is-apples-most-profitable/ Accessed April 28, 2016.

Chapter 2

Process-Flow Analysis

Manufacturing processes can be modeled as a sequence of process steps that are executed in a specific order. The steps and their sequence are referred to as a process flow. Process-flow modeling emulates a real manufacturing process.[1] This means that the process flow attempts to imitate the actual manufacturing process.

Process-flow modeling is generally thought of as a *bottom-up* approach to cost modeling. In a bottom-up model the overall response or characteristic of a product is determined by accumulating the properties (responses and characteristics) of each individual action that takes place in the course of manufacturing the product. The opposite of a bottom-up approach is the *top-down* method, in which high-level attributes are used to determine the responses or characteristics of the object without taking into account its constitute parts or the processes used to create it.

2.1 Process Steps and Process Flows

In process-flow models, an object accrues cost (and other properties) as it moves through the sequence of process steps, as in Figure 2.1.

Each process step starts with the state of the product after the preceding step ("Inputs"). The step then modifies the product and the output is a new state ("Outputs"), which forms the input to the process step that follows, and so on. Usually, process-flow models are constructed so that the form of the process step input matches the form of the output; this allows them to be readily sequenced together. Some types of process steps also provide

[1] Workflow modeling is also sometimes referred to as process-flow modeling. However, workflow modeling is a term usually ascribed to business processes rather than manufacturing processes.

a mechanism by which products can exit the process flow ("Fallout"). Objects that exit the process flow do not continue directly on to the next step in the sequence, although they may reenter the process flow at another point, either before or after the process step that removed them.

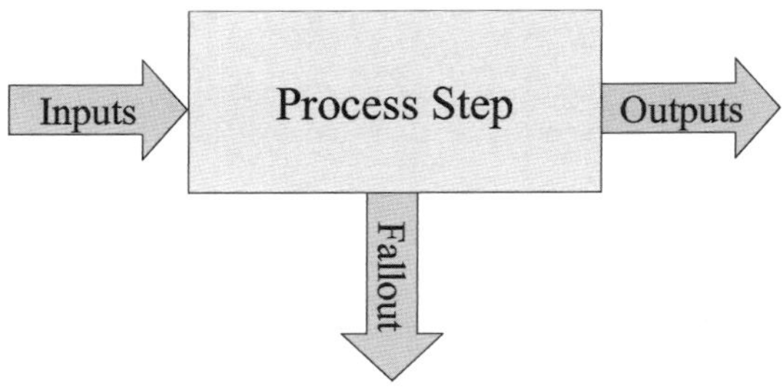

Fig. 2.1. Single process step.

When two or more process steps are sequenced together, a process flow is created. A linear sequence of process steps is called a "branch." The process flow for a complex manufacturing process could consist of one or more branches. Multiple branches imply that independent sub-processes are taking place that eventually merge together to form the complete product. A simple three-branch process flow is shown in Figure 2.2.

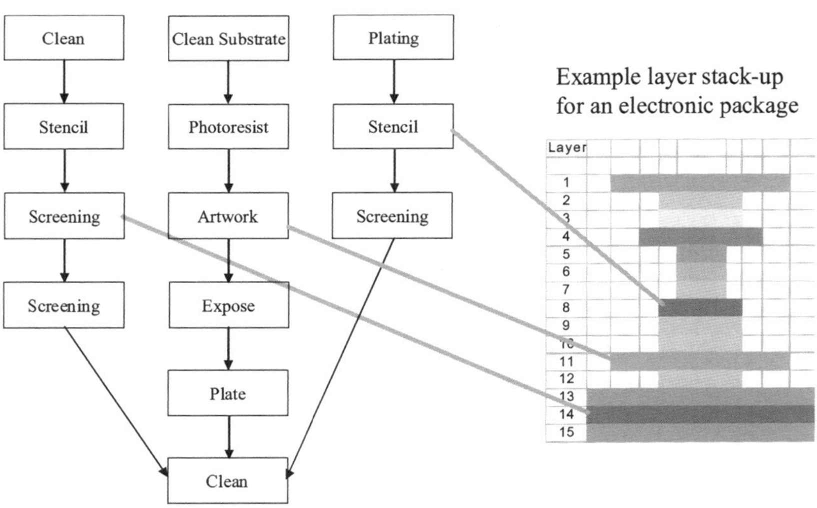

Fig. 2.2. A simple three-branch process flow for fabricating a multilayer electronic package. Each rectangle in the process flow on the left could represent a process step.

2.1.1 *Process-Step Sequence*

As mentioned above, a key attribute that differentiates process-flow modeling from other manufacturing cost analysis approaches is that it captures the order (or sequence) of the manufacturing activities. Sequence matters when product instances (units) can be removed at some intermediate point in a process — for example, by a test step. This is important because when an individual product is removed from the process (scrapped), the amount of money spent up to the point of removal must be known in order to properly allocate the scrapped value back into the product instances that remain in the process. If all the inspection/testing of a product occurred only after the completion of all manufacturing steps, then the sequence of those steps, while important to actually make the product, may not be important for modeling the manufacturing cost. However, if products are inspected and either repaired or scrapped at some interim point in the process, then the sequence is very important. Other methods capture the manufacturing activities, but do not readily capture the order in which the activities take place and are therefore less well suited for manufacturing processes that have significant in-process inspections, testing and rework — for example, electronics assembly processes.

2.1.2 *Process-Step Inputs and Outputs*

Numerous different product properties can be identified, modified and accumulated during the process steps. Obviously, for the purposes of cost modeling, we want to accumulate product cost through process steps; however, there are many other properties that may be useful to identify (and accumulate) and that may be required in order to accurately model the total cost of the product. Properties that may be used include:

- Cost – how much money has been spent (total and specific to particular cost categories – see Section 2.2).
- Time – how long it takes to perform the process step for a product. Actual elapsed time is useful for determining the throughput and

cycle time associated with the process. Touch time is associated with the labor content.

- Defects – the number of defects (total and of specific types) introduced by the process step.
- Mass – how much mass is added or subtracted from the product by the process step.
- Material content – inventory of all materials in the product.
- Material wasted – inventory of all materials in the waste stream for the product.
- Scrap – number of product instances scrapped.
- Energy – inventory of energy used (total and source specific).

These properties do not represent a comprehensive list; other properties may be useful to support other types of models and analyses.

2.2 Process-Step Calculations

Generally process steps can be divided into the following five types:

- Fabrication or assembly steps – These are the most general process steps.
- Test/inspection steps – These are unique because they can remove product instances from the process flow. (See Chapter 7 for a detailed discussion of test/inspection process steps.)
- Rework steps – These operate on product instances that have been removed from the process flow by a test or inspection step and can either permanently remove those units from the process flow (scrap them), or rework them and insert them back into the process flow. (See Chapter 8 for a detailed discussion of rework process steps.)
- Waste disposition steps – These operate on the waste inventoried during a process flow.
- Insertion steps – These allow objects to be inserted into process flows.

The commonality in the step types described above is that they each can contribute labor, materials, tooling, and equipment/capital costs. The following subsections describe the general calculation of these costs.

2.2.1 *Labor Costs*

Labor costs refer to the cost of the people required to perform specific activities. The labor cost of a process step associated with one product instance is determined from

$$C_L = \frac{U_L T L_R}{N_p} \tag{2.1}$$

where

U_L = the number of people associated with the activity (operator utilization); a value < 1 indicates that a person's time is divided between multiple process steps; a value > 1 indicates that more than one person is involved.

T = the length of time taken by the step (calendar time).

N_p = the number of product instances that can be treated simultaneously by the activity (note: this is a capacity, not a rate.)

L_R = the labor rate. If this is a burdened labor rate then the overhead is included in C_L; if it is not a burdened labor rate then overhead must be computed and added to the cost of the product separately.

The product $U_L T$ is sometimes referred to as the *touch time*. For example, if a process step takes five minutes to perform, and one person is sharing his or her time equally between this step and another step that takes five minutes to perform, then $U_L = 0.5$ and $T = 5$ minutes for a touch time of $U_L T = 2.5$ minutes. The throughput of the process step is given by the ratio N_p/T and the cycle time of the process step is the reciprocal of the throughput.

2.2.2 *Materials Costs*

The materials cost of a process step associated with one product instance is given by

$$C_M = U_M C_m \tag{2.2}$$

where

U_M = the quantity of the material consumed by one product instance, as described by its count, volume, area, or length.

C_m = the unit cost of the material per count, volume, area, or length.

Materials costs may include the purchase of more material than is used in the final product due to waste generated during the process, and it may include the purchase of consumable materials that are used and completely wasted during manufacturing, such as water (see [Ref. 2.1]).

2.2.3 *Tooling Costs*

Tooling costs are non-recurring costs associated with activities that occur only once or only a few times:

$$C_T = \frac{C_t N_t}{Q} \tag{2.3}$$

where

C_t = the cost of the tooling object or activity.

N_t = the number of tooling objects or activities necessary to make the total quantity, Q, of products.

Q = the quantity of products that will be made.

Examples of tooling costs are programming and calibration costs for manufacturing equipment, training people, and purchasing or manufacturing product-specific tools, jigs, stencils, fixtures, masks, and so on.

2.2.4 *Equipment/Capital Costs*

Capital costs are the costs of purchasing and maintaining the manufacturing equipment and facilities. In general, capital costs are determined from

$$C_C = \frac{C_e}{D_L}\left[\frac{T}{N_p T_{op}}\right] \tag{2.4}$$

where T and N_p are as defined in Equation (2.1), and

C_e = the purchase price of the capital equipment or facility.

T_{op} = the operational time per year of the equipment or facilities = (equipment operational time as a fraction) (hours/year).

D_L = the depreciation life in years. This equation assumes a "straight line" method is used to model depreciation; that is, depreciation is linearly proportional to the length of time of service.

The term in the brackets in Equation (2.4) is the fraction of the equipment's annual life consumed by producing one unit of the product. In some cases, the capital costs associated with a standard manufacturing process are incorporated into the overhead rate. Even if the capital costs are included in the overhead, Equation (2.4) may still be used to include the cost of unique equipment or facilities that must be created or purchased for a specific product.

2.2.5 *Total Cost*

The total manufacturing cost is the sum of the labor, material, tooling and equipment costs:

$$C_{manuf} = C_L + C_M + C_T + C_C + \left(C_{OH} + C_W\right) \tag{2.5}$$

where

C_{OH} = the overhead (indirect) cost allocated to each product instance (alternatively it may be included in C_L).

C_W = the waste disposition cost per product instance (management of hazardous and non-hazardous waste generated during the manufacturing process). This cost may be included in the process flow and be expressed as labor, material, tooling and capital costs.

Equation (2.5) represents the total manufacturing cost per unit manufactured. Many modifications can be made to Equations (2.1) through (2.5), including learning curves (see Chapter 10), volume-dependent pricing (e.g., for materials), and the inclusion of uncertainties (see Chapter 9).

2.2.6 *Capacity*

The labor and equipment/capital costs in Equations (2.1) and (2.4) depend on the number of product instances that can be concurrently processed by a given process step — that is, the capacity (N_p):

$$N_p = N_e N_u \tag{2.6}$$

where

N_e = the number of wafers or panels concurrently processed by the step.

N_u = the number-up (number of die or boards per wafer or panel).

In electronics, products are fabricated in formats that create many instances of the product concurrently, as shown in Figure 2.3. For integrated circuit manufacturing, individual die are fabricated on wafers of various diameters that may or may not have a flat edge.[2] In the case of printed circuit boards, the boards are fabricated on large (for example, 18 × 24 inch) rectangular panels. Algorithms that predict the number of die per wafer have been developed — for example, in [Ref. 2.2] and [Ref. 2.3]. An equation that gives the approximate number of die on a wafer, assuming that $F = 0$ and that each die is a square with a dimension of S, is given in [Ref. 2.2]:

[2] Generally wafers that are smaller than 200 mm diameter have one or possibly two flat edges. Larger wafers only have a “notch” to indicate orientation, as too much valuable area is taken up by flat edges on large wafers.

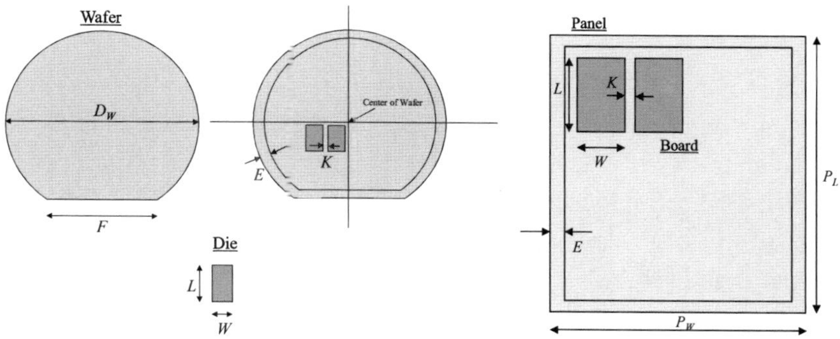

Fig. 2.3. Calculation of the number of die on a wafer or boards on a panel.

$$N_u = \left\lfloor \frac{\pi(0.5D_W - E)^2}{(S+K)^2} e^{-\left[\frac{S+K}{0.5D_W - E}\right]} \right\rfloor \tag{2.7}$$

where

D_W = wafer diameter.
E = the edge scrap (unusable wafer edge).
S = die dimension, $S = \sqrt{LW}$.
K = minimum spacing between die (kerf).
$\lfloor\ \rfloor$ = floor function (round down to the nearest integer).

Equation (2.7) works best when the die are small compared to the wafer.

Similarly, although considerably simpler because the panels are rectangular, the number of boards per panel can be found (see [Ref. 2.4]).

2.3 Process-Flow Examples

This section contains two process-flow analysis examples. The first example is a very simple two-step portion of a larger process. The second models a more extensive process that will be revisited in Chapters 3 and 7.

2.3.1 *Simple Pick & Place and Reflow Process*

Surface mount (SMT) assembly is often performed while the individual boards (or cards) are still on panels — that is, before the boards are singulated from the panel. In the following portion of a process flow (Figure 2.4), electronic parts are being assembled onto PCMCIA cards (52 × 82 mm) while the cards are still in a panel form. In this case there are 56 cards per panel (18 × 24 inch panel) and 42 parts per card with a cost of \$0.90 per part. Assuming 100,000 total cards will be manufactured, a labor rate of \$20/hour, a labor burden of 0.8, and 5-year straight-line depreciation on the equipment, what is the effective cost per card at the conclusion of the reflow process step?

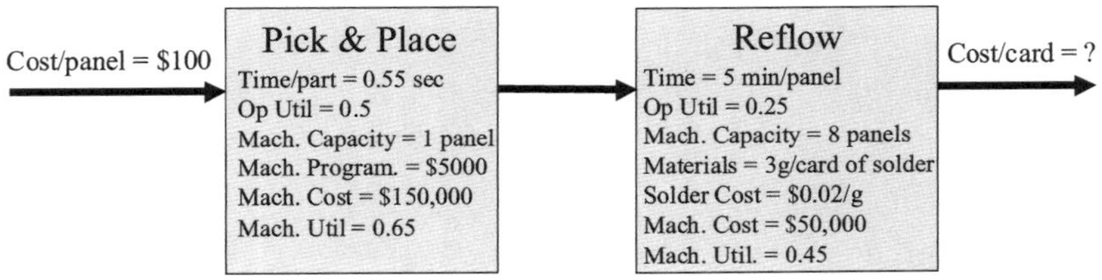

Fig. 2.4. Pick & Place and Reflow portion of a SMT assembly process.

Using the data describing the process steps in Figure 2.4 and noting that the panels have \$100 of accrued cost per panel prior to the portion of the process flow shown in Figure 2.4, the labor, materials, tooling and equipment costs associated with the pick & place step are given by:

$$\begin{aligned}
C_L &= \frac{(0.5)(0.55 \cdot 42 \cdot 56/60/60)(20 \cdot (1+0.8))}{(1)} = \$6.47/\text{panel} \\
C_M &= (42 \cdot 56)(0.90) = \$2116.80/\text{panel} \\
C_T &= \frac{(5000)}{(100{,}000/56)} = \$2.80/\text{panel} \\
C_C &= \frac{(150{,}000)}{(5)}\left[\frac{(0.55 \cdot 42 \cdot 56/60/60)}{(1)(0.65 \cdot 365 \cdot 24)}\right] = \$1.89/\text{panel} \\
C_{manuf} &= 100 + 6.47 + 2116.80 + 2.80 + 1.89 = \$2227.96/\text{panel}
\end{aligned} \tag{2.8}$$

where we have assumed that the \$5000 machine programming cost is a one-time cost. Note the cost of the parts is included as a material cost. The \$2227.96/panel becomes the input for the reflow process step. Using the data describing the process steps in Figure 2.4, the labor, materials, tooling and equipment costs associated with the reflow step are given by:

$$
\begin{aligned}
C_L &= \frac{(0.25)(5/60)(20 \cdot (1+0.8))}{(8)} = \$0.09/\text{panel} \\
C_M &= (3 \cdot 56)(0.02) = \$3.36/\text{panel} \\
C_T &= \$0.00/\text{panel} \\
C_C &= \frac{(50{,}000)}{(5)}\left[\frac{(5/60)}{(8)(0.45 \cdot 365 \cdot 24)}\right] = \$0.03/\text{panel} \\
C_{manuf} &= 2227.96 + 0.09 + 3.36 + 0.00 + 0.03 = \$2231.44/\text{panel}
\end{aligned} \tag{2.9}
$$

The effective cost per card after the reflow step is then \$2231.44/56 = \$39.85.

We have ignored a host of effects in this simple analysis. For one thing, we have not accounted for possible defects that could be introduced by either of these process steps (or that may be resident in the panels or the parts prior to these steps). This affects yield, which will be treated in Chapter 3; the processes associated with testing, diagnosing and potentially reworking the defective items will be addressed in Chapters 7 and 8. We have also assumed that the operators (labor) are fully utilized somewhere, even if they are not utilized on these process steps or for this product — that is, we are assuming that no idle time is unaccounted for. We have also assumed that the equipment will be used through its entire depreciation life, even if that life extends beyond the completion of the 100,000 cards fabricated in this example — that is, we are assuming that other products will use the equipment and that those products will pay for their use of the equipment.

2.3.2 *Multi-Step Process-Flow Example*

You are assigned to model a process that fabricates wafers containing integrated circuits (die). The process has the thirteen process steps performed in the order shown in Table 2.1.

All of the process steps apply to the whole wafer (<u>not</u> individual die). In addition, the parameters shown in Table 2.2 apply. What is the cost per die at the end of the thirteen-step process? The number of die per wafer in this case is exactly 528.

Table 2.1. Thirteen-Step Wafer Fabrication Process.

Step	Time (sec/wafer)	Op Util	Capacity (wafers)	Material Cost (per unit of material)	Units of Material (per wafer)	Tooling Cost	Tooling Life (number of wafers)	Equip Cost	Equip Operational Time (fraction)
A	10	1	1	0	0	0	100000	$150,000	0.6
B	60	2	1	3.2	1	0	100000	$20,000	0.6
C	30	0.5	12	0.1	4	1000	20000	$1,000,000	0.6
D	110	0.25	1	0	0	0	100000	$75,000	0.6
E	100	1	1	0	0	0	100000	$25,000	0.6
F	45	0.5	10	2	1	10000	100000	$10,000	0.6
G	14	1	2	0	0	5000	100000	$15,000	0.6
H	60	1	2	1	3	500	50000	$5,000	0.6
I	25	1.5	5	0.5	4	0	100000	$200,000	0.6
J	120	1	1	0.2	2	0	100000	$0	0.6
K	90	1	1	0.1	2	0	100000	$10,000	0.6
L	26	0.5	30	50	0.1	0	100000	$5,000	0.9
M	200	2	1	0	0	10000	1000	$5,000,000	0.5

Table 2.2. Parameters for the Wafer Process Example (the definitions of L, W, K, E, D_W and F are shown in Figure 2.3).

Labor rate (L_R)	22	$/hr
Labor burden (b)	0.8	
Years to depreciate (D_L)	5	years
Quantity	10000	wafers
Hours per year	8760	hours
Die dimension (L)	0.25	inches
Die dimension (W)	0.1	inches
Minimum spacing between die (K)	0.05	inches
Edge scrap width (E)	0.15	inches
Wafer diameter (D_W)	6	inches
Flat length (F)	2	inches

The process-flow cost model is easy to implement on a spreadsheet. Table 2.3 provides the results of applying Equations (2.1) through (2.4). The only challenge in the analysis is in the calculation of tooling costs. All of the tooling has to be paid for, whether it is used or not (there is no way to prorate the amount paid for tooling) and tooling is generally not transferrable between products. In this case Equation (2.3) becomes

$$C_T = \frac{C_t}{Q} N_t = \frac{C_t}{Q} \left\lceil \frac{Q}{Q_t} \right\rceil \qquad (2.10)$$

where Q_t is the number of objects that can be made for one tooling cost (C_t). The second term in Equation (2.10) is N_t and is calculated using a ceiling function; it rounds the ratio up. Equation (2.10) is relevant to calculating the tooling cost of Step *M* in Table 2.3.

Table 2.3. Thirteen-Step Wafer Fabrication Processes Cost Calculations (note, in some cases $C_L+C_M+C_T+C_C$ does not add up to exactly the Total Cost in the table due to round off in one or more of the numbers).

Step	Labor Cost (per wafer) C_L	Material Cost (per wafer) C_M	Tooling Cost (per wafer) C_T	Equip Cost (per wafer) C_C	Total Cost (per wafer) C_{manuf}	Accumulated Cost (per wafer)
A	\$0.11	\$0.00	\$0.00	\$0.02	\$0.13	\$0.13
B	\$1.32	\$3.20	\$0.00	\$0.01	\$4.53	\$4.66
C	\$0.01	\$0.40	\$0.10	\$0.03	\$0.54	\$5.20
D	\$0.30	\$0.00	\$0.00	\$0.09	\$0.39	\$5.59
E	\$1.10	\$0.00	\$0.00	\$0.03	\$1.13	\$6.71
F	\$0.02	\$2.00	\$1.00	\$0.00	\$3.03	\$9.74
G	\$0.08	\$0.00	\$0.50	\$0.00	\$0.58	\$10.32
H	\$0.33	\$3.00	\$0.05	\$0.00	\$3.38	\$13.70
I	\$0.08	\$2.00	\$0.00	\$0.01	\$2.09	\$15.79
J	\$1.32	\$0.40	\$0.00	\$0.00	\$1.72	\$17.51
K	\$0.99	\$0.20	\$0.00	\$0.01	\$1.20	\$18.71
L	\$0.00	\$5.00	\$0.00	\$0.00	\$5.00	\$23.72
M	\$4.40	\$0.00	\$10.00	\$12.68	\$27.08	\$50.80

The final cost per die is given by

$$\text{Cost per die} = \frac{\$50.80}{528} = \$0.10 \tag{2.11}$$

2.4 Technical Cost Modeling (TCM)

Technical cost modeling is a label used to describe the combination of traditional cost models and physical process models. Traditional cost models often fail to acknowledge direct connections between the labor, material, tooling and equipment requirements and the actual physical description of the product. In TCM, physical models are used to determine product technical characteristics, which are in turn used to compute costs [Ref. 2.5].

Algorithms describing the physical parameters associated with a process (temperature, pressure, flow rate, deposition rate, etc.) are used to predict values such as cycle time, power requirements, and materials consumption. In turn, these parameters are directly related to the costs of the materials, energy, equipment utilization and labor associated with the process. With this modeling approach, the cost of poorly understood processes can be estimated with some degree of certainty, and sensible technology development strategies for optimizing these processes can be devised.

TCM has been applied to a large cross-section of mechanical and electronic cost modeling problems, ranging from molding and casting to printed circuit board fabrication. TCM as a general concept, can be applied to any of the manufacturing cost modeling approaches discussed in Part I of this book. Many of the examples presented here (and problems that appear at the end of the chapters) represent TCM exercises in which the technical description of the product or system must be used to determine times and other attributes from which costs can be modeled.

2.5 Comments

Process-flow models are used to emulate manufacturing processes. They are particularly useful when the order in which activities happen is important. For example, if functional testing activities are included at points that are internal to a process, the sequence of steps is important and process-flow models are a good choice for modeling. However, process-flow models can often inhibit the ability to see the larger picture by focusing attention on detailed steps rather than the overall process.

References

2.1 Sandborn, P. A. and Murphy, C. F. (1998). Material-centric modeling of PWB fabrication: An economic and environmental comparison of conventional and photovia board fabrication processes, *IEEE Transactions on Components, Packaging, and Manufacturing Technology – Part C*, 21(2), pp. 97-110.

2.2 Ferris-Prabhu, A. V. (1989). An algebraic expression to count the number of chips on a wafer, *IEEE Circuits and Devices Magazine*, 5(January), pp. 37-39.

2.3 de Vries, D. K. (2005). Investigation of gross die per wafer formulas, *IEEE Transactions on Semiconductor Manufacturing,* 18(1), pp. 136-139.

2.4 Sandborn, P. A., Lott, J. W. and Murphy, C. F. (1997). Material-centric process flow modeling of PWB fabrication and waste disposal, *Proc. IPC Printed Circuits Expo.*, pp. S10-4-1 - S10-4-12.

2.5 Szekely, J., Busch, J. and Trapaga, G. (1996). The integration of process and cost modeling – A powerful tool for business planning, *Journal of the Minerals, Metals & Materials Society*, 48(12), pp. 43-47.

Problems

2.1 What properties would need to be accumulated by a process flow in order to support the analysis of disassemblability (i.e., to determine how much effort would be needed to disassemble a product)?

2.2 Formulate an algorithm that exactly determines the number of die that can fit on a wafer as a function of the parameters shown in Figure 2.3.

2.3 Compare the approximate number-up given by Equation (2.7) to the exact number-up calculated in Problem 2.2 (make a plot of the die area vs. number-up for square die).

2.4 Generally all the die on wafers and boards on panels are oriented the same direction when fabricated. Why? Note that the reason for maintaining the same orientation may be different for die on wafers than for boards on panels.

2.5 If the application described in Equations (2.8) and (2.9) could be manufactured in a smaller format, such that 72 cards could be fabricated on a panel, what would the effective cost per card be after the reflow step?

2.6 In the example given in Section 2.3.2, what is the cost per die at the end of the process if a step with the following characteristics is added between steps *G* and *H*: Time = 50 seconds, Op Util = 0.8, Capacity = 1 wafer, Material Cost = \$5/unit of material, Units of Material = 2/wafer, Tooling Cost = \$5000, Tooling Life = 1000 wafers, Equip Cost = \$150,000, and Equip Operational Time = 0.8?

2.7 Suppose that the final cost per die in the example in Section 2.3.2 is constrained to be no greater than \$0.094. The only parameter you can adjust is the material cost of step *L*. In this case the material cost can be lowered to any value (the tradeoff is the reliability of the product, which is outside the scope of this problem). What material cost of step *L* should you select?

2.8 Starting with the original example in Section 2.3.2, suppose that step *D* is replaced by the result of the parallel process as shown below. Now what is the final cost per die that result from the whole process? Assume that there are no tooling costs for D_1, D_2 and D_3. For D_1, D_2 and D_3 assume that the capacity of all the steps is 1 wafer, the equipment operational time is 0.75 for steps D_1, D_2 and D_3, and that there in 1 unit of material per wafer for all the steps. All other steps (except for *D*) are given in Table 2.1.

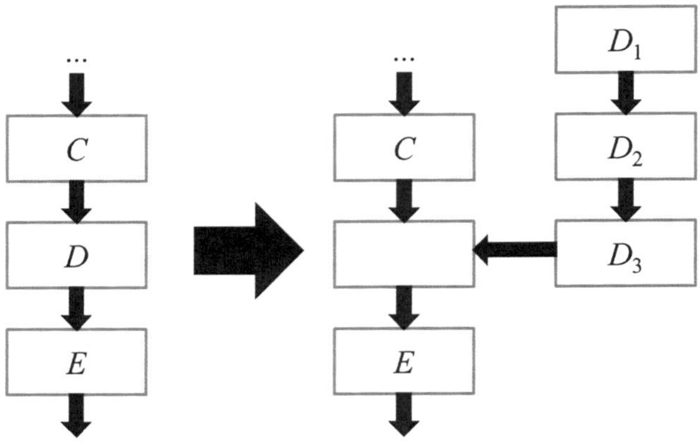

Step	Time (sec/wafer)	Operator Utilization	Material Cost (per wafer)	Equipment Cost
D_1	120	1	\$3.45	\$20,000
D_2	34	2	\$0	\$1,000,000
D_3	60	0.7	\$0.89	\$0

Chapter 3

Yield

Minimizing the manufacturing cost of a product is not sufficient to ensure that a product can be produced cost-effectively. The likelihood that a manufacturing process itself might introduce defects into the product being manufactured, with an associated cost for finding and correcting those defects, must be considered as well. For example, suppose process *A* manufactures a product for $50 per unit and introduces no defects; alternatively, process *B* manufactures the same product for $27 per unit but half of the products produced by process *B* are defective and must be discarded. For process *A*, the effective cost per good unit is $50 per unit, while for process *B* the effective cost per good unit is $27/0.5 = $54 per unit. This example makes it obvious that we must also consider the defects introduced into the manufacturing process in order to gain an accurate view of the effective cost of manufacturing a product.

According to the ISO 8402:1986 standard, quality is "the totality of features and characteristics of a product or service that bears its ability to satisfy stated or implied needs" [Ref. 3.1]. The cost of quality is defined as the cost incurred because less than 100% of the products produced can be sold [Ref. 3.2]. Generally, quality costs are composed of the following elements, [Ref. 3.2]:

- Prevention costs - the cost of preventing defects, including education, training, process adjustment, screening of incoming materials and components, supplier certification and audits, and so on.
- Appraisal costs - the costs of tests and inspections to assess if defects exist in manufactured or partially manufactured products.

- Internal failure costs - the costs of defects detected prior to delivery of the product to the customer.
- External failure costs - the costs of delivering defective products to the customer.

In this chapter, we will discuss internal failure costs through the introduction of the concepts of yield and yielded cost. Several other chapters in this book address quality costs as well: burn-in costs in Chapter 14 (prevention cost), functional testing in Chapter 7 (appraisal cost), diagnosis and rework in Chapter 8 (internal failure cost), sparing in Chapter 12 (external failure cost), and warranties in Chapter 13 (external failure cost).

Yield is defined as the probability that an item has no fatal defects. Non-fatal defects, like those that may cause a reduction in reliability, are not generally addressed in yield modeling. Restated, yield is the ratio of the number of items that are usable after the completion of a production process to the number of items that had the potential to be usable at the start of the process [Ref. 3.3]. Yield is an output, not an input. A process activity does not have a yield; it has a quality that results in a yield.

3.1 Defects

Defects occur in all types of manufacturing, including electronics manufacturing. According to Webster's Dictionary [Ref. 3.4], a defect is an imperfection; fault;[1] flaw; blemish; or deformity. There are several distinct types of defects. Firstly, there are gross defects that are large with respect to the size of the object being manufactured — for example, scratches, defects due to handling, or damage due to test probes. Gross defects generally result in catastrophic yield loss that causes products not to work at all. Secondly, there are parametric defects that may not result in any physically observable damage; however, they affect the object's performance. Parametric defects may be due to design flaws and often

[1] We will make a distinction between faults and defects when we discuss testing in Chapter 7. Generally, faults are defects that result in yield loss.

cause parts to “bin” lower,[2] or lead to reliability problems during field use. The third class of defects is random defects. Random defects that have a probability of occurrence are the focus of the remainder of the discussion in this chapter.

Depending on the extent and location of a defect, it affects either the yield or the reliability of the resulting electronic device. If the defect causes an immediate and obvious failure (a “fatal defect”) of the device prior to the completion of the manufacturing process, it is considered a yield problem. For example, missing metallization that causes an open circuit where two points on a signal line on a printed circuit board should have been connected will likely be detected as a yield problem. If the defect does not cause an immediate failure of the device, it is called a latent defect that may cause a failure of the device in the field that is perceived as a reliability problem. An example of a latent defect is a defect that reduces the thickness of a signal line in a printed circuit board that could become an open circuit after the device is used for several years.

Several metrics are used to measure defect levels. Defects can be measured in parts per million (ppm) defective. Defect density will be used in the discussion that follows, referring to defects per unit area, where the area is the area of a die (integrated circuit), wafer, board, or panel on which a board is fabricated. As mentioned, defects that result in yield loss are called faults or fatal defects. The likelihood that a random defect will become a fault is called the fault probability.

3.2 Yield Prediction

From a business perspective, the utility of accurately describing past yields and predicting the future yield of a product is obvious. Yield is arguably the single most influential metric upon which to gauge the financial success of a product, process, and manufacturer [Ref. 3.5]. Yield modeling

---

[2] Non-repairable items (such as integrated circuits) are often sorted by their final performance range at the end of their manufacturing process. Parts in different performance ranges (or “bins”) can be used for different applications and potentially are sold at different prices. An example of this is microprocessors, which may be binned by maximum clock frequency.

in electronics, specifically associated with the fabrication of semiconductor devices and later integrated circuits, has been performed since the 1960s; see [Ref. 3.6] for a review of the early history of yield modeling.

A simple definition of yield is

$$\text{Yield} = \frac{\text{Number of usable items after the process}}{\text{Total number of items}} \tag{3.1}$$

where the denominator of Equation (3.1) indicates two possibilities: if it refers to items that start the process, then this equation provides the process yield; if it refers to the items that complete the process, then Equation (3.1) gives the yield of the final product.

Mathematically, yield is the probability of obtaining an item with no (0) fatal defects, $\Pr(0,\lambda)$, where there are on average λ fatal defects per item. The essence of yield prediction is to obtain a numerical value of $\Pr(0,\lambda)$. The form of the equation for $\Pr(0,\lambda)$ depends on the spatial distribution of the fatal defects (distribution of defects over the physical area used to fabricate the items). The variable λ depends on the size distribution (distribution of defect physical sizes) of all potentially fatal defects.

The development of yield prediction relations is presented in the context of the fabrication of die (individual integrated circuits) on a wafer, as shown in Figure 3.1. However, the yield models developed are generally applicable to other physical items, such as printed circuit board fabricated on panels.

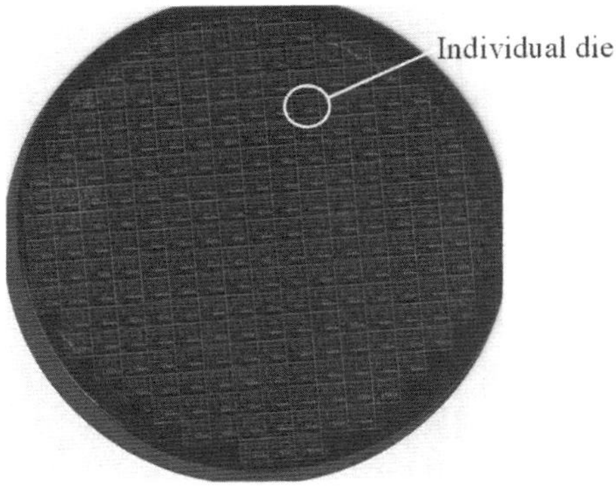

Fig. 3.1. Wafer containing individual die.

3.2.1 *The Poisson Approximation to the Binomial Distribution*

For die on a wafer, yield prediction requires calculating the probability of finding a particular state (a die with 0 faults) out of all possible states (die with 0, 1, 2 or more faults) when events (faults) are distributed over all states (die with 0, 1, 2 or more faults) according to some distribution law. In order to do this we need to use a counting technique (a method for determining the number of possible events) appropriate to the laws governing the way in which the events (faults) are distributed. On a die there are only two possible states (binomial): (1) the die has no faults, or (2) the die has one or more faults. Yield prediction is the determination of the probability of occurrence of the first case.

Consider the two states (just like heads and tails when flipping a coin):

$$p+q=1 \tag{3.2}$$

where p is the probability of getting a head and q is the probability of getting a tail when flipping a coin once. Now consider N coins (or the same coin flipped N times):

$$(p+q)^N=1 \tag{3.3}$$

Expanding $(p+q)^N$ using the Binomial Series,

$$(p+q)^N = p^N + Np^{N-1}q + \frac{N(N-1)}{2!}p^{N-2}q^2 + \ldots + q^N = \sum_{i=0}^{N}\binom{N}{i}p^i q^{N-i} \tag{3.4}$$

where N is an integer ≥ 1 and the binomial coefficient is given by

$$\binom{N}{i} = \frac{N!}{i!(N-i)!} \tag{3.5}$$

Equation (3.4) is known as the binomial distribution. Each term in the series given in this equation gives the probability that exactly i heads will be obtained when flipping the coin N times. The nth term in the series in Equation (3.4) is

$$\Pr(n;N,p) = \frac{N!}{n!(N-n)!}p^n(1-p)^{N-n} \tag{3.6}$$

$\Pr(n;N,p)$ is the probability of finding a state (n heads in N flips) when the events are distributed according to the binomial distribution. The probability of getting exactly no heads ($n = 0$) on N flips is

$$\Pr(0;N,p)=\frac{N!}{N!}(1-p)^N=(1-p)^N \tag{3.7}$$

Letting $\lambda = Np$ (λ is the mean of the binomial distribution), we get

$$\Pr(0;N,p)=\left(1-\frac{\lambda}{N}\right)^N \tag{3.8}$$

Taking the natural log of both sides of Equation (3.8) and using a Taylor series expansion,

$$\ln(1-x)=-x-\frac{x^2}{2}-\frac{x^3}{3}-\ldots-\frac{x^n}{n}-\ldots \tag{3.9}$$

we get

$$\ln\Pr(0;N,p)=N\left(-\frac{\lambda}{N}-\frac{\lambda^2}{2N^2}-\frac{\lambda^3}{3N^3}-\ldots\right)=-\lambda-\frac{\lambda^2}{2N}-\frac{\lambda^3}{3N^2}-\ldots \tag{3.10}$$

When N is large Equation (3.10) reduces to

$$\Pr(0;N,p)\cong e^{-\lambda} \tag{3.11}$$

Equation (3.11) is the probability of obtaining no heads when a coin is flipped N times (or N coins are flipped).

For our problem (faults in die), N is the number of possible faults in a die (not the number of unique faults) and p is the probability of one of the faults occurring (assuming all faults have the same probability of occurance).

We now wish to approximate the probability (in terms of λ) of obtaining an exact (n) number of events when N is large. Using the exact relation given in Equation (3.6), we can evaluate the following ratio:

$$\frac{P(n;N,p)}{P(n-1;N,p)}=\frac{N!}{n!(N-n)!}\frac{(n-1)!(N-n+1)!}{N!}\frac{p^n(1-p)^{N-n}}{p^{n-1}(1-p)^{N-n+1}} \tag{3.12}$$

When $N >> n \geq 1$ and $p << 1$, Equation (3.12) becomes

$$= \frac{(N-n+1)}{n}\frac{p}{(1-p)} \cong \frac{\lambda}{n} \qquad (3.13)$$

Using Equations (3.11) and (3.13), we can construct the following sequence of probabilities:

$$\begin{aligned} P(0;N,p) &\cong e^{-\lambda} \\ P(1;N,p) &\cong \lambda e^{-\lambda} \\ P(2;N,p) &\cong \frac{\lambda^2}{2}e^{-\lambda} \\ P(3;N,p) &\cong \frac{\lambda^3}{6}e^{-\lambda} \end{aligned} \qquad (3.14)$$

Generalizing the results, we obtain

$$P(n;N,p) \cong \frac{\lambda^n}{n!}e^{-\lambda} \qquad (3.15)$$

Equation (3.15) is the Poisson approximation to the binomial distribution and represents the probability of having a die with exactly n fatal defects. Observe that Equation (3.15) reduces to Equation (3.11) when $n = 0$. Equation (3.15) assumes that fatal defects are equally likely to occur in all die, which is not necessarily true; defects may be more likely in die at the edges of wafers than die in the center. It also assumes that the occurrence of a fatal defect is independent of whether a fatal defect has already occurred (which is also not necessarily true, since defects in wafers tend to cluster).

In Equation (3.15), λ is the mean number of occurrences of the event (faults) per die and is given by

$$\lambda = AD \qquad (3.16)$$

where A is the area of the die and D is the defect density (defects per unit area).

In general, D is not a constant over a wafer; rather, D is governed by its own probability distribution, $f(D)$. Using Equations (3.15) and (3.16) and summing over the distribution of defect densities, we obtain

$$P(n;AD) \cong \int_0^\infty \frac{(AD)^n e^{-AD}}{n!} f(D)dD \qquad (3.17)$$

Here, $f(D)$ is the distribution of defect densities (D) over the physical area in which the items are fabricated. Figure 3.2 shows an example of how $f(D)$ could be constructed for a wafer. The number of defects in each square in the grid are counted and divided by the area of the grid square to form a defect density (D) for each grid square. A histogram of the resulting values of D for all the grid squares can be created and fit with various mathematical distribution forms. The form of the defect density distributions distinguishes different yield models.

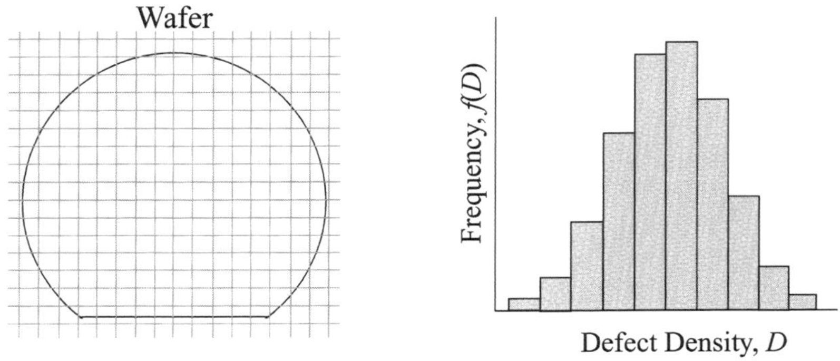

Fig. 3.2. Formation of defect density distributions.

Yield is the probability of obtaining a die with no faults ($n = 0$), with the assumption of a particular distribution of defect densities, $f(D)$:

$$Y = \Pr(0; AD) \cong \int_0^\infty e^{-AD} f(D)dD \tag{3.18}$$

3.2.2 *The Poisson Yield Model*

The Poisson yield model assumes that the defect density is constant — that is, that D is the same ($D = D_0$) in every grid square in Figure 3.2. This is represented as[3]

---

[3] δ is a Dirac delta function, which is defined by, $f(x) \equiv \int_{-\infty}^{\infty} f(y)\delta(y-x)dy$, in this case, the function only exists (is non-zero) at $y = x$. The Dirac delta function is a continuous analogue of the discrete Kronecker delta. In the context of signal processing it is often referred to as the unit impulse function.

$$f(D) = \delta(D - D_0) \tag{3.19}$$

Constant defect density means that the probability of obtaining a fatal defect is the same everywhere on the wafer. Using Equation (3.19) in Equation (3.18) we obtain

$$Y \cong \int_0^\infty e^{-AD} \delta(D - D_0) dD = e^{-AD_0} \tag{3.20}$$

Equation (3.20) is known as the Poisson yield equation, which predicts the yield of a die that has an area of A that is fabricated on a wafer with a constant defect density of D_0.

The Poisson yield equation generally predicts lower yield than what is actually observed. Why? The defect density is not really a constant. It varies from place to place on a wafer (and from wafer to wafer). For a constant number of defects, the Poisson yield equation predicts the worst-case situation. In reality, defects cluster and may be more likely at certain locations on the wafer. Consider the simple demonstration in Figure 3.3.

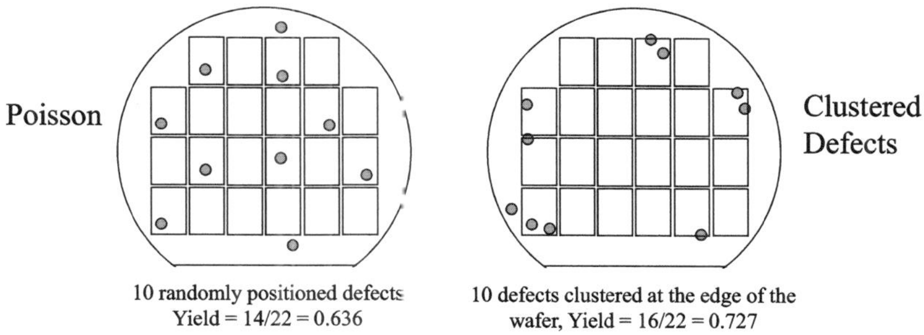

Fig. 3.3. Demonstration of the under-prediction of yield by the Poisson yield model.

3.2.3 *The Murphy Yield Model*

The Murphy yield model assumes defect density has a symmetric triangular distribution (Simpson distribution) defined by

$$f(D) = \frac{D}{D_0^2}, \quad 0 \le D \le D_0 \tag{3.21a}$$

$$f(D)=\frac{1}{D_0}\left(2-\frac{D}{D_0}\right),\ D_0 \le D \le 2D_0 \tag{3.21b}$$

and $f(D)=0,\ D \ge 2D_0$, which is shown in Figure 3.4. Substituting Equation (3.21) into Equation (3.18) gives

$$Y \cong \int_0^{D_0} e^{-AD}\frac{D}{D_0^2}dD+\int_{D_0}^{2D_0} e^{-AD}\frac{1}{D_0}\left(2-\frac{D}{D_0}\right)dD \tag{3.22}$$

Equation (3.22) becomes

$$Y \cong \frac{1}{D_0^2}\frac{e^{-AD}}{A^2}\left(-AD-1\right)\Bigg|_0^{D_0}+\frac{2}{D_0}\frac{e^{-AD}}{-A}\Bigg|_{D_0}^{2D_0}-\frac{1}{D_0^2}\frac{e^{-AD}}{A^2}\left(-AD-1\right)\Bigg|_{D_0}^{2D_0} \tag{3.23}$$

which reduces to

$$Y=\left[\frac{1-e^{-AD_0}}{AD_0}\right]^2 \tag{3.24}$$

Equation (3.24) is known as the Murphy yield model [Ref. 3.7]. For Equation (3.34), in the limit at D_0 approaches 0, Y approaches 1.

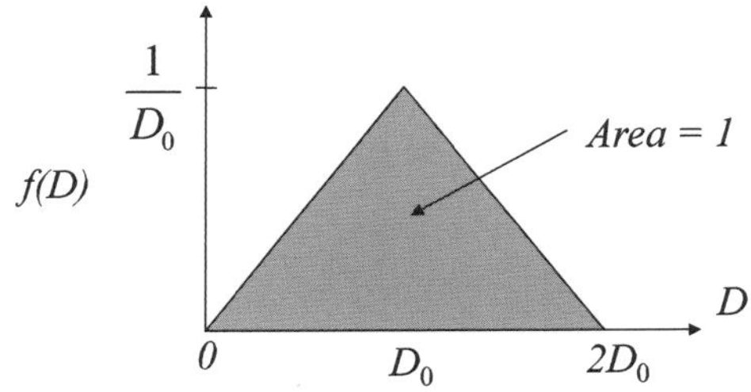

Fig. 3.4. Symmetric triangular defect density distribution.

3.2.4 *Other Yield Models*

Other yield model forms can be derived using alternative defect density distributions. These include:

Uniform: $f(D)=\frac{1}{2D_0},\ 0 \le D \le 2D_0$ resulting in $Y=\left(\frac{1-e^{-2AD_0}}{2AD_0}\right)$ (3.25)

Half Gaussian: $f(D)=\frac{2}{\pi D_0}e^{-\left(\frac{D}{D_0\sqrt{\pi}}\right)^2}$, $D\geq 0$ resulting in

$$Y=e^{\pi\left(\frac{AD_0}{2}\right)^2}erf\left(\sqrt{\pi}\frac{AD_0}{2}\right) \quad (3.26)$$

Exponential: $f(D)=\frac{e^{-D/D_0}}{D_0}$, $D\geq 0$ resulting in $Y=\frac{1}{(1+AD_0)}$ (3.27)

The half-Gaussian-based form is often referred to as the Stapper model; the exponential distribution-based form is referred to as the Price or Seeds model.[4] Other models exist based on the Erlang, Gamma, and Bose-Einstein distributions. Figure 3.5 shows a comparison of the yield models discussed so far. All the yield models predict approximately the same yield for small die and then diverge as die become larger. The Poisson model gives the most conservative estimate of yield.

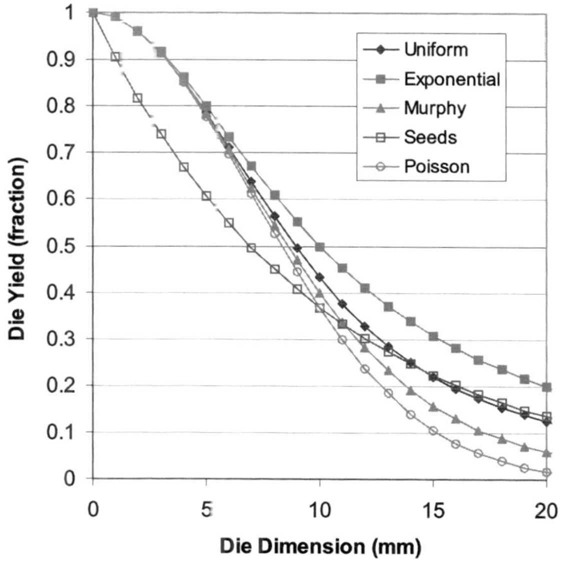

Fig. 3.5. Comparison of yield models. D_0 = 1 defect/cm², die dimension squared is the die area (A). The Seeds model referred to in this figure is given by $Y=e^{-\sqrt{AD}}$.

[4] Note $Y=e^{-\sqrt{AD}}$ is also referred to as the Seeds model.

State-of-the-art yield modeling often uses the negative binomial distribution [Ref. 3.8], which results in

$$Y = \left[1 + \frac{D_0 A}{\alpha}\right]^{-\alpha} \tag{3.28}$$

where α is a clustering parameter. The clustering parameter α ranges from 1 (highly clustered) to ∞ (no clustering or random). The negative binomial model assumes that the likelihood of a defect occurring at a given location increases linearly with the number of defects that have already occurred at that location. Several of the other yield models discussed in this chapter can be approximated through the appropriate choice of α. The negative binomial model makes no assumptions about the spatial independence of defects. The International Technology Roadmap for Semiconductors [Ref. 3.9] recommends using $\alpha = 2$.

3.3 Accumulated Yield

The yield models developed in this chapter can be used in several different ways. In a real item, there will be many different types of defects and each defect type can have its own unique defect density distribution that leads to its own unique yield (with respect to that defect type). The yields that are specific to a particular defect type may or may not be independent of each other. In the simplest approach, the defect density distribution can represent an aggregation of all the defect types; likewise, the yield is an aggregate yield from all relevant defect types.

Ferris-Prabhu [Ref. 3.3] characterizes the application of yield models as either composite or layered. This characterization is not based on aggregating the effect of defect types, but rather on distributing the yield contribution among multiple process steps (or in the case of integrated circuit manufacturing, different "layers"). In the composite applications, the yield models predict the yield of a die (or any other item) based on the average number of defects of all types over all process steps (or layers). In layered models, the yield of each individual layer (step in the manufacturing process) is determined, from which a composite yield can be formed.

Yield is the probability of no defects, and probabilities (if independent) can be accumulated by taking their product. In the case of a process flow where Y_i represents the aggregate yield of the ith process step, the accumulated yield is given by

$$Y = \prod_{i=1}^{n} Y_i \tag{3.29}$$

where n is the total number of process steps. If all the individual layer yields are modelled with the Poisson yield model, Equation (3.29) becomes

$$Y = \prod_{i=1}^{n} e^{-AD_i} = e^{-A\sum_{i=1}^{n} D_i} \tag{3.30}$$

Equation (3.30) implies that the sum of the defect densities across all the layers (process steps) equals the net effective defect density for the whole process. The only yield model for which this is mathematically true is the Poisson yield model.[5]

3.3.1 *Multi-Step Process-Flow Example*

As an example of accumulating yield through a process flow, we will return to the multi-step process-flow example presented in Section 2.3.2. If the individual process steps *A-M* introduce defects into a wafer with the defect densities given in the second column in Table 3.1, assuming that the Poisson yield model is applicable, what is the yield of die that result from this process?

The third column of Table 3.1 accumulates the defect densities through the steps. The fourth column of Table 3.1 is the yield of each individual process step calculated using Equation (3.20), where the area of the die is given by $A = LW$ from Table 2.2 (converted to cm). The fifth column of Table 3.1 is a the product of the individual step yields. The final yield of a single die from this process is 0.6834 (the last entry in the fifth column)

[5] The implications of this fact are discussed in detail in [Ref. 3.3]. The Poisson yield model is often used (with appropriate scaling — see [Ref. 3.3]) when yield is accumulated through a series of layers or process steps for this very reason, whereas other models are used for composite applications.

and can also be computed from the accumulated defect densities using Equation (3.30):

$$\text{Yield of a die} = e^{-(0.1613)(2.36)} = 0.6834 \tag{3.31}$$

where the area of the die is 0.1613 cm^2 = (0.25)(2.54)(0.1)(2.54). This result means that 68.34% of the die that result from this process will be defect-free.

Table 3.1. Thirteen-Step Wafer Process from Table 2.1 with Defect Densities Included (All of the process steps apply to the whole wafer, not individual die).

Step	Defect Density (defects/cm^2) D_i	Accumulated Defect Density (defects/cm^2)	Step Yield (per die) Y_i	Accumulated Yield (per die)
A	0.1	0.1	0.9840	0.9840
B	0.7	0.8	0.8932	0.8789
C	0.06	0.86	0.9904	0.8705
D	0.13	0.99	0.9793	0.8524
E	0.3	1.29	0.9528	0.8122
F	0.11	1.4	0.9824	0.7979
G	0.02	1.42	0.9968	0.7953
H	0.01	1.43	0.9984	0.7940
I	0.5	1.93	0.9225	0.7325
J	0.1	2.03	0.9840	0.7208
K	0	2.03	1.0000	0.7208
L	0.1	2.13	0.9840	0.7092
M	0.23	2.36	0.9636	0.6834

3.3.2 *The Known Good Die (KGD) Problem*

In the 1980s and 1990s, there was a lot of interest throughout the electronic packaging world in developing a technology called multichip modules (MCMs). An MCM is essentially the same as a printed circuit board with individual chips mounted on it, except that in MCMs the integrated circuits are not in their own packages (the single chip package) — they are just bare die mounted on an electronic interconnect. MCMs effectively eliminate one level in the packaging hierarchy. The benefits of omitting single chip packages include:

(1) Size/weight – Single chip packages take up a lot of space; systems can be made smaller and lighter if single chip packages are eliminated.

(2) Electrical performance – Single chip packages add lots of electrical parasitics, such as capacitance and inductance, which degrade the performance of a system.
(3) Reliability – Removal of single chip packages eliminates one source of potential interconnect reliability problems.

One of the most significant problems faced by MCM manufacturers is called known good die (KGD). In conventional electronic systems, die are packaged into single chip packages and then functionally tested prior to their sale and subsequent assembly onto boards. Unfortunately, bare die cannot be as easily tested prior to assembly. As a result, MCM manufacturers in the 1980s and 1990s could only functionally test the die in their systems after they were assembled into the MCM, rather than before assembly. The issues raised by the availability (or lack of availability) of functionally tested bare die is called *known good die* (KGD).

To illustrate the KGD problem consider the example shown in Figure 3.6. The first pass module yield is determined from

$$\text{First pass module yield} = (\text{Die Yield})^{\text{Number of die in the module}} \tag{3.32}$$

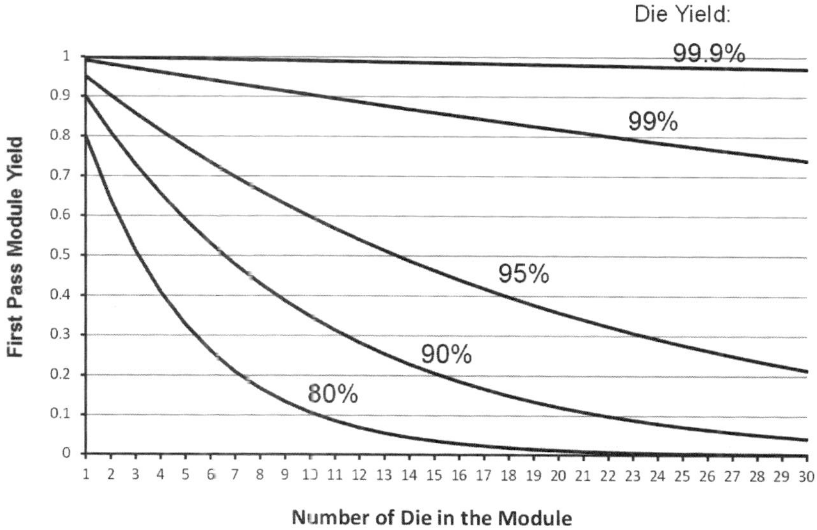

Fig. 3.6. First pass module yield that results from using the specified number of identical die with the indicated individual die yields.

Equation (3.32) assumes that all the die in the module have to be good in order for the module to be good. This example demonstrates that the use of multiple die with relatively high yields can result in low module yields.

Today, many integrated circuit manufactures can provide die that have been functionally tested at the wafer level. However, known good die (tested bare die) are often more expensive than chips (tested packaged die).

3.4 Yielded Cost

The ratio of the cost of a product to its yield is called yielded cost:

$$C_Y = \frac{Cost}{Yield} \tag{3.33}$$

We can appreciate the value of this definition by considering the example shown in Figure 3.7: if $C_{in} = 0$, $Y_{in} = 1.0$, and setting $C_i = 100$ and $Y_i = 0.9$ for each of the $m = 3$ steps, then $C_{out} = \$300$, $Y_{out} = 0.9^3 = 0.729$, and $C_Y = \$300/(0.9^3) = \$412$ per good assembly. The measurement of process-yielded cost (the yielded cost of a process) is valuable because it represents an effective cost per good assembly after a set of process steps, which potentially helps in evaluating the value of the process.

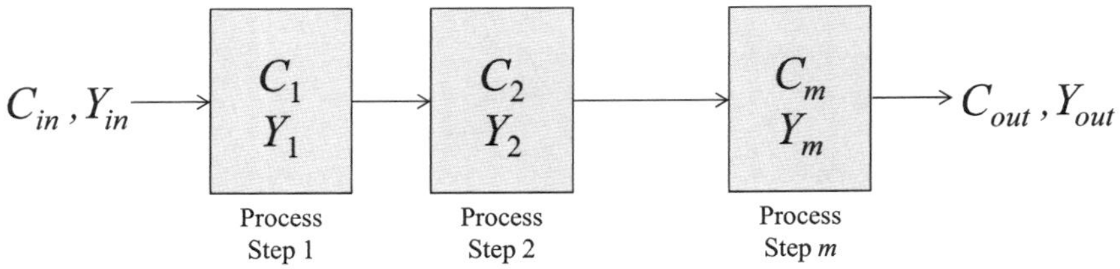

Fig. 3.7. A simple sequential process flow for illustrating yielded cost.

In general, for a sequential process flow, the final yielded cost of the items that result from the process is given by

$$C_{Y_{Final}} = \frac{C_{out}}{Y_{out}} = \frac{C_{in} + \sum_{i=1}^{m} C_i}{Y_{in} \prod_{i=1}^{m} Y_i} \tag{3.34}$$

While it is easy to evaluate the final yielded cost of a process flow, for example, using Equation (3.34); how can the yielded cost associated with a specific process step be evaluated? Step-yielded cost, C_{Ystep}, represents the true effective cost contribution of an individual step within the entire process. The criteria used for evaluating a model of step-yielded cost are [Ref. 3.10]:

(1) Individual step-yielded costs must be collected in some way to get the final yielded cost of the entire process.
(2) Step-yielded costs must account for upstream and downstream information for each step.
(3) Step-yielded costs must be independent of step order between steps that scrap items.

Collecting step-yielded costs is necessary because the accumulation of effective cost contributions should represent the effective cost of the entire process. Incorporating upstream and downstream information is necessary because step-yielded cost should account for both a step's affect on all other process steps, and all other process steps' affect on the step under consideration. Steps that scrap items through tests or inspections remove items from the process. The independence of step order for steps between those that scrap items is necessary because the contribution should be the same no matter where a step is in a process as long as items are not removed from the process.

Several approaches to calculating step-yielded cost have been used. The simplest model is called the *itemized* approach. The itemized approach defines $C_{Y_{Step}}$ as the cost of the step divided by the step's yield:

$$C_{Y_{Step}} = \frac{C_{Step}}{Y_{Step}} \tag{3.35}$$

In Figure 3.7, the itemized approach would give $C_{Y_{in}} = C_{in} / Y_{in}$ and $C_{Y_1} = C_1 / Y_1$. The total yielded cost after step 1 would then be $C_{in} / Y_{in} + C_1 / Y_1$. Since this is not, in general, equal to the actual process-yielded cost after step 1, which is, $(C_{in} - C_1) / Y_{in}Y_1$, this approach does not satisfy the first criteria ($C_{Y_{Step}}$ values cannot be collected to get $C_{Y_{Final}}$).

Several alternative methods of calculating step-yielded cost have been proposed (see [Ref. 3.10]). The most accurate method to measure the true effective cost contribution of a process step is the *omission* method [Ref. 3.10]. The omission approach calculates $C_{Y_{Step}}$ as the difference between $C_{Y_{Final}}$ computed with the step in the process flow, and $C_{Y_{Final}}$ computed without the step in the process flow. The step-yielded costs calculated with this method thus represent the change in $C_{Y_{Final}}$ by removing the step from the process flow. Under this definition, the yielded cost of the first step in Figure 3.8 would be

$$C_{Y_1} = \frac{C_{in} + C_1 + C_2}{Y_{in} Y_1 Y_2} - \frac{C_{in} + C_2}{Y_{in} Y_2} = \frac{C_{in}(1 - Y_1) + C_1 + C_2(1 - Y_1)}{Y_{in} Y_1 Y_2} \quad (3.36)$$

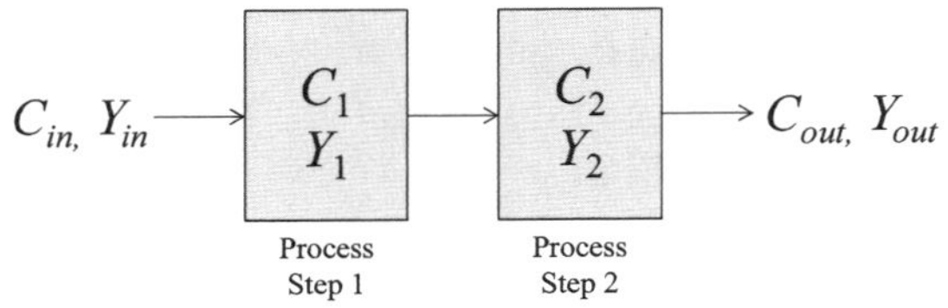

Fig. 3.8. Two-step process flow.

The omission method satisfies the three criteria given earlier in this section – the individual step-yielded costs can be collected to obtain the final yielded cost. If Equation (3.36) is separated into the sum of three terms, each term will have the process yield in the denominator and a step cost multiplied by a yield factor in the numerator. The second term is the cost of the first step divided by the process yield. This term represents the *base cost*, or the cost invested in the step. The first and third terms have a step cost multiplied by the fraction of assemblies made defective in the first step, all divided by the process yield. These terms represent *auxiliary* costs (wasted money on assemblies that will later be made defective or on assemblies that are already defective).

The $C_{Y_{Step}}$ value obtained with the omission approach represents the change in $C_{Y_{Total}}$ when removing the step from the process flow, and can be broken down into base cost and auxiliary cost components. Because the base costs and auxiliary costs are independent of step order, the step-yielded cost is also independent of step order.

The sum of all step-yielded costs for Figure 3.8 is

$$\begin{aligned}
&C_{Y_{in}} + C_{Y_1} + C_{Y_2} \\
&= \frac{C_{in} + (1-Y_{in})(C_1 + C_2)}{Y_{in}Y_1Y_2} + \frac{C_1 + (1-Y_1)(C_{in} + C_2)}{Y_{in}Y_1Y_2} + \frac{C_2 + (1-Y_2)(C_{in} + C_1)}{Y_{in}Y_1Y_2} \\
&= \frac{C_{in} + C_1 + C_2}{Y_{in}Y_1Y_2} + \frac{C_{in}(2 - Y_1 - Y_2)}{Y_{in}Y_1Y_2} + \frac{C_1(2 - Y_{in} - Y_2)}{Y_{in}Y_1Y_2} + \frac{C_2(2 - Y_{in} - Y_1)}{Y_{in}Y_1Y_2}
\end{aligned} \tag{3.37}$$

The sum of the base costs term $(C_{in} + C_1 + C_2) / Y_{in}Y_1Y_2$ equals the process-yielded cost, $C_{Y_{out}}$ from Figure 3.8. The additional terms in the last line of Equation (3.37) represent the sum of the auxiliary costs. Thus this method gives $C_{Y_{Step}}$ values that can be collected, according to the criteria set previously.

In addition, these $C_{Y_{Step}}$ values incorporate upstream and downstream information via the auxiliary costs. For example, in Equation (3.36), upstream information appears in the C_{in} term and downstream information appears in the C_2 term. The C_{in} term represents the incoming auxiliary cost on items to be made defective in the first step. That is, there will be some amount of cost invested into assemblies before they enter the first step. The assemblies made defective in the first step waste this cost by a factor of $(1-Y_1)$. Likewise, the C_2 term represents the auxiliary cost of the second step on assemblies made defective in the first step. Like the first case, there will be items made defective in the first step that will absorb cost from the second step. Thus the omission approach calculates $C_{Y_{Step}}$ values that incorporate upstream and downstream information with its auxiliary cost terms (the last three terms in Equation (3.37)). Furthermore, this approach defines $C_{Y_{Step}}$ values that are independent of step order. In Equation (3.36),

C_{Y_1} would not change if steps 1 and 2 were switched. This is because both the base cost and auxiliary cost terms are independent of step order. The base costs only depend on the cost of the base step and the process yield, $Y_{in}Y_1Y_2$, which remains the same during step switching. Likewise, both auxiliary cost terms have the same auxiliary yield factor, (1-Y_1), so switching step order will not affect the result. This is intuitive, because if cost is incurred before step 1, then the fraction (1-Y_1) of assemblies made defective in step 1 forces the loss of this incurred cost. Additionally, if cost is incurred after step 1, then these assemblies also absorb a fraction (1-Y_1) of this cost. Either cost is incurred on assemblies that are defective or on assemblies to be made defective and an amount C_{step}(1-Y_1) of cost is lost due to the defect generation in step 1. For these reasons, auxiliary costs, and thus, step-yielded costs, are independent of step order.

3.5 The Relationship Between Yield and Producibility

Producibility is the ability to reproduce units of a product identically and without waste, so that they satisfy all customer physical and functional requirements (quality, reliability, performance, availability and price) [Ref. 3.11]. Producibility is quantified using capability indices. Process capability is the ability of a process to produce output within specification limits and is measured using a capability index. An index value of a certain magnitude indicates the same performance of a process relative to specifications, regardless of the product. Capability index is defined as

$$\text{Capability Index} = \frac{\text{Product Requirements}}{\text{Process Capability}} \tag{3.38}$$

Several capability indices are used to quantify process capability, including the following:

$$C_p = \frac{HSL - LSL}{6\sigma} \tag{3.39}$$

$$C_{pk} = \frac{\min(HSL - \mu, \mu\text{-}LSL)}{3\sigma} \tag{3.40}$$

where *HSL* and *LSL* are the high and low specifications limits defined in Figure 3.9, μ is the mean of the process, and σ is the standard deviation of the process. For C_p and C_{pk}, bigger is better.

To explore the connection between yield and process capability, consider the three processes shown in Figure 3.10. The data describing the three processes is shown in Table 3.2. For the example shown in Figure 3.10, obviously process *A* would be preferred over process *C*; however, the C_p for both processes is the same, since they both have the same standard deviation. In the case shown in Figure 3.10, the C_{pk} of process *A* is larger than that of process *C*.

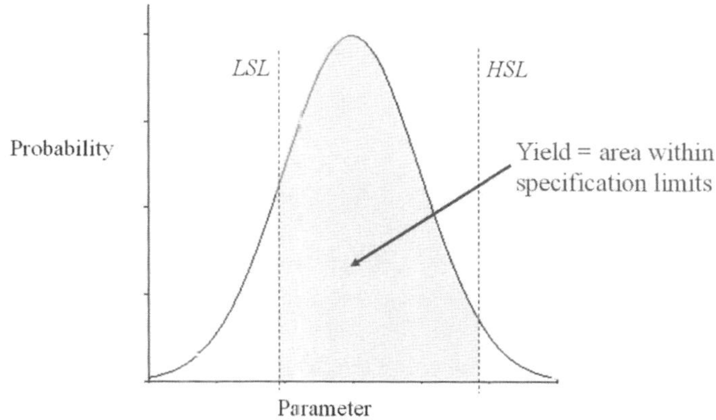

Fig. 3.9. Distribution of products produced by the process in terms of a critical parameter value. *HSL* and *LSL* are product-requirement specific.

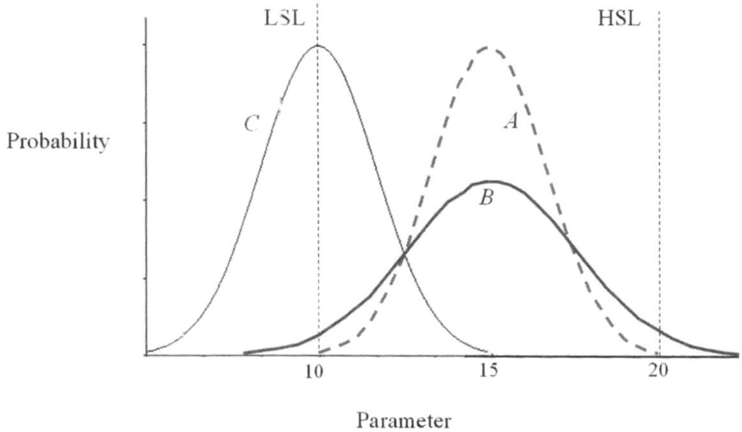

Fig. 3.10. Distribution of products produced by three processes.

Table 3.2. Data Describing the Three Processes in Figure 3.10.

Process	μ	σ	*HSL*	*LSL*	C_p	C_{pk}	Yield
A	15	3.54	20	10	0.47	0.47	0.84
B	15	4.95	20	10	0.34	0.34	0.69
C	10	3.54	20	10	0.47	0	< 0.50

From Table 3.2 we can see that a high C_p indicates high "quality" (repeatability) — that is, a small standard deviation. For processes with a constant standard deviation, C_{pk} can be used as an indicator of yield, but C_p cannot. See [Ref. 3.12] for additional discussion.

References

3.1 ISO (1986). *Quality-Vocabulary, ISO 8402*, (International Organization for Standardization, Geneva).

3.2 Sakurai, M. (1996). *Integrated Cost Management* (Productivity Press, Portland, OR).

3.3 Ferris-Prabhu, A. V. (1992). *Introduction to Semiconductor Device Yield Modeling* (Artech House, Norwood, MA).

3.4 Webster (1978). Webster's New Twentieth Century Dictionary of the English Language, Unabridged, 2nd Edition (William Collins+World Publishing Company).

3.5 Anderson, K. (2006). Innovative yield modeling using statistics, Proceedings of the SEMI/IEEE Advanced Semiconductor Manufacturing Conference.

3.6 Stapper, C. H. (1989). Fact and fiction in yield modeling, *Microelectronics Journal*, 20(1-2), pp. 129-151.

3.7 Murphy, B. T. (1964). Cost-size optima of monolithic integrated circuits, *Proceedings of the IEEE*, 52(12), pp. 1537-1545.

3.8 Stapper, C. H. (1975). On a composite model to the IC yield problem, *IEEE J. Solid-State Circuits*, SC-10(6), pp. 537-539.

3.9 *International Technology Roadmap for Semiconductors* (ITRS). http://www.itrs2.net/itrs-reports.html. Accessed May 5, 2016.

3.10 Becker, D. and Sandborn, P. (2001). On the use of yielded cost in modeling electronic assembly processes, *IEEE Transactions on Electronics Packaging Manufacturing*, 24(3), pp. 195-202.

3.11 Harry, M. J. and Lawson, J. R. (1992). *Six Sigma Producibility Analysis and Process Characterization*, (Addison-Wesley, Reading, MA).

3.12 Ramakrishnan, B., Sandborn, P. and Pecht, M. (2001). Process capability and product reliability, *Microelectronics Reliability*, 41(12), pp. 2067-2070.

Bibliography

In addition to the sources referenced in this chapter, there are several good sources of information on yield modeling, including:

Kuo, W. and Kim, T. (1999). An overview of manufacturing yield and reliability modeling for semiconductor products, *Proceedings of the IEEE*, 87(8), pp. 1329-1344.

Peters, L. (2000). Choosing the best yield model for your product, *Semiconductor International*, May 1.

IEEE Transactions on Semiconductor Manufacturing, February 1988 to present.

Problems

3.1 Would you expect the Poisson yield model to be more or less accurate as die sizes increase?

3.2 Derive Equation (3.28) Hint: the equation is derived by compounding the Poisson model with the gamma distribution, generating a "contagious" distribution.

3.3 Under what conditions does Equation (3.28) reduce to the Poisson yield model and the Seeds yield model given in Equation (3.27)?

3.4 How does the accumulated yield computed by summing defect densities compare with the accumulated yield found by multiplying probabilities for non-Poisson yield models? Is it always larger or smaller?

3.5 If the defect density introduced by Step *G* in Table 3.1 is changed to 0.25, what is the final yield per die for the entire process in Table 3.1? Make sure to express your yield calculations to at least 5 significant figures.

3.6 Assuming the use of a Poisson yield model is valid, under what conditions does the accumulation of defect densities for all process steps and the use of Equation (3.30) not work?

3.7 If a Murphy yield model is assumed (rather than a Poisson yield model), what is the final yield per die for the entire process in Table 3.1? Make sure to express your yield calculations to at least 5 significant figures.

3.8 What is the effective yielded cost per die at the end of the thirteen-step process given in Tables 2.1 and 3.1, assuming a Poisson yield model?

3.9 A round wafer (no flat) with a diameter of 150 mm has ten uniformly distributed defects on it. The die area is 1.2 cm^2. (a) What is the die yield? (b) Assume the wafer will go through eight additional process steps and the final target yield for die after all those additional steps is 75%. If all the steps introduce an equal number of uniformly distributed defects, how many total defects can each step contribute?

3.10 Using the omission method, what is the effective yielded cost of Step *H* in the process flow shown in Table 3.1? Does changing the cost of Step *B* affect the effective yielded cost of Step *H*? Why or why not? Does changing the cost of Step

K affect the effective yielded cost of Step *H*? Why or why not? Make sure to express your yielded cost calculations to at least 5 significant figures.

3.11 In the previous problem (Problem 3.10), if a zero cost test was added to the process flow between Steps *H* and *K* that removed all the defective wafers, would changing the cost of Step *K* affect the yielded cost of Step *H*? Why or why not?

3.12 You run a small company that applies a protective coating to electronic boards. It takes five minutes of labor and \$6 in materials to coat a single board. Your coating process has an 85% yield (assume that none of the defects introduced by your process are repairable). Assume that labor costs you \$35/hour (ignore overhead). If a prospective customer comes to you with a board to be coated, and you want to make a 10% profit on the job, how much should you charge the customer per board? Assume that the customer has \$1000 invested in each board before you get them for coating (and they are 90% yield when you get them).[6] The customer will reduce your payment by \$1000 for every good board that has one or more defects added to it by your process.

3.13 A semiconductor manufacturing facility has a yield that is controlled purely by random defects. The density of these random defects depends on the design rule used. More specifically, for a 1 μm design rule, the defect density is 0.5 defects/cm^2, while for a 0.5 μm design rule, the defect density is 2.0 defects/cm^2. (a) A die being fabricated has an area of 1 cm^2 and uses 1 μm design rules. Assume that the Poisson yield model is valid in each of the design rule regions on the die. Using the Poisson yield equation, estimate the yield of this die. (b) A die being fabricated has an area of 1 cm^2. 90% of this die area uses 1 μm design rules, while the rest uses 0.5 μm design rules. Using the Poisson yield equation estimate the yield of this die.

3.14 Assuming the number of particles of contamination on a wafer are distributed according to a Poisson distribution where there is a mean of 1.5 particles per square inch. Ignore the particle size. The process specification wafer state that there must be 12 or fewer particles in *each* of the six equal area sectors of the wafer. Assume a 6 inch diameter wafer with no flat edge ($F = 0$).

a) What is the expected yield from this process?

b) The manufacturer plans to migrate to an 8 inch diameter wafer (no flat edge). The same specification (12 or fewer particles in each of the six equal area sectors of the wafer) will be applied. What is the yield of the new wafers?

c) If we want to have a yield of 95% for the 8 inch diameter wafers, what should the mean number of particles per square inch be?

---

[6] You have no way of distinguishing the incoming good (non-defective) boards from the defective ones so you coat them all, but assume that the customer will be able to distinguish your defects from their original defects after you deliver the coated boards back to the customer.

3.15 You are using 200-mm diameter round wafers. You have been fabricating a particular 5 × 5 mm die and found that the yield of these die is 80%.

a) Using the simple Poisson model, find the defect density in the wafer.

b) Suppose that an alternative explanation of the observed 80% die yield is that some fraction of the wafer, f, is perfect and the rest of the wafer is totally dead (can never produce anything that is defect free). This would be called "perfect deterministic clustering of defects". What is f?

c) Let's consider a third explanation for the 80% observed die yield. In this case, assume that all the yield loss is due to a defect in one single structure on each die, i.e., only one thing can go wrong on each die and either it is non-defective or defective. In this case there is at most only one defect per die. This is not an unrealistic case for a MEMs fabrication, for example. What is the defect density that causes this case?

3.16 Why is the yield associated with Process C in Table 3.2 less than 0.5 rather than equal to exactly 0.5?

Chapter 4

Equipment/Facilities Cost of Ownership (COO)

Conventionally, equipment and facility purchase decisions have been based on initial purchase and installation costs. However, purchase costs do not consider the effect of equipment reliability and utilization, and the defects that equipment may introduce into products. Over the life of the production process, these factors may have a greater impact on cost of ownership than the initial purchase costs do. Cost of ownership (COO) is defined as the "total lifetime cost associated with acquisition, installation and operation of fabrication equipment" [Ref. 4.1]. SEMI E35 defines COO as the full cost of embedding, operating, and decommissioning, in a factory and laboratory environment, a system needed to accommodate a required volume [Ref. 4.2]. Cost of ownership relates the cost of acquiring and using a tool to the number of units produced over the life of the tool. Although "tool" traditionally refers to a single piece of production equipment, we can generalize "tool" to mean a specific machine, process, process step or facility.[1]

The concept of cost of ownership originated at Intel Corporation during an examination of the effective total cost of purchasing, operating, and maintaining, equipment for semiconductor device fabrication. COO matured and was introduced to the mainstream through SEMATECH in the 1990s [Ref. 4.3].

Cost of ownership is fundamentally different from process-flow-oriented cost modeling. In process-flow models, the actual path of a product through a fabrication or assembly process is emulated with an

[1] In the Part II introduction and Chapter 20, we will discuss a generalization of cost of ownership, as viewed by the customer, which will treat the complete cost of acquiring and using (and possibly disposing of) a product.

instance of a product accruing cost as it moves through a sequence of process steps. In a process-flow model, equipment and facilities costs are often lumped together into a single overhead rate, which in the case of traditional cost accounting is a multiplier of labor costs. In process-flow modeling a proportion of equipment costs can be charged to each instance (unit) of a product on a per-step basis. COO views the problem in a different way. In the COO approach, the sequence of process steps is not as important as the portion of the lifetime cost of a tool that is consumed by each specific instance of a product. Accumulating all of the fractional lifetime costs expended for all the equipment (i.e., tools) for one instance of a product provides an estimate of the cost of one instance of the product. In COO, the labor, materials and tooling costs are included within the lifetime cost of the particular piece of equipment (or tool).

Cost of ownership was originally developed for modeling integrated circuit fabrication costs. IC costs are dominated by equipment and facilities (labor, tooling and material contributions to the cost are small compared to the billion or more dollars required to construct and maintain an IC fabrication facility). The nature of COO makes it best suited for "equipment and facilities-centric products." Other types of electronic systems — for example, printed circuit board assembly, are far less dominated by equipment and facilities costs, and therefore are not as well suited for COO modeling.

4.1 The Cost of Ownership Algorithm

The fundamental cost-of-ownership algorithm is described by [Ref 4.4] as:

$$C_{ownership} = \frac{C_{fixed} + C_{variable} + C_{yield\ loss}}{(TPT)(Y)(U)} \tag{4.1}$$

where:

C_{fixed} = fixed cost: purchase, installation, etc.

$C_{variable}$ = variable cost: labor, material, utilities, overhead, etc.

$C_{yield\ loss}$ = cost due to yield loss: money invested into scrapped parts and production lost by producing defective parts.

TPT = Throughput.

Y = composite yield.

U = utilization: ratio of production time to total available time.

Equation (4.1) calculates a cost of ownership per instance of the product. The fixed costs include all the purchase, installation, and facilities costs (these costs are normally amortized over the lifetime of the tool). The variable costs are the costs incurred during normal tool operation, which include: material, labor, repair, utilities and applicable overhead costs. The throughput is defined by the time required to meet a process requirement or perform the required task. The composite yield is the operational yield of the tool, which includes breakage and processing errors caused by the tool. The utilization is the ratio of the production time to the total available time.

The yield-loss cost is the value of product that is lost due to operational losses and non-repairable defects caused by the tool. Yield models (Chapter 3) can be incorporated into COO models to estimate the yield loss caused by defects introduced by the tool.

COO models require information from many different sources. The Texas A&M Center of Excellence in Manufacturing Systems Research groups COO inputs for IC wafer processing into the following categories:

- equipment cost (fixed costs)
- annual operating cost (variable costs)
- process scrap yield
- die scrap yield
- downtime
- value of wafer at process step
- value of completed wafer.

In the above categories, process scrap yield (also known as mechanical throughput yield), is the operational yield of the tool, while die scrap yield is the defect-limited yield that is detected by wafer testing or probing (see Section 7.8.1). The downtime is the time that would not be used for production that is lost due to scheduled maintenance, calibration, standby, and repair.

4.2 Cost of Ownership Modeling

While Equation (4.1) is complete and captures the fundamental concepts of COO, actual implementation of a COO model is facilitated by dividing the contributions into capital, sustainment, and performance for each tool. In each of the following, the computed cost is the total cost per tool per unit time.

4.2.1 *Capital Costs*

Capital costs treat the costs to buy the machine, facilities, and/or process, how it depreciates, and what value it has at the end of the depreciation period. Assuming straight-line depreciation, the capital cost are given by

$$C_{cap} = \frac{P - R}{D_L} \tag{4.2}$$

where

P = the purchase price of the machine, facilities, and/or process and is assumed to include installation and any extra facilities needed to make it operational.

R = the residual value of the machine, facilities, and/or process at the end of the depreciation life.

D_L = the depreciation life.

4.2.2 *Sustainment Costs*

Sustainment costs treat all the costs required to keep the machine, facility and/or process operational. Both scheduled and unscheduled maintenance contribute to sustainment cost. The scheduled maintenance contribution (labor only) is given by

$$C_{sched\ maint} = N_{off} T_R L_R (1 + b) \tag{4.3}$$

where

N_{off} = the number of scheduled shutdowns for maintenance during off-production hours.

T_R = the time to perform scheduled maintenance activity (per scheduled maintenance instance).
L_R = the labor rate for maintenance activities.
b = the burden on the labor rate.

The unscheduled maintenance contribution (labor only) to sustainment cost is given by

$$C_{unsched\ maint} = N_{on}(MTTR)L_R(1+b) \tag{4.4}$$

where

N_{on} = the number of unscheduled shutdowns for maintenance during production hours = production time/MTBF, where *MTBF* is the mean time between failure for the machine, facility and/or process.
MTTR = the mean time to repair (per unscheduled maintenance instance).

Production time is the amount of time that production is taking place, e.g., hours or years. Note, as presented in Equations (4.3) and (4.4), $C_{sched\ maint}$ and $C_{unsched\ maint}$ only include the labor content; replacement parts and other materials may be included as well. In some cases all the maintenance costs may be subsumed by maintenance contracts, the cost of which can be substituted for $C_{sched\ maint}$ and/or $C_{unsched\ maint}$.

If unscheduled maintenance (or scheduled, for that matter) occurs during times when production would otherwise be occurring, the opportunity to produce profit-generating products is lost. The cost of the lost production is given by

$$C_{lp\text{-}maint} = V\frac{N_{on}\left(MTTR + T_{cool} + T_{start}\right)}{T_i} \tag{4.5}$$

where

T_{cool} = the time for the process (and/or the specific *tool*) to cool down before maintenance can begin.
T_{start} = the time for the process (and/or the specific *tool*) to warm up after the maintenance is completed.

T_i = the effective time interval between the completion of product instances by the process that the machine, facility or subprocess is associated with.[2]

V = the value of the product (profit that can be earned on one instance of the product).

4.2.3 *Performance Costs*

Performance costs measure the value (or lack thereof) of having the machine, facility or process included by accounting for change-overs, repairable and non-repairable defects, and the speed with which the process can produce products. The cost associated with change-overs is

$$C_{change-overs} = N_{co} T_{co} L_R (1+b) \tag{4.6}$$

where

N_{co} = the number of change-overs during production hours.

T_{co} = the time to perform a change-over (per change-over instance).

As with unscheduled maintenance, if change-overs occur during times when production would otherwise be occurring, the opportunity to produce profitable products is lost. The cost of the lost production is given by

$$C_{lp\text{ - }co} = V \frac{N_{co}\left(T_{co} + T_{cool} + T_{start}\right)}{T_i} \tag{4.7}$$

Also contributing to performance costs are repairable and non-repairable defects introduced by the machine, facility and/or process. The repairable defect cost is given by

$$C_{repairable\ defects} = D_r C_D \left(Production\ \ time\right) \tag{4.8}$$

where

D_r = the rate at which repairable defects are produced.

C_D = the cost of repairing one defect.

---

[2] This time could be characterized as the mean inter-arrival time to a process step after the end of the process flow of interest — that is, it is the average time between consecutive arrivals of product instances at the end of the process.

Non-repairable defects result in two cost contributions. First, the money spent on the product up to the point where it is scrapped must be included:

$$C_{scrap} = D_{nr} I (Production \;\; time) \tag{4.9}$$

where

D_{nr} = the rate at which non-repairable defects are produced.

I = the investment in the product up to the scrap point (i.e., how much has been spent on one product instance).

Second, non-repairable defects result in the loss of production time that could have been used to make product instances that could have been sold for a profit. The cost of the lost production is given by

$$C_{lp-s} = D_{nr} V (Production \;\; time) \tag{4.10}$$

The last performance cost is associated with effects on the number of units that can be produced by the process. The production-penalty cost is applicable to situations comparing alternate equipment or subprocesses. The penalty computes the effective cost of process time impacts due to the equipment or subprocess of interest:

$$C_{production\, penalty} = \left[\left(\frac{Production\, time}{T_i} \right)_{without} - \left(\frac{Production\, time}{T_i} \right)_{with} \right] V \tag{4.11}$$

The first term in Equation (4.11) is the number of product units made per year without the equipment or subprocess of interest in the overall process; the second term is the number of product instances made per year with the equipment or subprocess of interest in the overall process. If the rate at which the process can produce finished product instances is the same with and without the equipment or subprocess of interest, then there is no effective production penalty.

4.3 Using COO to Compare Two Machines

In this section cost-of-ownership is used to compare two pieces of manufacturing equipment. In this example, the objective is to determine which of the two machines should be purchased. The operational inputs governing the use of the chosen machine are given in Table 4.1.

Table 4.1. Operational Inputs.

Production hours per week	168
Production weeks per year	51
Hourly labor rate for maintenance (L_R)	\$20
Labor burden (b)	0.5
Estimated cost of repairing one defect caused by the machine (C_D)	\$20
Value of the product produced on the line (profit/product) (V)	\$25
Investment in the product prior to encountering this machine (I)	\$5.20

The capital cost inputs and computed per-week effective capital cost of each machine are shown in Table 4.2. The value in the last line in Table 4.2 for Machine *B* is computed using Equation (4.2):

$$C_{cap} = \frac{\$75{,}000 - \$10{,}000}{5}\frac{1}{51} = \$255/\text{week} \tag{4.12}$$

The quantity 1/51 appears in Equation (4.12) to convert the final value to cost per week.

Table 4.2. Capital Cost Inputs and Outputs.

	Machine *A*	Machine *B*
Capital cost of the machine (P)	\$70,000	\$75,000
Depreciation life (years) (D_L)	5	5
Residual sale (salvage) value of the machine (R)	\$10,000	\$10,000
Per-week capital cost (C_{cap})	**\$235**	**\$255**

The sustainment cost inputs and computed per-week effective sustainment cost of each machine are shown in Table 4.3. The values in the seventh, eighth and ninth rows in Table 4.3 for Machine *B* are computed using Equations (4.3) through (4.5):

$$C_{sched\ maint} = (4)(4)(\$20)(1+0.5) = \$480/\text{year} \tag{4.13}$$

$$C_{unsched\ maint} = \frac{(168)(51)}{2000}(0.5)(\$20)(1+0.5) = \$64/\text{year} \tag{4.14}$$

$$C_{lp\text{-}maint} = \$25\frac{\frac{(168)(51)}{2000}(0.5+1.5+1.5)}{110/60/60} = \$12{,}268/\text{year} \tag{4.15}$$

The product (168)(51) appearing in Equation (4.14) gives the production hours per year. The values computed by Equations (4.13) and (4.14) only account for labor costs. Finally, these three equations are used to determine the total sustainment cost for Machine *B*:

$$\text{Sustainment cost} = (\$480 + \$64 + \$12{,}268)\frac{1}{51} = \$251/\text{week} \qquad (4.16)$$

quantity 1/51 appears in Equation (4.16) to convert the final value to cost per week.

Table 4.3. Sustainment Cost Inputs and Outputs.

	Machine *A*	Machine *B*
Cool-down and start-up time (hours) (T_{cool} and T_{start})	2	1.5
Times per year the machine is down (scheduled maintenance, off production) (N_{off})	4	4
Hours of maintenance per scheduled down time (T_R)	4	4
Machine MTBF (hours)	2000	2000
Machine MTTR (hours)	0.5	0.5
Time interval between the completion of product instances including this machine (sec) $(T_i)_{\text{with}}$	120	110
Scheduled maintenance costs per year ($C_{sched\ maint}$)	$480	$480
Unscheduled maintenance and repair costs per year ($C_{unsched\ maint}$)	$64	$64
Lost production opportunity cost per year ($C_{lp\text{-}maint}$)	$14,459	$12,268
Per-week sustainment cost	**$294**	**$251**

The performance cost inputs and computed per-week effective sustainment cost of each machine are shown in Table 4.4. The values in the seventh through twelfth rows in Table 4.4 for Machine *B* are computed using Equations (4.6) through (4.11):

$$C_{change-overs} = (5)(51)\left(\frac{10}{60}\right)(\$20)(1 + 0.5) = \$1{,}275 \qquad (4.17)$$

$$C_{lp\text{ - }co} = \$25\frac{(5)(51)(10/60)}{110/60/60} = \$34{,}773 \qquad (4.18)$$

$$C_{repairable\ defects} = (0.5)(\$20)(168)(51) = \$85{,}680 \tag{4.19}$$

$$C_{scrap} = (1)(\$5.20)(51) = \$265 \tag{4.20}$$

$$C_{lp-s} = (1)(\$25)(51) = \$1{,}275 \tag{4.21}$$

$$C_{production\ penalty} = \left[\left(\frac{(168)(51)}{100/60/60}\right)_{\text{without}} - \left(\frac{(168)(51)}{110/60/60}\right)_{\text{with}}\right](\$25) = \$701{,}018 \tag{4.22}$$

Table 4.4. Performance Cost Inputs and Outputs.

	Machine A	Machine B
Change-over time (min) (T_{CO})	10	10
Change-overs per week (N_{CO})	5	5
Time interval between the completion of product instances excluding this machine (sec) $(T_i)_{\text{without}}$	100	100
Repairable defects produced by this machine per hour (D_r)	0.5	0.5
Number of assemblies per week scrapped due to defects caused by this machine (D_{nr})	1	1
Monthly consumable cost	\$4,834	\$3,427
Change-over costs per year (labor) ($C_{change\text{-}overs}$)	\$1,275	\$1,275
Lost production due to change-overs per year ($C_{lp\text{-}co}$)	\$31,875	\$34,773
Repairable defect costs per year ($C_{repairable\ defects}$)	\$85,680	\$85,680
Scrap costs per year (C_{scrap})	\$265	\$265
Lost production due to scrapped product per year ($C_{lp\text{-}s}$)	\$1,275	\$1,275
Production penalty per year ($C_{production\ penalty}$)	\$1,285,200	\$701,018
Per-week performance cost	**\$28,698**	**\$16,969**

Equation (4.18) assumes that the change-over can occur without incurring start-up or cool-down times (a "hot" change-over). Finally, Equations (4.17) through (4.22) are used to determine the total performance cost for Machine B:

$$\text{Performance cost} = \left[\begin{array}{c}\$1{,}275 + \$34{,}773 + \$85{,}680 + \$265 + \\ \$1{,}275 + \$701{,}018 + (\$3{,}427)(12)\end{array}\right]\frac{1}{51} = \$16{,}969/\text{week} \tag{4.23}$$

where \$3,427 is the monthly consumables cost and the value in Equation (4.23) is divided by 51 to convert the final value to cost per week.

The total cost of ownership per week of the machines is the sum of the last lines in Tables 4.2-4.4: $C_{ownership\ A}$ = \$29,227 and $C_{ownership\ B}$ = \$17,475. The results of this example demonstrate that even though Machine *B* was more expensive to purchase than Machine *A*, its cost of ownership is significantly less than that of Machine *A*'s.

4.4 Estimating Product Costs

The COO example considered in Section 4.3 compared two pieces of equipment. Ideally, to estimate a product's cost using COO, the fractional lifetime costs of all the equipment (tools) that an instance of a product encounters can be accumulated to estimate the cost of one instance of the product. This approach would be appropriate if the materials and recurring labor content in the product were negligible compared to the equipment and facilities contributions. However, in practice both the materials and recurring labor content have to be included within the lifetime cost of the equipment, or a hybrid model should be used that includes a COO treatment of the equipment and facilities costs and a treatment of materials and labor costs via a process flow or another approach.

Consider the inclusion of COO within the process-flow example provided in Section 2.3.2. Instead of using Equation (2.4) to compute the capital cost (C_C) associated with a process step, a COO model could be used. Consider Step *D* in Table 2.1. For this step, the equipment cost is \$75,000 and the computed effective capital cost per wafer, found from Equation (2.4) in Table 2.3 is \$0.09. The C_C in Table 2.3 is calculated as:

$$C_C = \frac{\$75{,}000}{5}\left[\frac{110/60/60}{(1)(0.6 \cdot 8760)}\right] = \$0.0872 \tag{4.24}$$

where C_e = \$75,000, D_L = 5 years, T = 110 sec/wafer, N_p = 1 and T_{op} = (0.6)(8760).

For illustration purposes, consider the piece of equipment in Step *D* to be Machine *B*, as discussed in Section 4.3. All the data for Machine *B* is consistent with the original assumptions about the equipment in Step *D* of the Section 2.3.2 example. The step time of 110 sec per wafer (capacity of

$N_p = 1$ wafer at a time) corresponds to the time interval between completed wafers of the process that includes Machine *B*. We will assume that the number of wafers that could be completed per week by the process that uses Machine *B* is (7)(24)(60)(60)/110 = 5498, resulting in an effective cost per wafer for Machine *B* of $16,966/5498 = $3.09, which is considerably larger than the effective capital cost in Step *D* of Section 2.3.2 given in Equation (4.24). The example in Section 2.3.2 could account for some of this discrepancy through the labor burden rate--that is, the maintenance of the equipment and facilities may be part of this overhead.[3] The example in Section 2.3.2 also includes a machine utilization of 0.6 that infers that the machine is non-operational 40% of the time (possibly down for maintenance). However, the calculation in Equation (4.24) does not account in any way for the lost production opportunities due to machine downtime or additional processing time created by the machine, which represent the majority of the effective cost of ownership of the machine.

References

4.1 Semiconductor Industry Association (1994). The National Technology Roadmap for Semiconductors, San Jose, CA, p. C-3.

4.2 SEMI (1995). E35: Cost of Ownership for Semiconductor Manufacturing Equipment Metrics, *Book of SEMI Standards*, Mt. View, CA.

4.3 LaFrance, R. L. and Westrate, S. B. (1993). Cost of ownership: The suppliers view, *Solid State Technology*, pp. 33-37.

4.4 Dance, D. and Jimenez, D. W. (1994). Applications of cost of ownership, *Semiconductor International*, pp. 6-7, September.

4.5 Sandborn, P. (2003). The economics of embedded passives, *Integrated Passive Component Technology*, Ulrich R. and Schaper L. editors, (Wiley-IEEE Press, Hoboken, NJ).

---

[3] The incorporation of various non-labor cost elements — for example, equipment and facilities maintenance — into a burden rate on the labor content associated with manufacturing a product is potentially problematic for products that are not labor-cost-dominated. This leads to inaccuracies in the allocation of overhead charges. Chapter 5 provides an introduction to activity-based costing, which is a methodology that attempts to accurately allocate overhead charges to products.

Bibliography

In addition to the sources referenced in this chapter, there are several good sources of information on equipment and facilities cost of ownership, including:

Dance, D. L. (1996). Modeling the cost of ownership of assembly and inspection, *IEEE Transactions on Components, Packaging, and Manufacturing Technology – Part C*, 19(1), pp. 57-60.

Nanez, R. and Iturralde, A. (1995). Development of cost of ownership modeling at a semiconductor production facility, *Proc. IEEE/SEMI Advanced Semiconductor Manufacturing Conference*, pp. 170-173.

Dance, D. and Jimenez, D. (2004). Lithography cost of ownership: revisited, *Semiconductor International*.

A bibliography of COO modeling literature can be found at: http://www.wwk.com/cost.html. Accessed April 28, 2016.

Problems

4.1 Rework the example in Section 4.3, assuming that change-overs require the machines considered in the example to be completely shut down and warmed back up--that is, include the cool-down and warm-up times.

4.2 In the example in Section 4.3, suppose you have the option of purchasing a Machine *C* that has a time interval between the completion of product instances of 108 seconds. How much more would you be willing to pay for Machine *C* than Machine *A*? All other properties of Machine *C* are identical to Machine *A*.

4.3 You are considering buying one of the following two machines for your printed wiring board fabrication facility. The use of the two machines is characterized by the data in Table 4.1 and the following:

Depreciation life (years)	5
Time interval between the completion of product instances without the machine (sec)	250

	Machine *A*	Machine *B*
Capital cost of the machine	$90,000	$75,000
Residual sale value of the machine	$12,000	$10,000
Time interval between the completion of product instances including the machine (sec)	252	251
Change over time (min)	10	8
Change overs per week	5	5

a) What are the capital costs (in $/week) for each machine?
b) What is the production-time penalty (in $/week) for each machine?
c) What is the cost of lost production (in $/week) due to change-overs for each machine?

4.4 Resistors can be fabricated inside of printed circuit boards; these are called embedded resistors [Ref. 4.5]. They are fabricated by printing or plating resistive materials on inner-layer pairs of the board. When the resistors are laid out on the inner layers they are sized to have lower resistance than required by the design. After the layer pair is fabricated, the resistors are trimmed to bring their resistance up to the required design value. You must purchase one of the following laser trimming machines. Using a cost-of-ownership model, which one is most cost effective?

Property	Laser trimmer #1	Laser trimmer #2
Capital cost	$200,000	$350,000
Operating cost	$2,000/year	$1,500/year
MTBF	300 hours	250 hours
MTTR	1.5 hours	2 hours
Warm-up time (min)	15	15
Cool-down time (min)	30	30
Average time per non-trimmed resistor (seconds)	0.03	0.03
Average time per trimmed resistor (seconds)	0.05	0.045
Depreciation life (years)	5	5
Residual value of the machine	$25,000	$35,000
Scheduled maintenance events/year	4	4
Hours to perform scheduled maintenance (per event)	4	4
Monthly consumable cost	$1000	$1000
Trimming defects (ppm)	37	40

ppm = parts per million (1 ppm = 1 defect in 1,000,000 tries).

In addition to the above information, assume that

- there are no change-overs.
- there are no repairable defects produced by either machine.
- the time interval between the completion of product instances time excluding trimming = 300 seconds/layer pair.
- 80 production hours per week.
- 50 production weeks per year.
- $30/hour labor rate for all maintenance.
- the burden rate (b) = 0.

- the effective value (profit) associated with one embedded resistor layer pair panel = $100.
- 97.7% of the fabricated resistors require trimming.
- 500 embedded resistors are on a board.
- 18 boards can be fabricated per layer pair panel.
- $500 has been invested in layer pairs prior to the trimming process.
- all trimming defects result in unusable board layer pairs (no rework is possible).

Layer pairs and panels are synonymous in this problem. Express your final numbers as cost of ownership per <u>week</u>.

Chapter 5

Activity-Based Costing (ABC)

Overhead costs are the portion of the costs of a product that cannot be clearly associated with particular operations, products, or projects and must be prorated among all the products made by an organization. Overhead costs include labor costs for persons who are not directly involved with a specific manufacturing process, such as managers and office workers; non-recurring costs necessary to design, test, and support products; facilities costs, such as utilities and mortgage payments on buildings; non-cash benefits provided to employees, such as health insurance, retirement contributions, and unemployment insurance; and other costs of running the business, such as accounting, taxes, furnishings, insurance, sick leave, and paid vacations. In traditional cost accounting, indirect or overhead costs are allocated to products and process steps based on their direct cost content — for example, via a labor burden rate that is a multiplier on labor costs (see Section 1.4).

Manufacturing organizations found that the traditional cost accounting treatment of overhead costs (allocation based on direct cost content) became increasingly inaccurate as the percentage of the overhead costs that made up a product's total cost rose. They found that it was not easy to correctly allocate overhead to products because while the same processes, equipment and facilities were used by multiple products, the overhead costs were not equally consumed by all the products. In one case a product might occupy more time on an expensive piece of equipment than another product, however, if the direct costs (labor and materials) are the same for both products the same overhead is allocated to both products, i.e., the additional cost for the use of the expensive piece of equipment is not taken into account when the direct costs are added to the products. As a consequence, when multiple products share common processes, equipment and facilities, there is a danger of one product effectively subsidizing other products.

In the early 1960s, General Electric's finance and accounting people noted that overhead costs are often the result of decisions that are made long before the costs are actually incurred [Ref. 5.1]. For example, engineering change orders (ECOs) can result in changes in the quantity of parts ordered, multiple machine change-overs, additional tooling costs, and part inventory cost changes. But traditional cost accounting mechanisms may not allow the cost ramifications of the ECOs to be communicated back to the engineering organization. GE's original work in this area forms the basis for "activity-based management" and activity-based costing.

In the early 1970s Staubus established a formal activity accounting system with guidelines on principles and practices [Ref. 5.2]. During the 1970s and 1980s, the Consortium for Advanced Management — International formalized the principles that have become known as activity-based costing (ABC) [Ref. 5.3]. Activity-based costing was first clearly defined in 1987 by Kaplan and Bruns [Ref. 5.4], who focused on the manufacturing industry.

5.1 The Activity-Based Cost Modeling Concept

While it is simple to accurately assign the direct labor and materials costs to products, it is more difficult to accurately allocate common resource costs to products. Any time multiple products share common resource costs, there is a danger of one product effectively subsidizing another — that is, one product is allocated too little of the common cost, and others are overburdened with too much of the common cost.

Activity-based costing is a method of assigning an organization's resource costs to the products and services it provides to its customers. In traditional cost accounting, overhead costs are most often allocated to products in proportion to labor hours and material costs (direct costs). In ABC, distinct activities associated with the manufacture of a product are identified and the primary cost drivers behind each of the activities are found. Once activities and their associated cost drivers are identified, an *activity rate* (in units of $/activity) is determined. If the number of times a particular product performs a particular activity is known, then the activity rate can be used to allocate costs associated with that activity to the product. The sum of all the costs associated with each activity is the overhead cost of the product.

The advantage of ABC models over other approaches is that they more accurately allocate overhead costs to products. Instead of using the direct cost as a basis to allocate common resource costs, ABC seeks to identify the actual cause-and-effect relationships and use it to assign costs. Once the costs of all the activities have been determined, the cost of each activity is attributed to each product based on the amount of the activity used by the product.

5.1.1 *Applicability of ABC to Cost Modeling*

Most frequently, activity-based costing is used as an accounting tool from an *ex post facto* (after the fact) perspective to assign known overhead costs from a previous period of time to processes and products. While ABC clearly has the potential to improve the accuracy of product cost estimates, it has been argued that ABC may not be appropriate for cost modeling because it is an accounting system designed primarily for external financial reporting.

So, what is ABC's applicability to cost modeling — that is, forecasting the costs of a product before it is manufactured? ABC can be used as a component of cost modeling when historical accounting data (tracking the costs associated with various activities over time) is available to calculate the activity rates and when those rates have predictive validity for future products. Like cost of ownership (Chapter 4), ABC is less likely to be used as an exclusive modeling approach, and more likely to be combined with other modeling approaches such as COO and process-flow modeling to form useful cost models for real products.

5.2 Formulation of Activity-Based Cost Models

This section develops the formulations necessary to perform activity-based cost modeling. However, first it is helpful to briefly review how traditional cost accounting handles overhead costs.

5.2.1 *Traditional Cost Accounting (TCA)*

In traditional cost accounting (TCA), the total cost of a product instance is the sum of the direct costs (labor and materials per product instance) and the indirect costs (overhead). The indirect or overhead costs are all the costs that are not directly identifiable with a single type of product, such as equipment, facilities, insurance, management, marketing, sales, and so on. Tooling costs can appear as either a direct or indirect cost. The overhead cost is computed for each product instance as a proportion of the direct costs, possibly as a "burden rate" on the labor or the sum of the labor and material costs. This assumes that overhead is directly related to the labor and material cost. Traditional cost accounting focuses on what it costs to *do* something — for example, drilling a through-hole in a printed circuit board; in addition, activity-based costing also accounts for the cost of *not doing* something, such as the cost of waiting for a required part. "Activity-based costing records the costs that traditional cost accounting does not do" [Ref. 5.5].

5.2.2 *Activity-Based Costing*

Traditional accounting systems allocate costs inaccurately. ABC doesn't eliminate or change any costs relative to traditional cost accounting, it simply determines more accurately how the costs are actually consumed. In order to correctly associate costs with products and services, ABC assigns cost to activities based on their use of resources.

The basic premises of ABC are the following:

(1) It focuses on indirect costs (overhead).
(2) Cost objects consume activities.
(3) Activities consume resources.
(4) The consumption of resources is what drives costs.

Understanding the relationship articulated in these basic premises is critical to successfully costing and managing product overhead. In contrast, in traditional cost accounting, costs are assumed to be consumed by products rather than activities.

The first step in ABC is to identify activities. Activities are all the actions performed by people and machines to design, manufacture and support a product. Next, the cost driver(s) associated with each activity must be identified. Activities use transactional drivers, such as the number of holes, number of layers, and so on, as opposed to labor hours, material cost, or machine hours. A cost driver is any factor that causes a change in the cost of an activity — cost drivers are the root cause of the work done in an activity. ABC assigns costs to cost objects based on their use or consumption of activities.

Once activities and their associated cost drivers are identified, an *activity rate,* A_R, (the units of A_R are \$/activity) is determined using

$$A_R = \frac{\text{Activity cost pool}}{\text{Activity base}} \tag{5.1}$$

where the *activity cost pool* is the total amount of overhead required by the activity (for all products) during some period of time. Cost pools are groups of individual costs. The *activity base* is the number of times the activity was performed on all products during the period of time.

The total cost of the ith activity for a product is determined from

$$C_{A_i} = A_{R_i} N_{A_i} \tag{5.2}$$

where N_{A_i} is the number of times the activity must be performed to manufacture a product. Equation (5.2) is the overhead allocated to the product by activity i. The sum of C_{A_i} over all activities associated with the product gives the total overhead cost of the product.

The overhead allocation to each instance of the product is given by

$$\text{Overhead allocation} = \frac{1}{N_{tp}} \sum_{i=1}^{\text{all activities}} C_{A_i} \tag{5.3}$$

where N_{tp} is the total number of instances (units) of the product manufactured.

The total cost of a product (per unit) is given by

$$\text{Total cost/unit} = \text{Overhead allocation} + C_L + C_M \tag{5.4}$$

where C_L is the labor cost per unit and C_M is the material cost per unit.

5.3 Activity-Based Cost Model Example

Consider the case shown in Table 5.1. Products *A* and *B* require different amounts of labor and different quantities of each product are produced. The assumed labor rate applicable to both products is L_R = \$20/hour and the total overhead to produce both products is \$100,000. Which product (*A* or *B*) is less expensive to produce?

Table 5.1. Product Comparison.

	Product *A*	Product *B*
Labor content (hours/unit)	1	2
Direct labor cost (\$/unit) (C_L)	\$20	\$40
Quantity required (N_{tp})	100	950

The direct labor cost in Table 5.1 is the product of the labor content and the labor rate.

The traditional cost accounting treatment of the products in Table 5.1 (assuming $C_M = 0$) is given in Table 5.2.

Table 5.2. Traditional Cost Accounting (TCA) Treatment of Products *A* and *B*.

	Product *A*	Product *B*
Overhead Allocation (\$/unit)	\$50	\$100
TCA Total (\$/unit)	**\$70**	**\$140**

The overhead allocation in Table 5.2 for Product *B* is computed using

$$C_{OH} = U_L T \frac{\text{Total Overhead}}{\text{Total Labor Hours}} = (2)\frac{\$100{,}000}{(1)(100)+(2)(950)} = \$100 \qquad (5.5)$$

where $U_L T$ is the number of labor hours per unit. The TCA total is the sum of the direct labor cost and the overhead allocation. Using the resulting TCA total from Table 5.2, the total TCA expenditure for both products is (100)(\$70)+(950)(\$140) = \$140,000.

Now let's calculate the costs of the two products using ABC. The total expenditure for both products using ABC will be the same as for TCA (\$140,000); ABC does not change the total expenditure, only how the costs are allocated among products. To perform ABC we need to identify the activities and their drivers, as in Table 5.3.

Table 5.3. ABC Activities and Drivers.

Activity	Cost ($)	Cost Driver	Product *A* (N_A)	Product *B* (N_A)	Activity Rate ($/cost driver item) ($A_R$)
Design and prototype	$30,000	Engineering hours	500	500	$30
Programming, setup and tooling	$10,000	Number of setups	1	3	$2,500
Fabrication	$40,000	Fabrication hours	100	1900	$20
Receiving	$10,000	Number of receipts	1	3	$2,500
Packing and shipping	$10,000	Number of customers	1	3	$2,500

The second column in Table 5.3 (cost) is the activity cost pool — the column sums to $100,000, the total overhead for both products. The third column is the cost driver associated with each particular activity. Activity usage quantities (N_A) are provided in the fourth and fifth columns — this is data collected or estimated for the specific products. For example, the activity rate is computed for the last activity ($i = 5$) using Equation (5.1):

$$A_{R_5} = \frac{\$10{,}000}{(1+3)} = \$2{,}500\,/\,\text{customer} \tag{5.6}$$

The ABC product costs are computed as shown in Table 5.4.

Table 5.4. ABC Product Costs.

	Product *A*	Product *B*
Design and prototype	$15,000	$15,000
Programming, setup and tooling	$2,500	$7,500
Fabrication	$2,000	$38,000
Receiving	$2,500	$7,500
Packing and shipping	$2,500	$7,500
Activity total ($)	**$24,500**	**$75,500**
Overhead allocation ($/unit)	**$245**	**$79.47**
ABC total ($/unit)	**$265**	**$119.47**

The costs in the first five rows of Table 5.4 are activity costs associated with each of the products, which are computed using Equation (5.2). For example, the activity cost associated with the fabrication step (the $i = 3$ activity) for Product B is given by

$$C_{A_3} = A_{R_3} N_{A_3} = (\$20)(1900) = \$38{,}000 \tag{5.7}$$

The overhead allocation for Product B is calculated using Equation (5.3):

Overhead allocation

$$= \frac{1}{N_{tp}} \sum_{i=1}^{5} C_{A_i}$$

$$= \frac{1}{950}(15{,}000 + 7{,}500 + 38{,}000 + 7{,}500 + 7{,}500) = \$79.47 \tag{5.8}$$

Finally, the total cost per unit is found for Product B using Equation (5.4):

$$\text{Total cost/unit} = \text{Overhead allocation} + C_L + C_M$$

$$= \$79.47 + \$40 = \$119.47 \tag{5.9}$$

For the example in the section, $C_M = 0$.

Using the resulting ABC total from Table 5.4, the total ABC expenditure for both products is (100)(\$265)+(950)(\$119.47) = \$140,000, which is the same total expenditure as found using the traditional cost accounting method. However, obviously, the results in Tables 5.2 and 5.4 show that the effective costs per unit are vastly different. If the manufacturing of Product A had been quoted to a customer for \$70/unit, as implied by TCA, significant money would have been lost, since its actual cost was \$265/unit.

5.4 Time-Driven Activity-Based Costing (TDABC)

Transaction drivers that are used to count the frequency of an activity or the number of times an activity is performed is only one way to address the problem. The problem can also be approached using "duration drivers" that represent the time required to perform an activity.

Duration drivers typically provide greater accuracy than transaction drivers when the time required per transaction is not the same for all products. The tradeoff is that duration drivers are generally more expensive to measure than transaction drivers.

Duration drivers measure the time it takes to perform an activity. The *capacity cost rate,* CC_R, (the units of CC_R are \$/unit time) is the "cost per time unit of capacity" determined using,

$$CC_R = \frac{\text{Activity cost pool}}{\text{Activity base time}} \tag{5.10}$$

where the *activity cost pool* is the total amount of cost or overhead[1] required by the activity for all products during some period of time. The *activity base time* in Equation (5.10) is the total time for the activity for all products during the specified time period.

Consider a simple example. Ten employees perform a set of tasks. The total annual cost of the ten employees is \$800,000. Each of the ten employees works 240 days per year and 8 hours per day. Deducting the time for breaks, training, etc., gives 375 minutes per day or 90,000 minutes of productive work per employee per year.[2]

The capacity cost rate is,

$$CC_R = \frac{800{,}000}{(10)(90{,}000)} = \$0.8889\,/\,\text{minute} \tag{5.11}$$

As an example consider the example provided in Table 5.5. In this example, the ten employees described above, perform three activates.

Table 5.5. ABC Analysis, Example Activities and Drivers.

Activity	Estimated Fraction of Total Time	Activity Cost Pool (\$/year)	Activity Cost Driver	Activity Base	Activity Rate (\$/cost driver item) (A_R)
Setups	0.65	\$520,000	Number of setups	400	\$1300
Receiving	0.15	\$120,000	Number of receipts	1300	\$92.31
Packing and shipping	0.2	\$160,000	Number of customers	2250	\$71.11

---

[1] We don't have to just use ABC for the overhead costs, it can be used to model all costs as is the case in the example in this section.

[2] In this case (240)(8)(60) = 115,200 minutes would be the theoretical capacity per year. 90,000 minutes is called the "practical capacity".

In Table 5.5, the Activity Cost Pool is the Estimated Fraction of the Total Time multiplied by the total annual cost ($800,000); the activity rates are calculated using Equation (5.1).

The data in Table 5.5 can also be approached using TDABC. In this case instead of determining the activity cost pool, we determine the actual unit time for each activity (i.e., the measured average time per unit). Table 5.6 shows the actual unit times; the total time for the activities is the product of the actual unit time and the activity base (in Table 5.5). The unit cost is CC_R calculated in Equation (5.11) multiplied by the actual unit times and the total cost is the product of the unit cost and the activity base in Table 5.5.

Table 5.6. TDABC Analysis.

Activity	Actual Unit Time (min)	Total Activity Time (min)	Unit Cost ($/unit)	Total Cost
Setups	1492	596,800	$1326.22	$530,489
Receiving	95	123,500	$84.44	$109,778
Packing and shipping	69	155,250	$61.33	$138,000
Total		875,550		

To understand the difference between ABC and TDABC, first observe that the analysis in Table 5.6 did not use either the estimated fraction of time per activity or the money spent on each activity (columns 2 and 3 in Table 5.5), rather it uses the actual unit times (column 2 in Table 5.6). The productive time can be calculated using,

$$\text{Productive time} = \frac{\sum_{i=1}^{\text{all activities}} \text{Total Activity Time}_i}{\text{Practical Capacity}} = \frac{875,550}{90,000} = 0.973 \quad (5.12)$$

where the numerator is the sum of column 3 in Table 5.6 and the 90,000 in the denominator is the practical capacity (from footnote 2). Equation (5.12) indicates that 97.3% of the practical capacity was actually used and as a result 97.3% of the total cost ($800,000) was allocated to customers. Also compare the ABC costs (column 3 in Table 5.4) to the TDABC costs (last column in Table 5.5). ABC bases its estimation of costs on its assumed distribution of effort, whereas TDABC uses the actual productive effort.

5.5 Summary and Discussion

The ABC example considered in Section 5.3 compared two existing products. How do we forecast costs for a new product using ABC? If activity rates corresponding to the various activities involved in a new product's manufacture can be determined from accounting data for previous products, then ABC can be used to establish the proper allocation of overhead costs for the new product.

The advantage of ABC models over other approaches is that they more accurately allocate overhead costs to products. The disadvantage is that historical accounting data (tracking total costs associated with various activities over time) is required to calculate the activity rates. ABC is relatively simple to perform once the information is obtained and it focuses attention on the causes (drivers) of costs. The criticisms of ABC are that one cost driver may not explain the behavior of all items in a cost pool and cost drivers might be difficult to identify. ABC is most appropriate when production overheads are high relative to direct costs and when there is a wide range of products, each of which uses different resources.

Like COO (Chapter 4), accounting for the sequence of activities — that is, the order in which the activities occur — is not straightforward using ABC. The difficulty is that the activity rate associated with an activity could depend on the order in which the activities occur. This could, of course, be resolved by defining multiple versions of an activity that depend on their location in the process flow; however, the possible sequences will most likely be limited to those that are accommodated by the activity set, resulting in a less general model.

References

5.1 Latshaw, C. A. and Cortese-Danile, T. M. (2002). Activity-based costing: usage and pitfalls, *Review of Business*, January 1. https://www.highbeam.com/doc/1G1-90192832.html. Accessed April 22, 2016.

5.2 Staubus, G. J. (1971). *Activity Costing and Input-Output Accounting*, (Richard D. Irwin, Inc., Homewood, IL).

5.3 Consortium for Advanced Manufacturing–International (CAM-I), http://www.cam-i.org/. Accessed April 22, 2016.

5.4 Kaplan, R. S. and Bruns, W., eds. (1987). *Accounting and Management: A Field Study Perspective*, (Harvard Business School Press, Boston, MA).
5.5 Drucker, P. F. (1999). *Management Challenges of the 21st Century*, (HarperCollins Publishers, New York, NY).

Bibliography

In addition to the sources referenced in this chapter, there are many books and other good sources of information on activity-based costing, including:

Emblemsvåg, J. (2003). *Life-Cycle Costing: Using Activity-Based Costing and Monte Carlo Methods to Manage Future Costs and Risks*, (John Wiley & Sons, Inc., Hoboken, NJ).

Kaplan, R. S. and Anderson, S. R. (2007). *Time-Driven Activity-Based Costing: A Simpler and More Powerful Path to Higher Profits*, (Harvard Business School Press, Boston, MA).

Lewis, R. J. (1993). *Activity-Based Costing for Marketing and Manufacturing*, (Quorum Books, Westport, CN).

Maher, M. W. (2005). Activity-based Costing and Management, *Handbook of Cost Management*, 2nd Edition, Weil, R. L. and Maher, M. W. eds., (John Wiley & Sons, Inc., Hoboken, NJ), pp. 217-241.

Van der Merwe, A. (2009). Debating the principles: ABC and its dominant principle of work, *Journal of Cost Management*, 23(5), pp. 1-9.

Problems

5.1 Define a "transactional driver."

5.2 What value of b (burden rate) does the example in Table 5.2 correspond to?

5.3 For the products described below, fill in the missing numbers in all the boxes.

Inputs	Product *A*	Product *B*	Product *C*	**Assumptions**	
Labor Content (hours/unit)	3.5	7		Labor Rate ($/hour)	35
Direct Labor Cost ($/unit)			52.548	Total Overhead ($)	
Quantity Required	1000	600		Total Labor Hours to Produce All Products	
				Total Overhead ($/labor hour)	22.018

Traditional Cost Accounting (TCA)	Product *A*	Product *B*	Product *C*
Overhead Allocation ($/unit)			
TCA Total ($/unit)			

Activity Based Cost (ABC)	Product *A*	Product *B*	Product *C*	
Design and Prototype	13333.333	22222.222		Activity Costs
Programming, Setup and Tooling	3000	9000		
Fabrication	2472.799	46983.185		
Receiving	3000	9000		
Packing and Shipping	2000	6000		
Activity Total ($)				
Overhead Allocation ($/unit)				
ABC Total ($/unit)				

Activity	**Cost ($)**	**Cost Driver**	**Product *A***	**Product *B***	**Product *C***	**Activity Rate**
Design and Prototype	90000	Engineering Hours	300	500		44.444
Programming, Setup and Tooling	15000	Number of Setups	1	3		3000
Fabrication	50000	Fabrication Hours	100	1900		24.728
Receiving	15000	Number of Receipts	1	3		3000
Packing and Shipping	16000	Number of Customers	1	3		2000

5.4 Based on the solution to Problem 5.3, if all these products were quoted to the customer based on the TCA estimation, which one would you make the largest profit on (in absolute dollars)?

5.5 Start with the ABC example in Section 5.3. For Product *A*, assume that the following activities are a function of quantity:

$$\text{Number of setups} = \left\lceil \frac{Quantity}{1000} \right\rceil$$

$$\text{Fabrication hours} = Quantity$$

$$\text{Number of receipts} = \left\lceil \frac{Quantity}{200} \right\rceil$$

Also assume that the activity rates for the following activities are constants (i.e., not derived):

Programming, setup and tooling, activity rate = $2500/setup
Fabrication, activity rate = $20/hour
Receiving, activity rate = $2500/receipt

If the manufacturer requires a 15% profit margin on all products,

a) What is the price versus quantity relationship for Product *A*? Plot it.

b) Is traditional cost accounting more accurate at high quantities or low quantities?

5.6 Acme electric manufactures circuit breaker boxes. The product manufacturing overheads for last year are known:

Utility costs (related to machine hours) = $298,000
Product setup costs = $189,200
Cost of ordering materials = $28,380
Cost of material requisitions = $52,030

Details of three product models are the relevant information for last year are:

	Model *1*	Model *2*	Model *3*
Number of production runs (setups)	26	37	27
Number of material orders	30	45	52
Number of material requisitions	45	150	105
Units produced	1000	2000	2500
Machine hours per unit	1.5	2.25	3
Direct labor hours per unit	0.5	1	2
Direct materials per unit	$15	$18	$23

Labor cost = $65/hour

a) Calculate the unit cost for each of the three products using traditional cost accounting (based on labor content)
b) Calculate the unit cost of each of the three products using ABC
c) Calculate the unit cost of each of the three products using traditional cost accounting (based on machine time content) – Hint: calculate the overhead allocation per machine hour (instead of per labor hour).

5.7 You run a manufacturing facility. Last year your facility manufactured 21 products with the following characteristics:

Products	Number of Parts in the Product	Quantity Manufactured	Fabrication Time (hours/part)	Design and Prototyping (Eng. hours)
1	13	100	120	14
2	10	234	98	8
3	34	1000	389	57
4	56	2000	600	110
5	112	9	1000	350
6	34	50	340	32
7	78	100	800	200
8	22	100	200	22
9	43	250	415	78
10	89	1000	900	300
11	6	50	60	4

Products	Number of Parts in the Product	Quantity Manufactured	Fabrication Time (hours/part)	Design and Prototyping (Eng. hours)
12	113	50	1150	400
13	212	50	2000	1000
14	19	1000	200	17
15	28	1245	300	30
16	111	20	1116	356
17	44	250	450	70
18	100	69	1000	347
19	55	345	567	86
20	34	25	335	40
21	12	500	123	12

In addition, the following data is known about last year:

- 1.1 million labor hours were used to build the 21 products (note, "labor hours" and "fabrication hours" are <u>not</u> the same)
- $37/hour labor rate
- Assume there is no inflation

Activity	Cost ($)	Cost Driver	Driver Quantity Data
Design and Prototype	$290,000	Engineering Hours	
Programming, Setup and Tooling	$150,000	Number of Setups	21
Fabrication	$70,000,000	Fabrication Hours	
Receiving	$150,000	Number of Receipts	312
Packing and Shipping	$150,090	Number of Customers	43

You are considering manufacturing the following 3 new products:

	Product *A*	Product *B*	Product *C*
Number of Parts in the Product	23	46	212
Number of Setups	1	1	1
Number of Receipts	12	3	32
Number of Customers	3	1	7
Quantity Required	25	154	1000

Use ABC to determine how much you should quote customers for each of the products (assume no profit in the quotes). Your answer should be based on last year's history (do not assume that products *A*, *B*, and *C* have or are necessarily going to be built)

Hints:

1) You will need to figure out the number of engineering hours and fabrication hours needed for the three new products (we did parametric modeling a couple of weeks ago – remember?)
2) You can figure out the labor hours associated with each new product from last year's ratio of labor hours to fabrication hours.

5.8 Using the example in Section 5.4, how much will a project that has 54 setups, 200 receiving activities, and 756 packing and shipping activities cost using ABC and TDABC?

Chapter 6

Parametric Cost Modeling

By definition, a *parametric* is a measurable or quantifiable characteristic of a system. Parametric equations are sets of equations that express a set of quantities as explicit functions of a number of independent variables, known as *parameters*.

A parametric cost estimation uses cost estimating relationships (CERs) to create cost estimates. A parametric cost estimating model is made up of one or more algorithms or CERs that describe the cost of a product or asset using technical and/or programmatic data (parameters). For example, if history has demonstrated that the cost of performing functional testing (the dependent variable) normally represents 50% of the manufacturing cost of an integrated circuit (the independent variable), then a parametric model for the test cost is simply 50% of the manufacturing cost.

Unfortunately, most parametric models are not this simple. CERs are commonly developed from regression analysis of historical costing information; however, other analytical methods, such as neural networks, can be used as well. Parametric models are especially useful for cost and value evaluations early in the product or system life cycle when detailed design information is not known. However, as we will discuss in Section 6.3, the scope of usage of parametric models is usually limited to certain ranges of parameter input values, due to the many assumptions built into the CERs.

Parametric cost estimation dates back to the 1930s. Statistical estimation of costs was suggested in 1936 by Wright [Ref. 6.1]. Wright developed equations that could be used to predict the cost of airplanes over long production runs, a theory that came to be known as the learning curve (see Chapter 10). In World War II, industrial engineers used Wright's learning curve model to predict the unit cost of airplanes. In 1948, the U.S.

Department of Defense established the Rand Corporation. In the mid 1950s, Rand developed the basis for parametric cost modeling called the cost estimating relationship (CER), see [Ref. 6.2]. Rand also formed the foundation for parametric aerospace estimating by merging the concept of the CER with the learning curves (see [Ref. 6.3]).

All of the methodologies considered in this book so far (process-flow modeling, cost of ownership, activity-based costing) are *bottom-up* approaches to cost modeling. In a bottom-up model the overall response or characteristic of a product is determined by accumulating the properties (response and characteristics) of the individual actions that take place to manufacture the product. This description does not apply to parametric cost modeling, which is a *top-down* approach in which high-level attributes are used to determine the response or characteristics of the object without a view to the constituent parts or the processes used to create the product.[1]

6.1 Cost Estimating Relationships (CERs)

To illustrate the parametric cost modeling concept, consider the following example. It has long been known that the cost of manufacturing aircraft can be correlated to the mass of the aircraft. Figure 6.1 shows historical data for commercial airliners and fighter jets. In this simple example the points on the graph in Figure 6.1 represent the relationship of price to mass for different aircraft. The lines traversing the data points represent a linear relationship determined using a simple least squares straight-line fit between the mass and the price, which is given by

$$\text{Commercial Airliners: } Price = 1.3212(OEW) + 33.6 \qquad (6.1a)$$

[1] The disadvantages of the top-down approaches are the advantages of the bottom-up approaches and vice versa. [Ref. 6.4]. Top-down models can underestimate the costs of solving difficult technical problems and there is no detailed justification of the final cost estimate. By contrast, bottom-up models produce a justification. However, bottom-up approaches are more likely to underestimate the costs of system activities such as integration. Bottom-up modeling is also more expensive and time consuming.

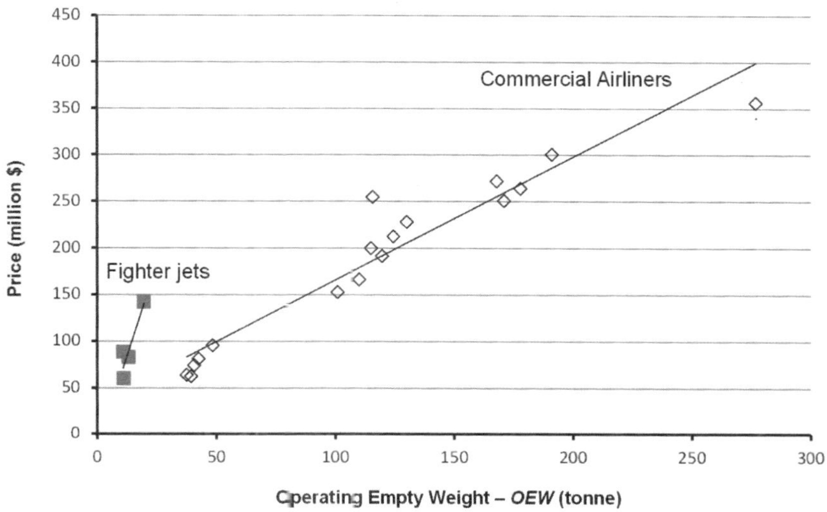

Fig. 6.1. Historical data for purchase price versus operating empty weight for fighter jets and Boeing and Airbus commercial airliners [Ref. 6.5].

$$\text{Jet Fighters: } Price = 7.9124(OEW) + 15.62 \tag{6.1b}$$

where *OEW* is the operating empty weight in tonnes and price is the purchase price of the aircraft in millions of dollars ($US). Using Equation (6.1), it is possible to predict the future price of a commercial airliner or a jet fighter based only on its mass. Equation (6.1) is a cost estimating relationship (CER).

In the case of aircraft we did not consider any of the details of how the aircraft are manufactured; we only identified one factor that has a correlation to the final price of the airplane and used it to construct a predictive model. The example provided in Figure 6.1 and Equation (6.1) is simple, but nonetheless represents an illustration of the principles of parametric cost estimating. Variations of this approach are widely used in industry to predict the cost of products under development and their subsequent life cycles.

A cost estimating relationship (CER) is an algorithm used to estimate a particular cost or price using an established relationship with an independent variable [Ref. 6.6]. If you can identify one or more independent variables (drivers) that demonstrate a measurable correlation with the cost or price of a product, system or service, you can develop a CER. The CER you develop may be simple (e.g., a ratio, or a curve fit, as in the example in this section) or it may involve a more complex mathematical expression or a system of equations.

6.1.1 *Developing CERs*

The following steps represent the CER development process [Ref. 6.6]:

Step 1. Define the dependent variable that the CER will estimate. The CER could be used to estimate price, cost, labor hours, material cost, or some other relevant measure. The more detailed the definition of the dependent variable, the easier it will be to gather the data needed for CER development.

Step 2. Select the independent variables to be tested. Independent variables for CER development can be identified from experience and/or published sources of information. The selected variables should be quantitatively measurable and have available historical data. If historical data does not exist, it will be impossible to use the variable for prediction. Because performance characteristics are often known (from system requirements) before design characteristics, it is better to develop CERs based on performance, as opposed to design characteristics.

Step 3. Collect data. Information should be collected at as low a level of detail as possible — information can always be aggregated later. Multiple sources of data are rarely comparable (or combinable) without manipulation or normalization. For example, the data in Figure 6.1 was collected from different sources and the items included in an aircraft's prices may not have been consistent from one source to another. Possible adjustments to data include timing (inflation, cost of money), cost scope (elements included or not included in the costs), learning curves (Chapter 10), and production volume.

Step 4. Explore the relationship between the dependent and independent variables. The degree of correlation (if any) between the independent and dependent variables must be determined. This can be accomplished using analytical techniques that range from simple graphical analysis and curve fitting to complex mathematical analysis — for example, ratio analysis, moving averages, and various types of regression analyses.

Step 5. Select the relationship that best predicts the dependent variable. After exploring the possible relationships, select the one that is the best predictor of the dependent variable. A high degree of correlation between an independent variable and the dependent variable can be a good indicator that the independent variable represents a good predictor. The selected estimate should also be checked for reasonableness (e.g., see Problem 6.7).

Step 6. Document. Documentation of the CER is an important step that permits others to understand how the CER can be used. Documentation needs to include details about the data used (what it was and where it came from), the time period that the data represents, and adjustments that were made to the data.

6.2 A Simple Parametric Cost Modeling Example

In this section we develop a simple parametric cost model relevant to electronic systems. Assume that your organization has had 16 ASICs (application specific integrated circuits) manufactured during some period in the past. All use 0.35 μm CMOS technology, and were produced on 300 mm wafers ($E = 2$ mm, $K = 0.3$ mm as defined in Figure 2.3) that cost C_w = \$5000/wafer to process.[2] You wish to develop a CER that can be used to estimate the recurring die cost (C_{die}), given a gate count (N_G) of ASICs you may manufacture in the future using the same process. The data you have is shown in Table 6.1.

---

[2] A detailed discussion of ASIC costs can be found in [Ref. 6.7] and [Ref. 6.8].

Table 6.1. ASIC Die Cost versus Gate Count Data.

Die Size (square inches) - A_{die}	Available Gates - N_G
0.5	5,000,000
0.32	2,000,000
0.16	400,000
0.1	180,000
0.08	100,000
0.02	10,000
0.05	50,000
0.04	25,000
0.12	300,000
0.33	1,000,000
0.2	1,000,000
0.25	900,000
0.075	90,000
0.065	92,000
0.03	12,000
0.035	20,000

First, the usable wafer area (the area in which die can be fabricated) is given by

$$\text{Usable Wafer Area} = \pi\left(\frac{D_W}{2} - E\right)^2 \tag{6.2}$$

where D_W is the diameter of the wafer and E is the edge scrap allowance (see Figure 2.3). The effective die area (the wafer area occupied by one die assuming the die are square) is given by

$$\text{Effective Die Area} = \left(\sqrt{A_{die}} + K\right)^2 \tag{6.3}$$

where K is the scribe street or kerf (minimum distance between adjacent die). The number-up (number of die on the wafer) can be estimated as

$$N_u = \frac{\text{Usable Wafer Area}}{\text{Effective Die Area}} = \frac{\pi\left(\frac{D_W}{2} - E\right)^2}{\left(\sqrt{A_{die}} + K\right)^2} \tag{6.4}$$

Equation (6.4) is an overestimation of the number of die that can fit on a wafer (see Section 2.2.6). The cost per die is then given by

$$C_{die} = \frac{C_w}{N_u} \tag{6.5}$$

where C_w is the cost of processing one wafer. Now we need to relate the number of gates to the die area using the historical data in Table 6.1. Plotting the data in Table 6.1, we obtain Figure 6.2. A logarithmic fit of the data in Figure 6.2 gives

$$N_G = 2\text{x}10^7 A_{die}^{1.9572} \tag{6.6}$$

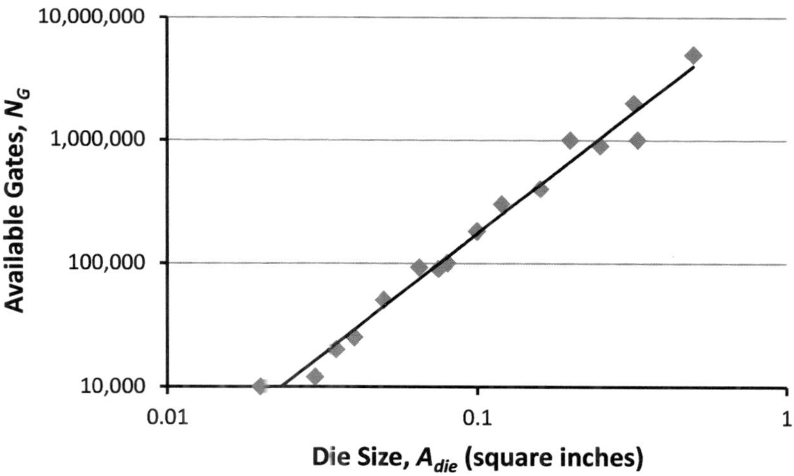

Fig. 6.2. Historical ASIC data.

Finally, combining Equations (6.4) through (6.6), we obtain

$$C_{die} = \frac{C_w\left[\left(\frac{N_G}{2\text{x}10^7}\right)^{0.2555} + K\right]^2}{\pi\left(\frac{D_W}{2} - E\right)^2} \tag{6.7}$$

Substituting for known quantities, Equation (6.7) can be reduced to

$$C_{die} = 0.07266\left[0.01363N_G^{0.2555} + 0.3\right]^2 \tag{6.8}$$

Equation (6.8) is potentially a valuable model for the recurring cost per die of fabricating ASICs. Note that this equation does not include the NRE (non-recurring) costs of designing the ASIC, testing the ASIC (see Chapter 7), or packaging the finished die into a chip.

Equation (6.8) is simple to use and accurately reflects your organization's history of having ASICs fabricated.

6.3 Limitations of CERs

The widespread use of CERs in the form of simple cost factors, equations, curves, and rules of thumb clearly establishes that there is value in CERs and that there are a wide variety of situations in which they can be used. However, if an unknown source provided you with Equation (6.8), would you know how to use it? Would you know the circumstances under which it is valid and when it is not? Would you know that it is only valid for 300 mm wafers?

In this section we discuss the limitations of CERs. Due to these limitations and constraints, it is incumbent upon the user to thoroughly understand the basis of a parametric model before using it.

6.3.1 *Bounds of the Data*

Strictly speaking, CERs are only relevant for forecasting costs of items that are within the bounds of the sample (the database) on which the development of the CER was based. Although the validity of extrapolation beyond the sample is statistically questionable, it is often practiced by users of CERs because, in many instances, the products and systems of interest are outside the range of the sample. The question is whether or not the CER is relevant if it is extrapolated — for example, is Equation (6.8) accurate for a 10-million-gate ASIC when the highest gate count included in the database used to develop the CER was 5 million gates?

6.3.2 *Scope of the Data*

In cost estimating, there are rarely large, directly applicable databases, and the source data has to be evaluated to determine if it can be applied to the desired estimate. For example, if we only knew the relationship between the price of commercial airliners and *OEW* (Equation (6.1a)), could we apply it to fighter aircraft? The answer is no — fighter aircraft are not within the scope of commercial airliners.[3] Similarly, Equation (6.8) was developed for 0.35 μm minimum feature size ASICs; can we use it for 0.15 μm ASICs? While Equation (6.8) only corresponds to 300 mm diameter wafers, is Equation (6.7) valid for 200 mm wafers (assuming that C_w is updated for 200 mm wafers)?

CER development is not necessarily limited to only developing extremely specific CERs, as in Equation (6.8). Use of more comprehensive databases and more sophisticated mathematical modeling allows the development of parametric models that relate cost based on a more generic system descriptions and complexity.

6.3.3 *Overfitting*

Overfitting occurs when a model inadvertently describes random error or noise in the data instead of, or in addition to, describing the underlying relationship it is targeting. Overfitting occurs when a mathematical model is created that is excessively complex, i.e., when it has too many parameters (or is higher order than it needs to be) for the number of observations that actually exist. Overfitting means that you are fitting both the predictable component of the data and the noisy part. An overfit model will generally have poor predictive performance, because it exaggerates minor fluctuations (noise) in the data. With a small sample, it is often

[3] This points out a common problem with CERs. If the CER is not sufficiently documented (Step 6 in Section 6.1.1), it could easily be misused. For example, what if Equation (6.1a) was provided and we knew it corresponded to airplanes, but did not know what kind of airplanes?

possible to write an equation that fits the data perfectly, but the equation is completely useless outside the range of the sample.[4]

As an example, consider the commercial airline data used in Section 6.1. Figure 6.3 shows the same data fit with a straight line and with a 6th order polynomial. The 6th order polynomial fit has a better correlation coefficient (i.e., coefficient of determination, R^2). Does that mean that it is a more meaningful curve fit to the data? Obviously not — the straight line fit provides a much better forecast of commercial airline prices, even though the 6th order polynomial fits the data set better.

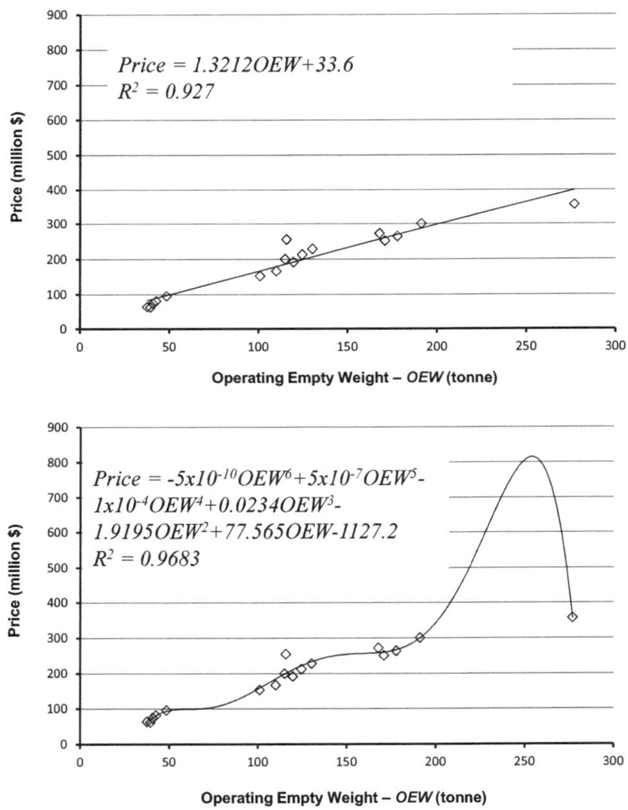

Fig. 6.3. Example of overfit data.

[4] Enrico Fermi recalled the following: "I remember my friend Johnny von Neumann used to say, 'with four parameters I can fit an elephant and with five I can make him wiggle his trunk.'" [Ref. 6.9].

6.3.4 *Don't Force a Correlation When One Does Not Exist*

If there is no discernible correlation between an independent variable and the dependent variable, then a parametric model that includes the independent variable should not be used (see Figure 6.4). For parametric models to be valuable, they should only include independent variables that have some effect on the dependent variable. A line of best fit could be drawn through the data in Figure 6.4, but a more accurate conclusion might be that there isn't a correlation between procurement life and introduction date for EPROM parts.

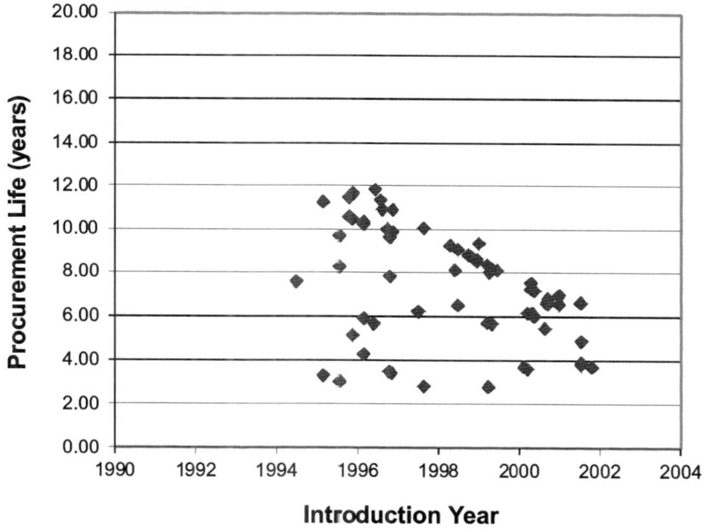

Fig. 6.4. Procurement life versus introduction date for EPROM memory devices. Procurement life is defined in [Ref. 6.10]. EPROM stands for Erasable Programmable Read Only Memory.

6.3.5 *Historical Data*

A statistical CER can be derived from information on past occurrences, but there is no guarantee that the past is a reliable guide to the future. An estimate based on past performance may be wrong if the technology or the world changes in some fundamental way. This is not meant to imply that the occurrence of "disruptive" technologies automatically makes CERs invalid.[5] Some CERs transcend the disruption, or even anticipate that disruptive technologies will occur and their impact on cost even though they cannot predict what the technologies are.

6.4 Other Parametric Cost Modeling/Estimation Approaches

Parametric cost modeling approaches appear in many contexts and are used for many different applications. All share the common attribute of being based on the use of historical data to infer the cost of future products and systems.

6.4.1 *Feature-Based Costing (FBC)*

Parametric cost models that are applied to the determination of the cost of mechanical and solid objects is usually referred to as feature-based cost modeling. Feature-based cost modeling involves the identification of a product's cost-driving features, such as the number of holes, edges, folds, or corners, and the determination of the costs associated with each of these features [Ref. 6.12].

Feature-based cost models have become popular for use in the design of mechanical systems because they can readily be incorporated into CAD systems to automatically estimate manufacturing costs of objects based on their features concurrent with their design. Feature-based cost modeling first appeared in the 1950s when Boeing estimated the cost of various casting processes — sand casting, die casting, investment casting and permanent mold casting as a function of a single casting feature, casting volume [Ref. 6.13].

The fundamental idea behind feature-based costing is that products can be described as a collection of associated features — holes, flat faces, edges, folds, etc. It then follows that each product feature has a cost

[5] Disruptive technologies are defined as technologies that fundamentally change an existing market. The term was first used by Bower and Christensen in 1995 [Ref. 6.11] and is used in business and technology to describe innovations that improve a product or service in ways that the market does not expect, typically by lowering price, improving performance or functionality, or allowing introduction of the product or service to a different set of consumers.

implication during production [Ref. 6.14]. The assumption is that because the same features appear in many different parts and products, the cost information determined for a class of features can be reused for multiple products. Although feature-based costing has gained popularity, there is no accepted consensus across disciplines and organizations on what a feature is. Therefore, organizations must create their own feature definitions.

6.4.2 *Neural Network Based Cost Estimation*

Neural network based cost estimation is an extension of parametric modeling that can potentially represent more complex relationships between process and product design parameters than the simple CERs used in most parametric approaches [Refs. 6.15, 6.16]. An artificial neural network (ANN), or simulated neural network (SNN), is a group of interconnected artificial neurons that makes use of a mathematical model to perform information processing. In most cases, an ANN is an adaptive system that changes its structure based on external or internal information that flows through the network.

For cost estimating purposes, the fundamental idea is to make a computer program learn the correlation between product-related attributes and cost — that is, to provide attribute data (and corresponding costs) to a computer such that it learns which product attributes influence the final cost and how much influence they have [Ref. 6.12]. The ANN approximates the functional relationship between the attribute values and the cost using past examples. Once the computer program is trained, the attribute values of a new product can be provided to the network that then applies the function relationship obtained via training to the new attributes and computes a cost. The network (functional relationship) created is a CER.

It has been demonstrated that neural networks can produce better cost predictions than conventional regression methods [Ref. 6.16]. In cases where an appropriate CER can be identified, regression models have significant advantages in terms of accuracy, variability, model creation and model examination [Ref. 6.16]. The advantage that neural networks have over regression-analysis-type parametric costing is that they are able to detect hidden relationships among data. However, to be effective, neural networks require large databases of similar products, which is problematic for industries that have limited product offerings. The artificial neural network also, unfortunately, becomes a "black box" CER that cannot produce a detailed list of the reasons and assumptions behind the cost estimate.

6.4.3 *Costing by Analogy*

Analogy estimates cost based on historical data for analogous systems or subsystems [Ref. 6.17]. In costing by analogy, a current product or system, similar to the new product or system, is used as a cost basis. The cost of a proposed new product or system is estimated by adjusting the cost of a known system to account for differences between the systems. Adjustments are made using scaling parameters that account for differences in size, performance, technology, and complexity. Quantitative data based adjustments are generally preferable to adjustments based on qualitative judgments from subject-matter experts. Analogy estimates typically use a single historical data point as the basis for the estimate [Ref. 6.18].

6.5 Summary and Discussion

Many of the most accurate cost estimation and quoting models in the world are based on parametric cost models. Parametric costing is relevant when a new product or service is similar to products and services that have been previously provided and there is a sufficiently large and detailed historical database of the previously provided products and services.

Parametric models can be very accurate for well known and well defined products. For example, the most accurate cost models for fabricating printed circuit boards are parametric models. However, parametric models represent a top-down modeling approach and are only valid when used to determine the cost of products that fall within the scope of the original data used to create the model; problems occur when a complete picture of this scope is not available.

CERs can be developed and used for estimating all stages of a product life cycle, provided applicable data is available. Three additional topics in this book discuss applications of parametric models: learning curves (Chapter 10), service costing (Chapter 18) and software development costing (Chapter 19). The determination of CERs is a highly developed science and many publications provide more detail than the introduction provided in this chapter (see the bibliography for relevant sources).

References

6.1 Wright, T. P. (1936). Factors affecting the cost of airplanes, *Journal of Aeronautical Science*, 3(2), pp. 122-128.

6.2 Levenson, G. S., Boren Jr., H. E., Tihansky, D. P. and Timson, F. (1972). Cost-Estimating Relationships for Aircraft Airframes, Rand Corporation Report, R-761-PR. http://www.rand.org/pubs/reports/2007/R761.1.pdf. Accessed April 22, 2016.

6.3 Stuparu, D. and Vasile, T. (2009). Elementary statistical techniques used in cost estimating relationships (CER's) development, *Annals. Economic Science Series* XV, pp. 392-399.

6.4 Sommerville, I. (2007). Chapter 26 – Software cost estimation, *Software Engineering*, 7th Edition (Addison-Wesley, Harlow, England).

6.5 Irastorza, J. (2010). An aircraft worth its weight in gold? March 13, 2010, Available at: http://theblogbyjavier.wordpress.com/2010/03/13/an-aircraft-worth-its-weight-in-gold/. Accessed April 22, 2016.

6.6 Chapter 4 - Developing and using cost estimating relationships, Volume 2 – Quantitative Techniques for Contract Pricing, Contract Pricing Reference Guides, Defense Procurement and Acquisition Policy, Available at: https://acc.dau.mil/CommunityBrowser.aspx?id=379490. Accessed April 22, 2016.

6.7 ASIC Outlook 1998. An application specific report and directory, "Chapter 5 – ASIC Cost Effectiveness," ASIC Outlook 1998, An Application Specific Report and Directory, Integrated Circuit Engineering Corporation, 1998. Available from http://smithsonianchips.si.edu/ice/cd/ASIC98/SECTION5.PDF. Accessed April 22, 2016.

6.8 Liu, J. (1995). Detailed model shows FPGAs' true cost, *Electronics Design, Strategy, News,* pp. 153-158, May 11, 1995. Available from: http://www.edn.com/design/systems-design/4348855/EDN-Access--05-11-95-Detailed-model-shows-FPGAs-true-cost. Accessed on April 22, 2016.

6.9 Dyson, F. (2004). Turning points: A meeting with Enrico Fermi, *Nature*, 427, p. 297.

6.10 Sandborn, P., Prabhakar, V. and Ahmad, O. (2011). Forecasting technology procurement lifetimes for use in managing DMSMS obsolescence, *Microelectronics Reliability*, 51, pp. 392-399.
6.11 Bower, J. L. and Christensen, C. M. (1995). Disruptive technologies: catching the wave, *Harvard Business Review*, pp. 43-53, January-February.
6.12 Rush, C. and Roy, R. (2000). Analysis of cost estimating used within a concurrent engineering environment throughout a product life cycle, *Proceedings of the 7th ISPE International Conference on Concurrent Engineering: Research and Applications*, Lyon, France, pp. 58-67.
6.13 Creese, R. C. and Patrawala, T. B. (1998). The return of feature based cost modelling, *Proceedings of the SPIE Conference on Intelligent Systems in Design and Manufacturing*, Vol. 3517, Boston, MA, pp. 172-182.
6.14 Brimson, J. A. (1998). Feature costing: beyond ABC, *Journal of Cost Management*, pp. 6-12.
6.15 Bode, J. (1998). Neural networks for cost estimation, *Cost Engineering*, 40(1), pp. 25-30.
6.16 Smith, A. E. and Mason, A. K. (1997). Cost estimation predictive modelling: Regression versus neural network, *Engineering Economist*, 42(2), pp. 137-162.
6.17 Chapter 3 – Affordability and life-cycle resource estimates, Defense Acquisition Guidebook, Defense Acquisition University, Available at: https://acc.dau.mil/CommunityBrowser.aspx?id=488329. Accessed April 22, 2016.
6.18 Dysert, L. R. (2005). So you think you're an estimator? *Cost Engineering*, 47(9), pp. 30-35.
6.19 Chapter 18, Use of cost estimating relationships, DOE G 413.3-4, U.S. Department of Energy Technology Readiness Assessment Guide, March 28, 1997. Available at: https://www.directives.doe.gov/directives-documents/400-series/0430.1-EGuide-1-Chp18/@@download/file. Accessed April 22, 2016.

Bibliography

In addition to the sources referenced in this chapter, there are many books and other good sources of information on parametric costing, including the following:

Parametric Cost Estimating Handbook, Fall 1995, which can be accessed at: https://acc.dau.mil/CommunityBrowser.aspx?id=322656. Accessed April 22, 2016.
The International Society of Parametric Analysts (ISPA) (http://www.ispa-cost.org/) has several resources for the development and use of CERs including the ISPA Parametric Estimating Handbook: http://www.ispa-cost.org/ISPA_PE_Hdbk_4thED.pdf. Accessed April 22, 2016.
Journal of Cost Analysis and Parametrics

Problems

6.1 The manufacturers of a particular electronic product have observed that the cost of a completed instance of the product varies directly with the number of chips (integrated circuit parts) it contains. Thus, the sum of the number of chips in a specific product's design can serve as an independent variable (cost driver) in a CER to predict the cost of the completed product. Assume an analysis of the product indicates that each instance of the product is allocated \$5.23 of non-recurring and overhead cost, and an additional cost of \$1.10 per chip is required. Write the CER for the product cost. If a product is to contain 30 chips, what is the estimated cost of the product using your CER?

6.2 Based on its formulation (not the data from which it is formulated), is Equation (6.8) likely to be an overestimation or underestimation of the cost per die? Provide specific reasons for your answer.

6.3 Assuming that the cost of processing a 300 mm wafer was \$5000/wafer in 2002, but has decreased by 5% per year since then, formulate a version of Equation (6.8) that depends on the year in which the ASIC is fabricated.

6.4 Assuming a Poisson yield model, re-derive Equation (6.8) to be the effective cost per good (non-defective) die. Assume that the defect density of the process is $D =$ 1 defect/cm^2 and that individual defective die are disposed of—that is, they have no salvage value.

6.5 Assuming all the die in the ASIC example in Section 6.2 have an aspect ratio of 2:1 (the example in Section 6.2 assumes that they are square, which corresponds to an aspect ratio of 1:1). Write a new CER that relates gate count to the die cost. Hint: a number-up calculation is discussed in Section 2.2.6 and Problem 2.2.

6.6 The data given in the table below was observed for a specific type of test. Create a CER for the effective cost per part that is passed by the test step (your CER should be in terms of fault coverage, incoming cost and incoming yield, which are the inputs to the test operation defined in Section 7.4). If for some later part, $C_{test} =$ \$500, what fault coverage (f_c) does your CER tell you this corresponds to? Is this a reasonable result, why or why not?

Fault Coverage, f_c (fraction)	Test Cost, C_{test} ($/part tested)
0.05	50
0.14	51
0.157	51.3
0.21	51.2
0.23	51
0.3	56
0.33	55
0.45	78
0.56	105
0.8	170
0.9	190
0.94	230

6.7 Data on hazardous waste disposal costs has been collected and the following CER has been determined (from [Ref. 6.19]),

$$C_{disposal} = 200 + 275D_r - 0.19M_l \tag{6.9}$$

where

$C_{disposal}$ = the cost to dispose of drummed hazardous waste.
D_r = the number of drums.
M_l = the number of miles between the location that generated the waste and the hazardous waste disposal facility.

The CER in Equation (6.9) has been checked and the parameters are within acceptable tolerances. Equation (6.9) also fits the known data well. Unfortunately, this is not a reasonable CER. Why not? Is there anything that is intuitively unreasonable about this CER?

6.8 You work for a company that builds environmentally controlled inventory storage facilities for electronic parts. All the facilities you have built in the past are listed in the table below. Assuming no inflation, write an equation that predicts the total cost of one of your storage facilities. The objective is to produce a reasonable[6] model that fits the existing data with an $R^2 > 0.95$.

[6] "Reasonable" in this case excludes anything greater than a 3rd order polynomial.

Floors	Gross Floor Area (ft^2)	Perimeter (ft)	Total Cost
2	600	200	$2,084,440
3	500	103	$1,703,173.5
1	1000	800	$3,659,600
4	1435	450	$6,158,784
1	2000	179	$5,341,878.5
2	600	98	$1,800,574
3	780	74	$2,295,105
4	1400	500	$6,347,960
1	600	196	$1,800,574
2	3000	219	$8,248,677
3	600	600	$4,032,540
4	4000	800	$14,638,400
1	600	100	$1,666,990
2	400	234	$1,669,782
3	2540	700	$9,390,006
4	600	500	$4,310,840

Chapter 7

Test Economics

For many electronic systems, testing[1] is an important driver that significantly affects the total cost of manufacturing. In some cases, more than 60% of a product's recurring cost can be attributed to testing costs [Ref. 7.1]; for integrated circuits, testing costs approach 50% of the total product cost [Ref. 7.2]. When the products that result from a manufacturing process are imperfect, four costs are potentially involved:

- the cost of determining whether a given instance of the product is good or bad (testing);
- the cost of determining what defect caused the faulty product and where it is located (diagnosis);
- fixing the defect (rework); and
- eliminating the causes of the defect(s) (continuous improvement).

Depending on the maturity of the product, its placement in the market, and the profit associated with selling it, all, some or none of these cost activities may be performed. Understanding the test/diagnosis/rework costs may determine the extent to which the system designer can control and optimize the manufacturing cost, and the extent to which it makes sense to do so.

The ultimate goal of any functional test strategy is to answer the following questions:

(1) When should a system be tested? At what point(s) in the manufacturing process?

---

[1] In this chapter we are concerned with recurring functional (pass/fail) and diagnostic testing. This chapter does not treat environmental testing — i.e., qualification. A discussion of qualification is included in Section 11.3.

(2) How much testing should be done? How thorough should the testing be?
(3) What steps should be taken to make the system more testable?

The answers to these questions would be easy with unlimited time, resources, and money. We could stop after every step in the manufacturing process and perform a full function test, and add structures to the system such that every circuit could be accessed and tested. These measures, unfortunately, are far from practical, so engineers are usually faced with determining how to obtain the best test coverage possible for the least cost.

The specific goal of test economics is to minimize the cost of discarding good products and the cost of shipping bad ones. This goal is enabled through the development of models that allow the yield and cost of products that pass through test operations to be predicted as a function of both the properties of the product entering the test and the characteristics of the test operation (its cost, yield, and ability to detect faults in the product it is testing).

7.1 Defects and Faults

A *defect* is a flaw that causes a system not to work under certain conditions, where the conditions under which the defect appears are relevant to the specified operational conditions of the product. A *fault* is the effect of a defect on the system. Test equipment (testers) measure or detect faults. For example, a defect in an electronic system might be a broken wirebond. The fault detected by the tester due to this defect would be an electrical open circuit (where a short circuit was expected). A diagnosis activity isolates the fault and relates it to an actual defect — that is, diagnosis determines where the open circuit is and that a broken wirebond caused it.

Two other definitions occur in testing discussions. An *error* is the manifestation of a fault that results in an incorrect system output or state (it may occur some distance from the actual fault site). *Failure* is the deviation of a system's specified behavior, caused by an error. In general, faults may cause errors that in turn cause failure; however, the terms fault, failure and error have often been used interchangeably.

In order to develop a basis for understanding test economics, we must first relate defects to faults. Once we have a basis for mapping defects to faults, we can address the concepts of defect coverage and fault coverage, followed by a derivation of the yield after a test operation as a function of the fault coverage associated with the test.

7.1.1 *Relating Defects to Faults*

Most tests (and testers) are designed to detect specific types of faults. Generally, a defect cannot be measured directly and there is not a one-to-one mapping between defects and faults — that is, a given type of defect can appear as several different types of faults and a particular fault type may be the result of more than one type of defect.

A *fault spectrum* is defined as the fault rate per fault type, or the number of occurrences of a particular type of fault in the device under test. Fault types for electronic components include opens, shorts, static faults, dynamic faults, voltage faults, temperature faults, and many others [Ref. 7.3]. The fault spectrum can be determined from similar previously manufactured products. Using a previous product's fault spectrum has several inherent problems [Ref. 7.4]. First, the measured fault spectrum depends on the fault coverage of the tests, and second, there is no basis for predicting a fault spectrum for fundamentally new products that use new technologies.

Another approach to determining the fault spectrum is by relating it to the defect spectrum [Ref. 7.4]. The defect spectrum describes the average number of defects per device under test per defect type. The total number of defects per defect type (a defect spectrum element) can be calculated using

$$d_j = \frac{dpm_j n_e}{10^6} \tag{7.1}$$

where

d_j = the number of defects of defect type j in the device under test.

dpm_j = the number of defects of defect type j per million elements (ppm).

n_e = the number of elements in the device under test.

Assume in Equation (7.1) that the device under test is a packaged chip; the element is a wirebond from the bare die to the leadframe in the package; and defect type j is a broken wirebond. If the defect level for wirebonding is 100 ppm and there are 200 I/Os to be wirebonded to the leadframe in order to package the die, then the total number of defects of type "broken wirebond" is 0.02 broken wirebonds in one chip.

The defect spectrum is related to the fault spectrum by a conversion matrix. Where the conversion matrix defines how a defect is distributed (statistically) among fault types, then

$$f = Cd \tag{7.2}$$

where f is the fault spectrum (vector of fault types), d is the defect spectrum (vector of defect types), and C is the conversion matrix. To understand the conversion matrix, consider Figure 7.1.

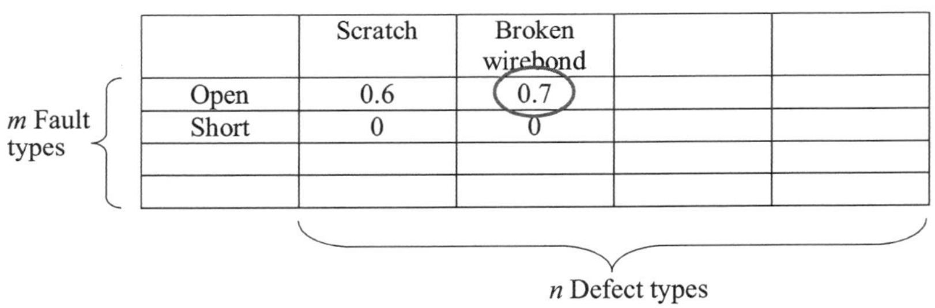

Fig. 7.1. Interpretation of the conversion matrix.

The circled quantity in Figure 7.1 represents the fraction of defects of defect type 2 (broken wirebond) that appear as faults of fault type 1 (open circuit); this would be the C_{12} element of the conversion matrix. In general, $n \neq m$ — the number of fault types does not equal the number of defect types. Ideally the sum of each column of C is equal to 1 — that is, every defect appears as a fault of some type that the testing can find (however, this is usually not the case). If the columns add to 1, it is called "conservation of defects."

As an example of the formation of a conversion matrix element, consider a hypothetical die wirebonded to a leadframe. First, break wirebond #1. Does the open circuit test detect the problem? If the wirebond is one of many ground I/Os on the die, the open circuit test may not detect the problem. Then re-bond wirebond #1. Repeat the process for all the bonds between the die and the leadframe. When all wirebonds have been successively tested, the matrix element is given by the following ratio[2]:

$$C_{12} = \frac{\text{Number of broken wirebonds successfully detected by the open circuit test}}{\text{Total number of wirebonds on the die}} \tag{7.3}$$

We have denoted the matrix element in this case as C_{12}, indicating that it relates fault type 1 (open circuit) to defect type 2 (broken wirebond).

Expanding and generalizing Equation (7.2), we obtain

$$\begin{pmatrix} f_1 \\ f_2 \\ \vdots \\ f_m \end{pmatrix} = \begin{pmatrix} C_{11} & C_{12} & \cdots & C_{1n} \\ C_{21} & C_{22} & \cdots & \\ \vdots & \ddots & & \\ C_{m1} & \cdots & \cdots & C_{mn} \end{pmatrix} \begin{pmatrix} d_1 \\ d_2 \\ \vdots \\ d_n \end{pmatrix} \tag{7.4}$$

The fraction of devices under test that are faulty due to fault type *i* from Equation (7.4) is given by

$$f_i = C_{i1}d_1 + C_{i2}d_2 + \ldots + C_{in}d_n = \sum_{j=1}^{n} C_{ij}d_j = \sum_{j=1}^{n} f_{ij} \tag{7.5}$$

where $f_{ij} = C_{ij}d_j$ is the fraction of devices under test that are faulty due to fault type *i*, which is related to defect type *j*.[3] Consider the following example numbers:

$C_{12} = 0.7$	70% of broken wirebond defects (defect type 2) appear as open circuit (fault type 1) faults
$d_2 = 0.2$	20% of devices under test are defective due to broken wirebond defects (defect type 2)

---

[2] Note that this simple example assumes that all wirebonds between the die and leadframe are equally likely to be defective (broken), which is generally not the case.

[3] f_{ij} is a useful quantity because it is the same for all test methods. It is the relationship between faults of fault type *i* and defects of defect type *j* before testing has been done.

$f_{12} = C_{12}\, d_2 = (0.7)(0.2) = 0.14$ 14% of devices under test that are faulty due to open circuit faults (fault type 1) can be related to broken wirebond defects (defect type 2)

Consider an expanded example, in which we define the conversion matrix as

$$\begin{array}{cc} \text{placement error } (j=1) & \text{broken wirebond } (j=2) \end{array} \quad n = 2$$

$$C = \begin{pmatrix} 0.1 & 0.7 \\ 0.8 & 0 \\ 0.1 & 0.3 \end{pmatrix} \begin{array}{l} \text{open } (i=1) \\ \text{short } (i=2) \\ \text{other } (i=3) \end{array} \quad m = 3$$

$$\begin{array}{ccl} 1.0 & 1.0 & \text{sum of the columns equals 1} \end{array}$$

If the fraction of devices under test that are defective due to placement errors ($j = 1$) is given by

$$d_1 = \frac{(1000)(10)}{10^6} = 0.01 \tag{7.6}$$

where placement is a 1000 ppm process and there are 10 placements per board; thus the boards have a 99% yield with respect to placement defects. Similarly, if the fraction of devices under test that are defective due to broken wirebonds ($j = 2$) is given by

$$d_2 = \frac{(100)(4300)}{10^6} = 0.43 \tag{7.7}$$

where wirebonding is a 100 ppm process and there are 4300 wirebonds per board, thus the boards have a 57% yield with respect to wirebond defects. Note, in this case, the overall board yield (if the only defects were placement errors and broken wirebonds) would be

$$\text{overall board yield} = 1 - \sum_{j=1}^{n} d_j = 1 - (0.01 + 0.43) = 0.56 \tag{7.8}$$

or 56%. (Note that we would have also arrived at the value of 0.56 by taking the product of 0.99 and 0.57).[4] Using the values of the elements of the defect spectrum computed in Equations (7.6) and (7.7), the values of f_{ij} for $j = 2$ are

$$f_{12} = (0.7)(0.43) = 0.301$$

$$f_{22} = (0)(0.43) = 0$$

$$f_{32} = (0.3)(0.43) = 0.129$$

The value of 0.301 computed for f_{12} means that 30.1% of the boards that are faulty due to $i = 1$ (open circuit) faults are related to $j = 2$ (broken wirebonds). The relationship between the fault spectrum and the defect spectrum for this example is given by Equation (7.4) as

$$\begin{pmatrix} f_1 \\ f_2 \\ f_3 \end{pmatrix} = \begin{pmatrix} 0.1 & 0.7 \\ 0.8 & 0 \\ 0.1 & 0.3 \end{pmatrix} \begin{pmatrix} 0.01 \\ 0.43 \end{pmatrix} = \begin{pmatrix} 0.302 \\ 0.008 \\ 0.130 \end{pmatrix} \tag{7.9}$$

For example, we can see from Equation (7.9) that 30.2% of the boards are faulty due to open circuit faults. Note that the sum of the fault spectrum elements is 0.44 and 1-0.44 = 0.56 or a 56% yield, which agrees with Equation (7.8).

One additional check can be performed using this example. Computing the additional f_{ij} terms for $j = 1$,

$$f_{11} = (0.1)(0.01) = 0.001$$

$$f_{21} = (0.8)(0.01) = 0.008$$

$$f_{31} = (0.1)(0.01) = 0.001$$

Using the computed values of f_{ij},

$$\sum_{i=1}^{m} \sum_{j=1}^{n} f_{ij} = \sum_{i=1}^{m} f_i = 0.44 \tag{7.10}$$

---

[4] The product of 0.99 and 0.57 is actually 0.5643, not 0.56. Equation (7.8) determines yield by summing the defects, giving the worst possible case, whereas multiplying yields is an average case (a higher yield). Note that $1-(d_1+d_2-d_1d_2) = 0.5643$.

For the conversion matrix used in this example, defects are conserved, and therefore, the sum in Equation (7.10) results in the total defect fraction, $\sum_{j=1}^{n} d_j$.

7.2 Defect and Fault Coverage

Defect coverage is the fraction of defects present that are detected by a test; fault coverage is the fraction of total possible faults that could be present that are detected by a test activity[5]:

$$\text{Fault Coverage} = \frac{\text{Number of detected faults}}{\text{Number of total possible faults}} \tag{7.11}$$

Fault coverage is a measure of the ability of a set of tests (a collection of test vectors) to detect a given class of faults that may occur in a device under test. Fault coverage has also been referred to as fault cover, test coverage, and test efficiency; however, the term test coverage is usually used in reference to software as opposed to hardware. In this section we relate the fault coverage to the detectable defects. Section 7.3 discusses relating the fault coverage to the yield of units passed by the test.

The defect spectrum of the defects detected (the number of defects per defect type) can be determined from the fault spectrum of faults detected using the following relation:

$$d_{cover_j} = \sum_{i=1}^{m} \left(\frac{f_{cover_i}}{f_i} \right) f_{ij} \tag{7.12}$$

---

[5] This definition is sometimes referred as "raw coverage." Related metrics that could also be defined include:

$$\text{Testable Coverage} = \frac{\text{Number of detected faults}}{\text{Number of total faults} - \text{Number of untestable faults}}$$

$$\text{Fault Efficiency} = \frac{\text{Number of detected faults} - \text{Number of untestable faults}}{\text{Number of total faults}}$$

Here, d_{cover_j} is the fraction of all devices under test with detected defects of defect type j; f_{cover_i} is the fraction of all devices under test with detected faults of fault type i. Dividing the result of Equation (7.12) by the fraction of devices under test that are actually defective due to defects of defect type j (d_j) gives the defect coverage of the test for defect type j. The ratio appearing in Equation (7.12) is the fault coverage for fault type i — that is, the fraction of existing faults detected by the test:

$$f_{c_i} = \frac{f_{cover_i}}{f_i} \tag{7.13}$$

To explore how Equation (7.12) works, consider a few trivial cases. If f_{c_i} = 1 for all i, then the equation reduces to d_j, which implies a defect coverage of 1. When $f_{c_i} = 0$ for all i, then it gives 0 for all j, which implies a defect coverage of 0. Using the example generated in Section 7.1, we can compute the defect coverage for different types of defects (e.g., with $f_{c_1} = 0.5$, $f_{c_2} = f_{c_3} = 1$) as

$$\frac{d_{cover_1}}{d_1} = \frac{(0.5)(0.001)+(1.0)(0.008)+(1.0)(0.001)}{0.01} = 0.95$$

$$\frac{d_{cover_2}}{d_2} = \frac{(0.5)(0.301)+(1.0)(0.0)+(1.0)(0.129)}{0.43} = 0.65$$

This result predicts that 95% of the defects of defect type 1 and 65% of the defects of defect type 2 will be detected by the test with the specified fault coverages.

For analog and digital circuits, fault coverages are usually determined through fault simulation. Fault simulation analyzes the operation of a circuit under various fault conditions (a collection of test patterns) to determine the extent to which the given test patterns detect a specific type of fault. For more information on fault simulation see [Ref. 7.5].

Now that we have a description of fault coverage, we need to relate the fault coverage of a test operation to the yield of units being tested and to the resulting yield after the test operation has identified faults.

7.3 Relating Fault Coverage to Yield

Let's next define a *test step*. Test steps have all the same attributes as other types of process steps — namely, labor, material, tooling, and equipment contributions, and the introduction of their own defects. In addition to these characteristics, test steps can also remove products from the process (scrapping). The first attribute of a test step to consider is the outgoing yield. A basic test step is shown in Figure 7.2.

Let's determine the number of units that pass the test step (M) and the outgoing yield (Y_{out}). Note that testing does not improve the yield of a process — rather, it provides a method by which good and bad units can be segregated. (If the test step does not introduce any new defects, the net yield out (passed and scrapped) is the same as the yield in).

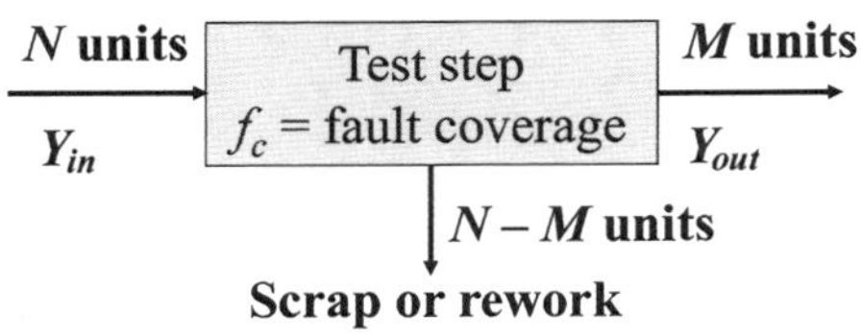

Fig. 7.2. Basic test step.

7.3.1 *A Tempting (but Incorrect) Derivation of Outgoing Yield*

Consider the following example. In Figure 7.2, let $N = 100$ units and the incoming yield be $Y_{in} = 90\%$ (0.9). This data implies that there will be (100)(0.9) = 90 good (non-defective) units and (100)(1-0.9) = 10 bad units (one or more defects) entering the test step. The fault coverage of the test step is $f_c = 80\%$ (0.8), assuming for simplicity that there is only a single fault type. In this case there will be 90 good units leaving the test (assuming the test step does not introduce any new defects and that there are no false positives — see Section 7.5).

It is tempting to claim that the number of bad units that are scrapped by the test is (0.8)(10) = 8, i.e., 80% of the bad units are correctly detected by the test step. If this were the case, (1-0.8)(10) = 2 bad units would be missed by the test and not be scrapped. So, $M = 90 + 2 = 92$ units are passed by the test step (90 good units and 2 bad units). In this case the outgoing yield would be given by

$$Y_{out} = 1 - \frac{2}{92} = 0.9783$$

Fortunately, this yield is too small and M is too large — that is, the test step actually does a better job than this. Why?

7.3.2 *A Correct Interpretation of Fault Coverage*

To illustrate the error in the example in Section 7.3.1, consider the situation shown in Figure 7.3.

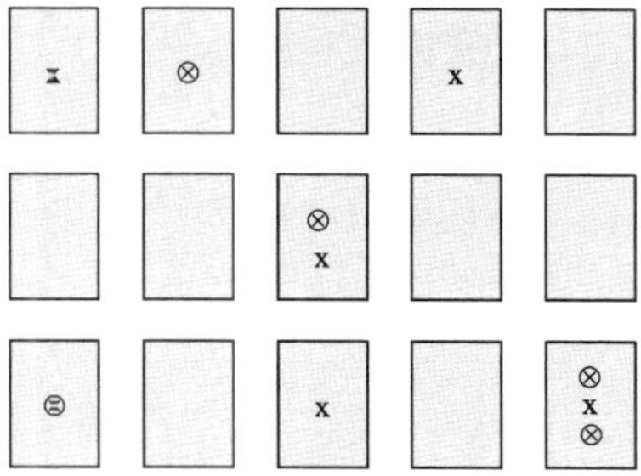

Fig. 7.3. 15 units, with 10 defects (x) subjected to a test step with a fault coverage of 0.5.

In Figure 7.3 exactly half the defects are detected by the test (every other defect is circled as an example of this). Counting units, we can see that there are N = 15 total units going into the test activity; 8 are good (without defects), 7 are bad and the incoming yield is equal to, $Y_{in} = 8/15 = 0.5333$. Treating this case like the previous example, we would have predicted that the number of units passed by the test would be $M = 8 + (1-0.5)(7) = 11.5$, giving an outgoing yield of $Y_{out} = 8/11.5 = 0.6958$.[6] In reality the number of units passed by the step (simply counting the units with no circled x's in Figure 7.3 is $M = 8 + 3 = 11$, giving an outgoing yield of $Y_{out} = 8/11 = 0.7273$).

[6] Don't be too concerned about that fact that we are dealing with fractions of units and not rounding them to whole units. If you are uncomfortable with this, multiply all the quantities we are working with by 10 or 100.

The original calculation of Y_{out} would have been correct if the fault coverage represented the fraction of faulty units detected by the test; however, fault coverage is the fraction of faults detected, not the fraction of faulty units detected. The original calculation of Y_{out} would still be correct if the maximum number of faults per unit was one, but in the example shown in Figure 7.3 this is obviously not the case. The reason that real test steps perform better (in the sense that they detect and scrap a larger portion of the defective units) than the results with the misinterpreted fault coverage is that a defective unit may have more than one defect in it; but the test only needs to successfully detect one fault to remove the unit from the process.

7.3.3 *A Derivation of Outgoing Yield (Y_{out})*

This section derives a general relationship for Y_{out} in terms of Y_{in} and fault coverage (the fraction of faults detected by the test), following the derivation of Williams and Brown [Ref. 7.6].[7]

To start the derivation we first need to review some results from probability theory. The binomial probability mass function is given by

$$\Pr(k;n,p) = \frac{n!}{k!(n-k)!} p^k (1-p)^{n-k} \tag{7.14}$$

$\Pr(k;n,p)$ is the probability of obtaining exactly k successes in n independent Bernoulli trials.[8] In our context, Equation (7.14) will be the probability of exactly k faults in a space where n faults are possible (all faults equally likely) and the probability of a single fault occurring is p.

---

[7] Note, a similar derivation and result to that in Williams and Brown's work appeared at approximately the same time in Agrawal *et al.* [Ref. 7.7], see Section 7.3.4.

[8] Equation (7.14) is derived in every introductory text on probability. The simplest application of it is flipping coins, where $\Pr(k;n,p)$ is the probability of obtaining exactly k heads when flipping the coin n times (or flipping n coins), where the probability of obtaining a head on a single flip is p. The equation assumes only two states are possible (heads or tails) — that is, it is binomial. Equations (7.14) and (7.15) are the same as Equations (3.6) and (3.7) in Section 3.2.1.

The yield (the probability of all possible faults being absent) in this case is given by

$$Y = \Pr(0;n,p) = (1-p)^n \tag{7.15}$$

Another basic concept from probability theory that we need for our development is sampling without replacement. Consider a box containing *n* things, *k* of which are defective. We draw one thing out at random. The probability of getting a defective thing is *k/n* (on the first draw or with replacement), so drawing out *m* things (without replacement, i.e., not replacing each thing after it is drawn) is the probability that exactly *x* of the *m* things drawn out are defective:[9]

$$f(x) = \frac{\begin{pmatrix} k \\ x \end{pmatrix}\begin{pmatrix} n-k \\ m-x \end{pmatrix}}{\begin{pmatrix} n \\ m \end{pmatrix}} \tag{7.16}$$

Equation (7.16) is known as the hypergeometric distribution (or hypergeometric probability mass function).

The problem is to determine the probability of a test activity <u>not</u> finding any faults ($x = 0$), when *k* faults are actually present, given that the test activity can see *m* faults out of *n* possible faults (*n-m* faults cannot be seen by the test). Note that *m/n* is the fault coverage. Another way of stating the problem is: What is the probability of testing for *m* faults out of *n* possible faults, when the device under test has *k* faults and none of the *m* faults that the test activity can detect are part of the *k* faults that are present ($x = 0$)?

As an example of using the hypergeometric distribution, consider the simple example shown in Figure 7.4. In the figure, there are $n = 8$ possible faults (*n* things), $k = 3$ faults are actually present, and $m = 4$ of the possible faults can be detected with the test (*m* things are drawn out).

---

[9] We have used the following notation:

$$\begin{pmatrix} k \\ x \end{pmatrix} = \frac{k!}{x!(k-x)!}$$

This is known as the binomial coefficient — "*k* choose *x*," the number of combinations of *k* distinguishable things taken *x* at a time.

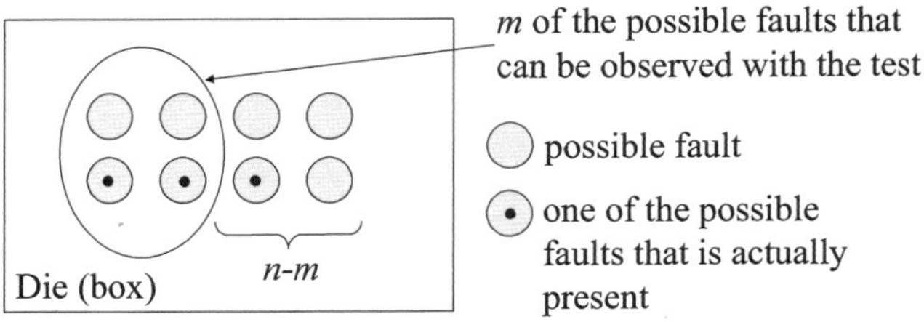

Fig. 7.4. Die as a box example.

What is the probability that the test activity won't uncover (i.e., won't draw out) any ($x = 0$) of the exactly k faults that are present? Substituting $x = 0$ into Equation (7.16),

$$f(x=0) = \frac{\binom{k}{0}\binom{n-k}{m-0}}{\binom{n}{m}} = \frac{\binom{n-k}{m}}{\binom{n}{m}} \tag{7.17}$$

The probability of accepting (passing) a die with exactly k faults (when m out of the n possible faults are tested for) is given by

$$P_k = \Pr(k;n,p)\frac{\binom{n-k}{m}}{\binom{n}{m}} = \binom{n}{k}p^k(1-p)^{n-k}\frac{\binom{n-k}{m}}{\binom{n}{m}} \tag{7.18}$$

Reducing the binomial coefficient terms we obtain:

$$\frac{\binom{n}{k}\binom{n-k}{m}}{\binom{n}{m}} = \frac{(n-m)!}{k!(n-m-k)!} = \binom{n-m}{k} \tag{7.19}$$

To get the probability of accepting a die with one or more faults, we must sum P_k over all k from 1 to n-m (the maximum number of faults is n-m; the rest are detectable using the test):

$$P_{bad} = \sum_{k=1}^{n-m}\binom{n\text{-}m}{k}p^k(1-p)^{n-k} \tag{7.20}$$

Equation (7.20) can be reduced to the following quantity (see Problem 7.6):

$$P_{bad} = (1-p)^m - (1-p)^n \tag{7.21}$$

The defect level is given by

$$\text{defect level} = \frac{\text{Probability that a bad die is accepted}\,(P_{bad})}{P_{bad} + \text{Probability that a good die is accepted}} \tag{7.22}$$

Note the denominator of Equation (7.22) is not 1.0; rather, it is only the probability that a die (good or bad) is accepted — that is, the pass fraction (introduced in Section 7.4). The second term in the denominator is the yield (if there are no false positives). Substituting from Equations (7.15) and (7.21) we obtain

$$\text{defect level} = \frac{(1-p)^m - (1-p)^n}{(1-p)^m - (1-p)^n + (1-p)^n} = 1-(1-p)^{n-m} \tag{7.23}$$

Further manipulating Equation (7.23) and substituting and rewriting it in terms of yield,

$$\text{defect level} = 1-(1-p)^{n-m} = 1-\left[(1-p)^n\right]^{\frac{n-m}{n}} = 1-Y^{\frac{n-m}{n}} \tag{7.24}$$

Realizing that m/n is the fault coverage (f_c) and that the yield out of the test is 1 minus the defect level,

$$Y_{out} = 1 - \text{defect level} = Y_{in}^{1-f_c} \tag{7.25}$$

where Y_{in} is the yield of units entering the test activity, Y_{out} is the yield of units that have been passed by the test activity and f_c is the fault coverage associated with the test activity. Equation (7.25) is the fundamental result from Williams and Brown [Ref 7.6] that forms the basis for much of test economics and the modeling of test process steps.

We can gain some intuitive understanding of Equation (7.25) by constructing a plot. Figure 7.5 shows the outgoing yield versus fault coverage for various values of incoming yield.

In Figure 7.5, as fault coverage approaches 100%, outgoing yield is 100% independent of the incoming yield. This makes sense because at

100% fault coverage the test step successfully scraps every defective unit (regardless of the fraction of units that are defective coming into the test), only letting good units pass. When fault coverage drops to 0, the outgoing yield should equal the incoming yield (the test is not doing anything). When the incoming yield is 100%, every incoming unit is good and therefore every outgoing unit is also good, regardless of fault coverage. As the incoming yield becomes small, the output yield is also small for all but fault coverages that approach 100%.

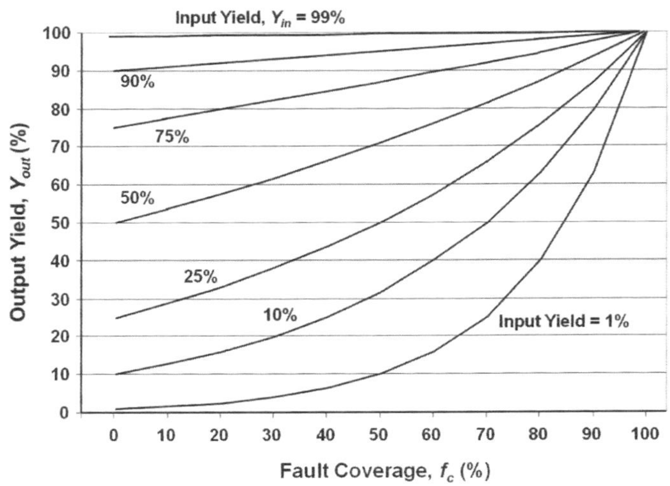

Fig. 7.5. Outgoing yield versus fault coverage from Equation (7.25).

Returning to the simple example in Section 7.3.1, let $N = 100$ units and the incoming yield, Y_{in} = 90% (0.9). This implies that there will be (100)(0.9) = 90 good (non-defective) units and (100)(1-0.9) = 10 bad units (one or more defects) entering the test step. If the fault coverage of the test step is f_c = 80% (0.8). In this case there will be 90 good units leaving the test and the outgoing yield is given by (7.25) as

$$Y_{out} = (0.9)^{1-0.8} = 0.9791$$

which is larger than the 0.9783 that resulted from the incorrect interpretation of fault coverage.

7.3.4 *An Alternative Outgoing Yield Formulation*

While the Brown and Williams result in Equation (7.25) is simple and widely used, it suffers from a potential problem that limits its accurate application to some types of testing [Ref. 7.8]. The model disregards defect clustering, assuming a Poisson distribution of defects (this assumption is embedded in Equation (7.15)), whereas the distribution when defects are clustered tends to be negative binomial. Agrawal *et al.* [Ref. 7.7] proposed an alternative model that includes clustering. In this model the outgoing yield is given by

$$Y_{out} = 1 - \frac{Y_{bg}}{Y_{in} + Y_{bg}} \tag{7.26}$$

where, Y_{bg} is the probability (or yield) of a bad unit being tested as good. This is given by

$$Y_{bg} = (1 - f_c)(1 - Y_{in})e^{-(n_o - 1)f_c}$$

where n_o is the average number of defects per unit. The derivation of Equation (7.26) is virtually identical to that of Equation (7.25), except that Pr($k;n,p$) is given by a negative binomial distribution that assumes that the likelihood of an event occurring at a given location increases linearly with the number of events that have already occurred at that location (clustering) [Ref. 7.9].

7.4 A Test Step Process Model

The results developed in Section 7.3 allow us to determine the yield of units that pass test steps. In this section we will complete the process step model for a test activity. The usefulness of such a model should be apparent. It can be used in sequence with other fabrication and assembly process steps as part of a larger process-flow model and in conjunction with rework models (see Chapter 8). Figure 7.6 shows the fundamental test step that we wish to formulate. In Figure 7.6, C_{test} is the cost of performing the test per unit (product instance) tested, S is the fraction of the incoming product scrapped by the test step, and the functional form of

Y_{out} has been given in Equation (7.25).[10] We wish to determine the functional form of C_{out} and S in terms of C_{in}, Y_{in}, C_{test}, and f_c.

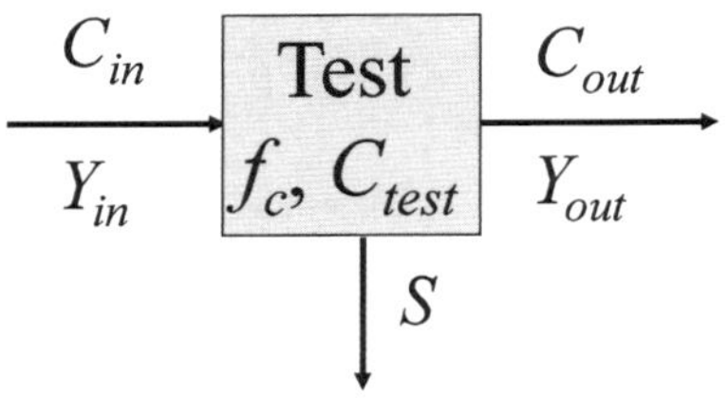

Fig. 7.6. Fundamental test step.

Our first guess at a value of the resulting outgoing cost might be $C_{out} = C_{in} + C_{test}$. This is in fact the actual money spent on the units that pass the test. But what about the units that do not pass the test (scrapped units)? $C_{in} + C_{test}$ has also been expended on each scrapped unit. The money spent on the scrapped units cannot be ignored; it is not reimbursed when the units reach the scrap heap. The effective cost of each passed unit, including an allocation of the money spent on the scrapped units, is given by

$$C_{out} = C_{in} + C_{test} + \frac{N_S\left(C_{in} + C_{test}\right)}{N_P} \tag{7.27}$$

where N_S is the number of units scrapped and N_P is the number of units passed. Note that we would expect C_{out} to reduce to $C_{in} + C_{test}$ if the scrap equaled zero (implying that $N_S = 0$) due to either an input yield of 100% or a fault coverage (f_c) of zero.

In order to rewrite Equation (7.27) in terms of C_{in}, Y_{in}, C_{test}, and f_c, we must analyze the number of units moving through the test step, Figure 7.7. Units are conserved by the process step, therefore

$$N_G + N_B = N_S + N_P \tag{7.28}$$

---

[10] The remaining development in this chapter uses Williams and Brown Equation (7.25) result; however, it could also be performed using the Agrawal *et al.* result in Equation (7.26).

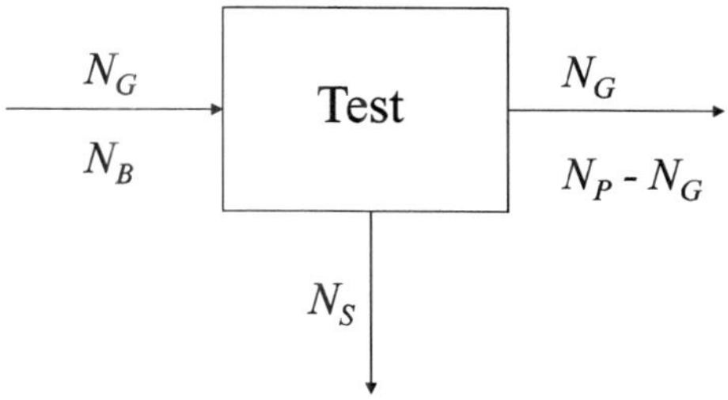

Fig. 7.7. Number of units moving through a test step. N_G = number of good units entering the test step, N_B = number of bad (defective) units entering the test step, N_P = total number of units passed by the test step, and N_S = total number of units scrapped by the test step.

Using the definition of yield out, $Y_{out} = \frac{N_G}{N_P}$, the number of units scrapped is given by

$$N_S = N_G + N_B - \frac{N_G}{Y_{out}} \tag{7.29}$$

By definition, the scrap fraction (S) is given by

$$S = \frac{N_S}{N_G + N_B} \tag{7.30}$$

and the pass fraction is

$$P = 1\text{-}S \text{ or } P = \frac{N_P}{N_G + N_B} \tag{7.31}$$

Substituting $Y_{out} = N_G/N_P$ into Equation (7.31) we obtain

$$P = \frac{N_G}{Y_{out}(N_G + N_B)} \tag{7.32}$$

Realizing that $Y_{in} = \frac{N_G}{N_G + N_B}$ and using Equation (7.25) we obtain

$$P = Y_{in}^{f_c} \text{ and } S = 1\text{-}Y_{in}^{f_c} \tag{7.33}$$

Substituting Equations (7.30), (7.31), and (7.33) into Equation (7.27), we obtain

$$C_{out} = C_{in} + C_{test} + \frac{1\text{-}Y_{in}^{f_c}}{Y_{in}^{f_c}}(C_{in} + C_{test}) \tag{7.34}$$

which, when reduced, becomes

$$C_{out} = \frac{C_{in} + C_{test}}{Y_{in}^{f_c}} \tag{7.35}$$

Equation (7.35) is the final form of C_{out} that we will use in test step process modeling.

7.4.1 *Test Escapes*

Test escapes are the bad units that are passed by the test step. Test engineers would define this as a Type II tester error [Ref. 7.10]. The number of test escapes can be seen in Figure 7.7 (N_P-N_G). A more useful general measure of test escapes is the escape fraction (E). The escape fraction is given by

$$E = \frac{N_P - N_G}{N_G + N_B} = \frac{N_P - N_G}{N_G} Y_{in} \tag{7.36}$$

Rearranging terms we obtain

$$E = \frac{N_G}{Y_{out} N_G} Y_{in} - \frac{N_G}{N_G} Y_{in} = \frac{Y_{in}}{Y_{out}} - Y_{in}$$

where we have used the fact that $N_P = N_G/Y_{out}$. Finally using Equation (7.25), we obtain

$$E = Y_{in}^{f_c} - Y_{in} \tag{7.37}$$

7.4.2 *Defects Introduced by Test Steps*

Test steps, like all other types of process steps, can introduce their own defects. For example, probes used to contact test pads on boards can damage the pads or the underlying circuitry, or defects can be introduced through handling when loading or unloading a sample into a tester.

If the defects (characterized by Y_{test}) are introduced on the way into the test activity prior to the application of the test, then we can simply replace all instances of Y_{in} with $Y_{in}Y_{test}$ in Equations (7.25), (7.35) and (7.33):

$$Y_{out} = \left(Y_{in} Y_{test}\right)^{1-f_c} \tag{7.38a}$$

$$C_{out} = \frac{C_{in} + C_{test}}{\left(Y_{in} Y_{test}\right)^{f_c}} \tag{7.38b}$$

$$S = 1\text{-}\left(Y_{in} Y_{test}\right)^{f_c} \tag{7.38c}$$

Similar relations can be found for the pass fraction and escape fraction.

Alternatively, if the defects are introduced on the way out of the test activity (after the actual application of the test), then the relations for C_{out} and S are unchanged and only Y_{out} is modified:

$$Y_{out} = Y_{in}^{1\text{-}f_c} Y_{test} \tag{7.39}$$

7.5 False Positives

A *false positive* is defined as a positive test result in subjects that do not possess the attribute for which the test is conducted. Test engineers would define false positives as a Type I tester error [Ref. 7.10]. In testing, this means that a test step will erroneously identify good units as bad at some non-negligible rate. In fact, data at the board and system level has shown that as many as 46% of all identified failures are not actually failures, but false positives [Ref. 7.11]. Recall from the introduction to this chapter that one of the goals of test economics is to "minimize the cost of discarding good products"; false positives are the dominant mechanism by which good products are discarded.

False positives may occur for many reasons, including intermittent contact of test pins, operator error, misinterpretation of data, poor design of load boards, or poor characterization of the automatic test equipment [Ref. 7.11]. A study of the economic impact of false positives using actual Honeywell data is provided in [Ref. 7.11].

The treatment of false positives affects both the number of units moving through the process and the yield of those units. The test step is characterized by both fault coverage and false positives, where f_p is the probability of testing a good unit as bad. (This should not be confused with the escape fraction, E, which is the probability of testing bad units as good). Parameter f_p is a function of the tester quality, not the fault coverage.

Let the number of units that come into the test affected by the false positives be N_{in} and the yield coming in be Y_{in}. Let the number of units going out (after false positives are created) be N_{out} and their yield be Y_{out}. These units consist of both good (g) and bad (b) units such that $N_{in}=N_{ing}+N_{inb}$ and $N_{out}=N_{outg}+N_{outb}$ (Figure 7.8).

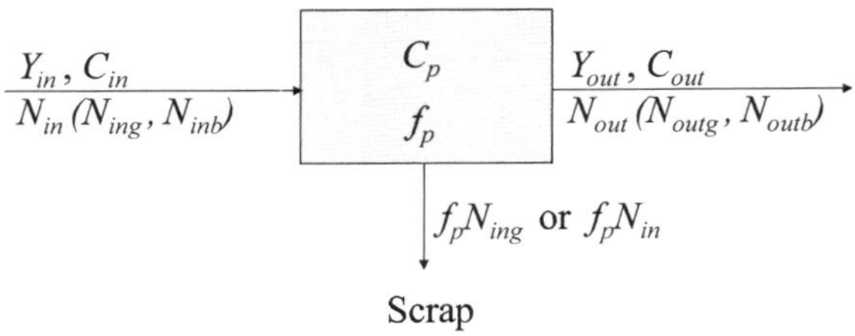

Fig. 7.8. Notation for false positive formulations.

In Figure 7.8, C_p is the portion of the test cost incurred to create false positives. There are several approaches to modeling the effect of the false positives. If we assume that the number of false positives sent to scrap by the test step will be f_pN_{ing}, based on the assumption that false positives only act on good units. The false positive fraction is given by

$$f_p = \frac{N_{ing} - N_{outg}}{N_{ing}} \tag{7.40a}$$

The cost, yield and scrap are modified as follows:

$$Y_{out} = \frac{N_{outg}}{N_{out}} = \frac{(1-f_p)N_{ing}}{(N_{in} - f_pN_{ing})} = \frac{(1-f_p)Y_{in}}{1-f_pY_{in}} \tag{7.41a}$$

$$C_{out} = \frac{C_{in}+C_p}{P} = (C_{in}+C_p)\frac{N_{in}}{N_{out}} = (C_{in}+C_p)\frac{N_{in}}{N_{in}-f_pN_{ing}} = \frac{C_{in}+C_p}{1-f_pY_{in}} \tag{7.42a}$$

$$S = \frac{f_pN_{ing}}{N_{in}} = f_pY_{in} \tag{7.43a}$$

Note that we are only considering the false positives portion of the test activity here (not the fault coverage portion). An alternative assumption is that the number of false positives sent to diagnosis by the test step will be

f_pN_{in}, based on the assumption that false positives act on all units.[11] The false positive fraction is given by

$$f_p = \frac{N_{in} - N_{out}}{N_{in}} \tag{7.40b}$$

and the cost, yield and scrap are modified as follows:

$$Y_{out} = \frac{N_{outg}}{N_{out}} = \frac{(1-f_p)N_{ing}}{(1-f_p)N_{in}} = \frac{N_{ing}}{N_{in}} = Y_{in} \tag{7.41b}$$

$$C_{out} = \frac{C_{in}+C_p}{P} = (C_{in}+C_p)\frac{N_{in}}{N_{out}} = (C_{in}+C_p)\frac{N_{in}}{N_{in}-f_pN_{in}} = \frac{C_{in}+C_p}{1-f_p} \tag{7.42b}$$

$$S = \frac{f_pN_{in}}{N_{in}} = f_p \tag{7.43b}$$

In other words, f_p in this case reduces the good and bad units proportionately, thus leaving the yield unchanged.

7.5.1 *A Test Step with False Positives*

Let's include the notion of false positives within the test step developed in Section 7.4. To construct the formulation we must first make an assumption about when the false positives occur relative to the fault coverage portion of the test step. Let's assume that the false positives are introduced prior to the fault coverage (Figure 7.9).

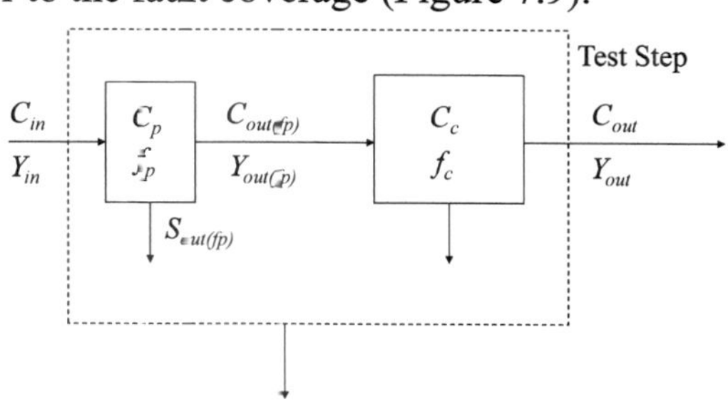

Fig. 7.9. Test step with false positives introduced prior to fault coverage, where $C_p + C_c = C_{test}$.

[11] In this case, the false positives can be created from already defective units — defective units detected as defective by the test step for the wrong reasons.

In Figure 7.9, $C_{out(fp)}$, $Y_{out(fp)}$ and $S_{out(fp)}$ are derived from Equations (7.41) through (7.43). Applying Equations (7.25) and (7.35) to the process in Figure 7.9 gives

$$Y_{out} = Y_{out(fp)}^{1-f_c} \tag{7.44}$$

$$C_{out} = \frac{C_{out(fp)} + C_c}{Y_{out(fp)}^{f_c}} \tag{7.45}$$

The net scrap from the test step is a bit more complicated to formulate. The total scrap is the scrap from the false positives portion of the step added to the scrap from the fault coverage portion of the step, as follows (see Section 7.6 for more discussion on computing S for cascaded process steps):

$$S = S_{out(fp)} + \left(1 - S_{out(fp)}\right)\left(1 - Y_{out(fp)}^{f_c}\right) \tag{7.46}$$

As an example, assume that f_p represents the false positives on all units (good and bad). In this case, Equations (7.44) through (7.46) reduce to

$$Y_{out} = Y_{in}^{1-f_c} \tag{7.47}$$

$$C_{out} = \frac{\dfrac{C_{in} + C_p}{1 - f_p} + C_c}{Y_{in}^{f_c}} = \frac{C_{in} + C_p + \left(1 - f_p\right)C_c}{\left(1 - f_p\right)Y_{in}^{f_c}} \tag{7.48}$$

$$S = f_p + \left(1 - f_p\right)\left(1 - Y_{in}^{f_c}\right) \tag{7.49}$$

It is easy to check some limiting cases of this solution. If $f_p = 0$ (no false positives), then Equations (7.47) through (7.49) reduce to Equations (7.25), (7.35) and (7.33). If $f_p = 1$ (every device under test is identified as a false positive), then $S = 1$ (everything is scrapped).

Assuming, alternatively, that the false positives affect the test after the fault coverage and that f_p represents the probability of a false positive in a good unit only, then Equation (7.41a) results in

$$Y_{out} = \frac{\left(1 - f_p\right)Y_{in}^{1-f_c}}{1 - f_p Y_{in}^{1-f_c}} \tag{7.50}$$

which is equivalent to the false positives result derived in [Ref. 7.12].

7.5.2 *Yield of the Bonepile*

The yield (fraction of good units) in the set of units scrapped by the test activity is called the bonepile yield [Ref. 7.12]. In the case where f_p represents the fraction of false positives on just good units,

$$Y_{BP} = \frac{f_p Y_{in}}{f_p Y_{in} + (1 - f_p Y_{in})\left[1 - \left(\frac{(1 - f_p) Y_{in}}{1 - f_p Y_{in}}\right)^{f_c}\right]} \tag{7.51a}$$

In Equation (7.51a), Y_{BP} is the number of good units scrapped (N_{in} multiplied by Equation (7.43a)) divided by the total number of units scrapped (N_{in} multiplied by Equation (7.46). using Equation (7.41a)). Trivial cases of Equation (7.51a) can be checked if $f_c = 0$, $Y_{BP} = 1$ and, $f_p = 0$, $Y_{BP} = 0$. Similarly, in the case where f_p represents the fraction of false positives on all units,

$$Y_{BP} = \frac{f_p Y_{in}}{f_p + (1 - f_p Y_{in})(1 - Y_{in}^{f_c})} \tag{7.51b}$$

7.6 Multiple Test Steps

It usually makes sense to test at more than one point in a process. If a process step that inserts a large number of defects into a product has just been completed, it may be prudent to test before continuing to spend money processing a defective product. Alternatively, before starting a process step that is going to cost a lot, it may be advisable to test so that the expensive processing is not wasted on an already defective product. Either way, the decision to test comes down to a tradeoff between using resources to perform a test and the possibility of wasting resources on processing a product that is already defective. Multiple test steps are also a method of modeling the details of different aspects of a single test activity — test activities that treat more than one fault type where the fault types treated have different fault coverages.

7.6.1 *Cascading Test Steps*

Figure 7.10 shows a pair of cascaded test steps. The formulation in this case is relatively straightforward except for the treatment of the scrap, since it is calculated as a fraction of the units that start the entire process.

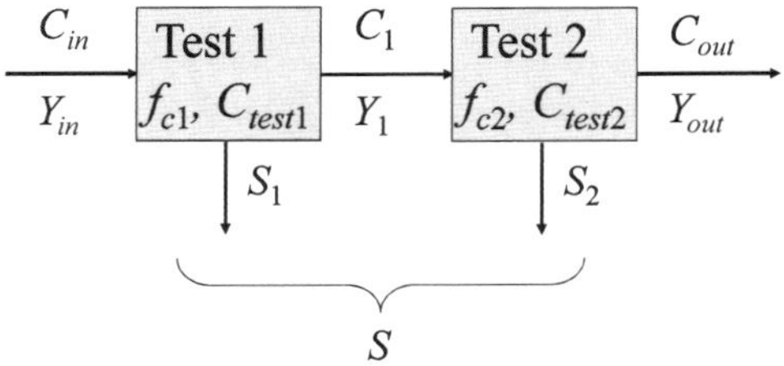

Fig. 7.10. Cascaded test steps.

Y_1, C_1, and S are computed from Equations (7.25), (7.35) and (7.33) or variations thereof, as discussed in the preceding sections. Y_1 and C_1 then replace Y_{in} and C_{in} in Equations (7.25) and (7.35) to compute the final outgoing cost and yield. However, the calculation of the total scrap (S) is a bit more complicated because S is a fraction of the quantity of units that start the process (but S_2 is a fraction of only the quantity of units that start the Test 2 step). For the case shown in Figure 7.10, the total scrap fraction is given by

$$S = \left(1 - Y_{in}^{f_{c1}}\right) + Y_{in}^{f_{c1}}\left(1 - Y_1^{f_{c2}}\right) \tag{7.52}$$

The first term in Equation (7.52) is S_1 and the second term is the product of the pass fraction from Test 1 and the scrap fraction S_2. Reducing Equation (7.52) and using $Y_1 = Y_{in}^{1-f_{c1}}$, we obtain

$$S = 1 - Y_{in}^{f_{c1}} Y_{in}^{f_{c2}(1-f_{c1})} \tag{7.53}$$

7.6.2 *Parallel Test Steps*

Figure 7.11 shows a pair of parallel test steps. In the figure, $Y_{in} = Y_{in1}Y_{in2}$ where Y_{in1} and Y_{in2} could represent the product yield with respect to different independent defect mechanisms. If this is the case, then

$$Y_{out} = Y_1 Y_2 = Y_{in1}^{1-f_{c1}} Y_{in2}^{1-f_{c2}} \quad (7.54)$$

$$C_{out} = \frac{C_{in} + C_{test1}}{Y_{in1}^{f_{c1}}} + \frac{C_{in} + C_{test2}}{Y_{in2}^{f_{c2}}} \quad (7.55)$$

$$S = S_1 + S_2 = \left(1 - Y_{in1}^{f_{c1}}\right) + \left(1 - Y_{in2}^{f_{c2}}\right) \quad (7.56)$$

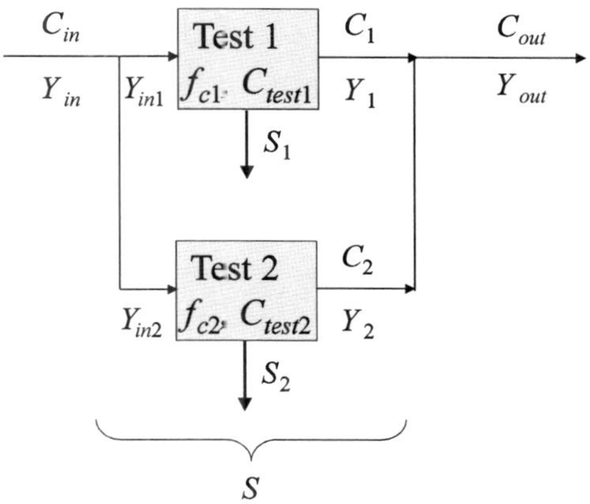

Fig. 7.11. Parallel test steps.

7.7 Financial Models of Testing

Sections 7.2 – 7.6 of this chapter treat the fundamental defining attribute of a test activity — namely, its ability to identify and scrap defective units. Beyond this unique ability, test steps have properties in common with all other types of process steps (equipment, tooling/programming, recurring labor, design/development and material costs).

A complete picture of test cost consists of several components, as shown in Figure 7.12. The test cost is a sum of the costs of these components [Ref. 7.13]. Test preparation includes the fixed costs associated with test generation, test program creation, and any design effort for incorporating test-related features. Test execution includes the costs of all the test hardware (hardware tooling) and the cost of the tester itself (including the capital investment, its maintenance, and facilities).

Test-related silicon captures the cost of incorporating specific design for test (DFT) features into the integrated circuits (see Section 7.8.3 for a discussion of DFT). Finally, imperfect test quality includes the effects of test escapes and defects introduced by the testing activity.

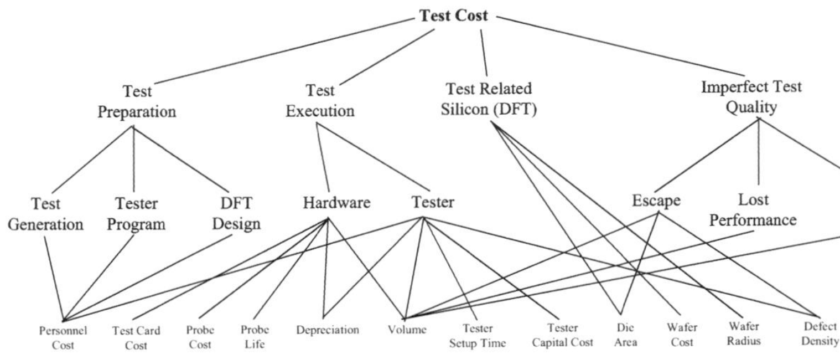

Fig. 7.12. Test cost dependency tree for an integrated circuit [Ref. 7.13].

The majority of the elements that appear in Figure 7.12 can be treated using the general methods developed previously in this book, including process-flow modeling (Chapter 2) and cost-of-ownership modeling (Chapter 4). Several detailed financial models have appeared in the literature that implement all or a portion of the dependencies shown in Figure 7.12. These include: Nag *et al.* [Ref. 7.13] and Volkerink *et al.* [Ref. 7.14]. In [Ref. 7.14], the effects of time-to-market delays that may be associated with test development are also included.

7.8 Other Test Economics Topics

There are many other topics within functional testing that have an economic impact on the system being fabricated. In this section we briefly introduce several of these topics.

7.8.1 *Wafer Probe (Wafer Sort)*

In the context of this chapter, wafer probing represents a test activity with a delayed ability to scrap identified defective units. Generally speaking, wafer probing or testing would be the first time that die fabricated on a wafer are functionally tested. There are three basic elements involved in the wafer probing operation. First, the wafer prober is a material handling system that takes wafers from their carriers, loads them into a flat chuck, and aligns and positions them precisely under a set of fine contacts on a probe card. Mostly, this test is performed at room temperature, but the prober may also be required to heat or cool the wafer during the test. Secondly, each input/output or power pad on the die must be contacted by a fine electrical probe. This is done with a probe card, whose job is to translate the small individual die-pad features into connections to the tester. Thirdly, the functional tester or automatic test equipment (ATE) must be capable of functionally exercising the chip's designed features under software control. Any failure to meet the published specifications is identified by the tester and the device is catalogued as a reject. The tester/probe card combination may be able to contact and test more than one die at a time on the wafer. This parallel test capability enhances the productivity of the wafer probe.

Die (individual unpackaged chips) that are catalogued as rejects are marked (traditionally using a drop of ink) or by digitally registering the location of individual defective die. Since the die are part of a larger wafer with many die on it, and it probably is not practical to immediately separate them from the wafer, the rejected die must continue in the process and be scrapped later (see Figure 7.13).[12]

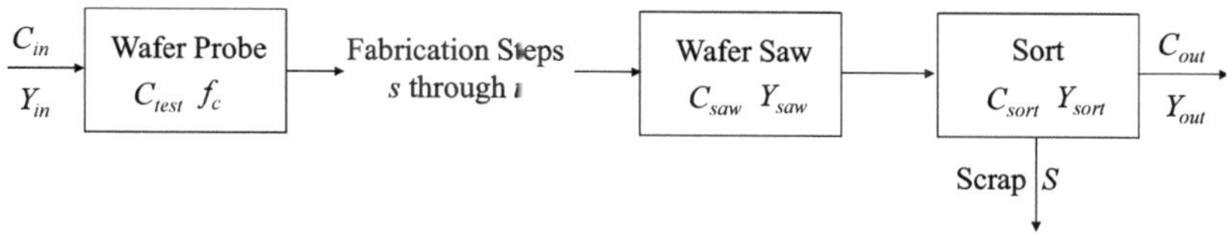

Fig. 7.13. Testing during wafer fabrication.

The important attribute is that the outgoing cost of a wafer probe test step is simply $C_{in} + C_{test}$ (since no die are actually scrapped at the test step). The defective die continue to be processed until after the die are singulated from the wafer and a "sorting" step is encountered. At the sorting step, the

[12] This applies unless enough die on the wafer are defective to make it more economical to scrap the entire wafer than to continue processing it.

marked die are finally scrapped. General relations for the cost and yield of individual die in a wafer probing situation are,

$$C_{out\ per\ die} = \frac{C_{in} + C_{test} + \sum_{k=s}^{t} C_{step_k} + C_{saw} + C_{sort}}{N_u Y_{in}^{f_c}} \tag{7.57}$$

$$Y_{out} = Y_{in}^{1\text{-}f_c}\left[\prod_{k=s}^{t} Y_k\right] Y_{saw} Y_{sort} \tag{7.58}$$

$$S = 1 - Y_{in}^{f_c} \tag{7.59}$$

where N_u (number up) is the number of die on the wafer, and C_{in}, C_{test}, C_{stepk}, C_{saw} and C_{sort} are assumed to be wafer costs while Y_{in}, Y_k, Y_{saw}, and Y_{sort} are assumed to be die yields.

Boards, which are fabricated on panels are subject to the same model as die on wafers.

7.8.2 *Test Throughput*

A key economic contributor to the recurring cost of testing is throughput. The process of performing a functional test on a complex system can be long [see Ref. 7.15]. Functional testing can be a bottleneck in the production process for ICs, boards, and systems. In general, the test throughput rate (units/time) is given by

$$TPT_t = \frac{1}{T_p Y_{in} + T_f\left(1 - Y_{in}\right) + T_h + T_t} \tag{7.60}$$

where

Y_{in} = the incoming yield.
T_p = the average pass time.
T_f = the average fail time.
T_h = the handling time (loading the tester).
T_t = the dead time (between samples).

Equation (7.60) assumes a single tester in the process sequence. Note that the times for passing good units and failing bad units can be different. This is because, in general, it takes substantially longer to pass a good unit than to fail a bad unit because testing can stop when the first fault is found (there is no need for the tester to find all the faults unless a rework activity is planned). Consequently, tests are organized to look for the most common fault first and the least common fault last. Alternatively, every test vector must be applied to determine that a good unit is in fact good.

7.8.3 *Design for Test (DFT)*

The semiconductor industry has been very successful in satisfying Moore's Law over the last twenty years.[13] One of the by-products of the increasing technological ability of the semiconductor industry has been a steadily decreasing cost per transistor. Unfortunately, the cost of functional testing per transistor has not followed the same relation.

The reason for the cost trend shown in Figure 7.14 is that the performance of today's circuits is approaching and surpassing that of the automatic test equipment. Thus it is becoming increasingly difficult and expensive to accurately test devices and circuits. The relationship shown in Figure 7.14 indicates that in about 2015 it will be less expensive to make a transistor than to test one. One of the implications of this trend is that it is becoming more economical to use expensive IC real estate to fabricate special circuitry that enables faster, less expensive functional testing than to perform functional testing at the board level. The technologies associated with creating special circuitry on the IC or board are known as design for test (DFT).

Design for test can take two different forms, ad-hoc and structured. Ad-hoc DFT is based on the use of "good" design practices. Structured DFT usually takes the form of built-in self test (BIST) or scan. BIST involves the inclusion of a BIST controller that generates test patterns, controls the clock of the circuit under test and collects and analyzes the responses. The focus of the scan is to obtain control and observability for flip-flops by adding a test mode to the circuit, such that when the circuit is in test mode, all flip-flops functionally form one or more shift registers. The inputs and outputs of these shift registers (scan registers) are made into the primary

[13] Moore's Law says that the density of ICs doubles every 18 months.

inputs and outputs. This type of scan is referred to as full scan, but other variations exist. Both BIST and scan increase the size of the system — either a larger chip area and/or a larger board area.

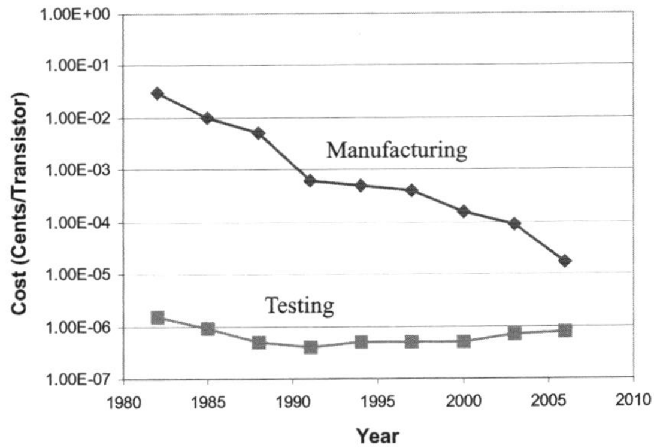

Fig. 7.14. Trends in automatic testing of ICs: Costs of manufacturing and testing transistors in the high-performance microprocessor product segment [Ref. 7.16].

The economic tradeoffs associated with structured DFT are complex. On one hand, DFT has the following potential benefits:

- better test access (higher fault coverage and better diagnostic resolution);
- higher test throughput (decrease in test time);
- more practical at-speed testing;
- less expensive test equipment;
- less time and effort needed for test tooling and programming; and
- shorter time to market (for systems that include ICs with DFT structures).

On the other hand, structured DFT does not come for free. Costs include

- more expensive and larger area ICs, and
- larger area boards with higher assembly costs.

As an example of the economic tradeoff problem associated with DFT, consider a 1 GHz microprocessor chip with 400 I/Os (pins). In order to obtain reliable results, testing should be performed at the rated clock speed of the chip. Assume that the tester costs \$6000/pin (1 GHz testers are expensive), or \$2.4M to perform this test. Alternatively, we could design and fabricate a version of the 1 GHz microprocessor chip with BIST. In this case, we will only need a tester to provide DC command signals to the microprocessor to perform the required BIST, then to read out the result from the microprocessor. In this case a 20 MHz tester that costs \$391/pin will do, so our tester cost is \$156,400, or a tester savings of \$2,243,600. So is our conclusion that using DFT is always preferable to not using DFT correct? In fact, some of the economic arguments for DFT do stop at this point. But, unfortunately, there are several other effects in play here, and we know from our knowledge of cost of ownership (Chapter 4) that high equipment costs are not always the primary driver behind a product's cost. Let's extend our economic analysis of DFT one more step (although this will still be a very rough approximation).

The first thing we need to consider is the fact that the area of the die increases when we include BIST. A die area increase translates into fewer die fabricated on a wafer, which in turn means a higher die cost. Die size increases for adding BIST range from 3% [Ref. 7.17] to 13% [Ref. 7.13], for this case we will use 5%. If the original chip (no BIST) had an area of $A_{noDFT} = 1$ cm^2, then the new die has $A_{DFT} = 1.05$ cm^2. This assumes a Seeds yield model that gives the die yield as

$$Y = \frac{1}{1 + AD} \tag{7.61}$$

where D is the defect density (assumed to be 0.222 defects/cm^2). The yields of the two die are $Y_{noDFT} = 0.818$ and $Y_{DFT} = 0.811$, the yield of the larger die being slightly lower. A rough approximation of the fabrication cost of a good die (yielded cost) is given by [Ref. 7.13]:

$$C_{fab} = \frac{Q_{wafer}}{\pi R_{wafer}^2 B_{waf\_die}} \left(\frac{A}{Y} \right) \tag{7.62}$$

where

Q_{wafer} = the fabricated wafer cost (\$1300/wafer).
R_{wafer} = the radius of the wafer (100 mm).

B_{waf_die} = the die tiling fraction that accounts for wafer edge scrap, scribe streets between die and the fact that rectangular die cannot be perfectly fit into a circular wafer. We will use 0.9.

Using Equation (7.62), the cost of fabricating a non-DFT die is \$5.62/die and a DFT die is \$5.95/die.

We also have to consider the design cost associated with the DFT die. Using a simple assumption that it costs \$500,000/cm^2 to design a die, the design costs (C_{design}) are \$500,000 for the non-DFT die and \$525,000 for the die with DFT.

We now need to take care of the tester cost. It is not realistic (at least for small volumes) to assume that a tester is purchased for only this die. Therefore, we will compute the portion of the tester cost that should be allocated to each die that is tested as

$$C_{tester} = C_{equip} \frac{T_{die}}{T_{op} D_L} \quad (7.63)$$

where

C_{equip} = the cost of purchasing the tester, facilities needed by the tester, and maintenance of the tester minus the residual value of the tester at the end of its depreciation life.

T_{die} = the effective time to load, unload, and test one die (6 seconds/die).

T_{op} = the effective operational time of the tester per year (10,512,000 seconds/year).

D_L = the depreciation life of the tester in years (4 years).

Equation (7.63) assumes that the tester is fully utilized testing something else when it is not testing the die we are concerned with. Using this equation, the effective tester cost per non-DFT die is \$0.342/die and for die with DFT is \$0.022/die. You should already be able to see that the tester cost difference of \$0.32/die is mitigated by the die fabrication cost difference of \$0.33/die.

One more non-recurring cost is the cost of a probe card to actually contact the wafer to test the die. Assuming that a probe card for the non-DFT die costs \$1000 (C_{probe}) and can test 100,000 die before needing to be replaced, the probe card of the die with DFT is simpler and only costs \$100.

Let's put it all together. The total effective cost per die in our simple model is given by

$$C = C_{fab} + C_{tester} + \frac{C_{design}}{N_D} + \frac{C_{probe}}{N_D}\left(\frac{N_D}{100{,}000}\right) \tag{7.64}$$

where N_D is the quantity of die to be fabricated.

Plotting ΔC ($C_{no\text{-}DFT} - C_{DFT}$) versus N_D we obtain the result in Figure 7.15. Figure 7.15 shows that for our simple example and assumptions, for quantities below ~3000, the inclusion of DFT is economically advantageous; for quantities between 3000 and 1,000,000 non-DFT should be used, and for quantities above 1,000,000 it doesn't make much difference.

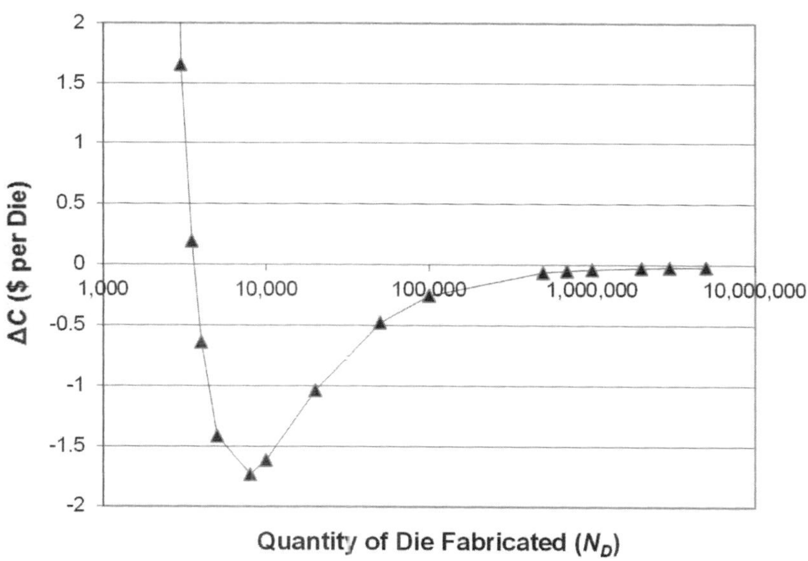

Fig. 7.15. Difference in cost between non-DFT die and die containing DFT as a function of the quantity of die fabricated. This result was computed using the simple demonstration model developed in this section.

It should be stressed that the simple model developed in this section is only for demonstration purposes and should not be used to draw any general conclusions. In fact, the model ignores many additional critical effects that will affect the applicability of DFT, including test generation costs, tester programming costs, variation in testing times, test quality (i.e., fault coverage), time-to-market costs, and yield learning. For models that include these and other effects, readers are encouraged to see Nag *et al.* [Ref. 7.13] and Ungar and Ambler [Ref. 7.18] for more detailed models that treat the application-specific tradeoffs associated with DFT.

A more general result from a more detailed model is shown in Figure 7.16. The uncertainty region in Figure 7.16 envelops the majority of the application-specific inputs. However, even the model used to create Figure 7.16 does not include time-to-market effects and assumes a very simplified number-up calculation (as in Equation (7.62)).

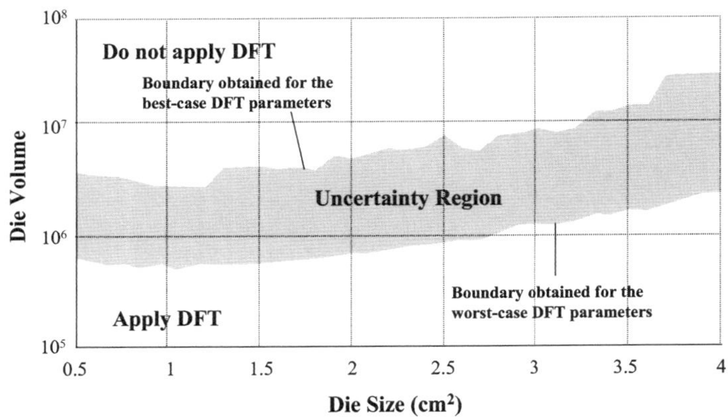

Fig. 7.16. DFT and non-DFT domains as a function of die size and production volume [Ref. 7.13].

Design for test is fundamentally a cost-avoidance proposition (see Section II.2). Traditionally, cost avoidance is a more difficult sell to customers and management than more direct returns on investment. The historical difficulty with DFT is that management often views the investment as a tradeoff between spending the money on improving the process yield or improving the detection of flaws caused by imperfect process yield. Stated in this way, management will often choose to focus company dollars on yield improvement rather than on DFT.

7.8.4 *Automated Test Equipment Costs*

The automated test equipment (ATE) cost is traditionally expressed as cost per digital pin. For example, the price of a functional tester ranged from \$8000-\$10,000 per pin in 2002. The actual price of a high-end VLSI logic tester has increased twenty-five times over the last two decades from ~\$400,000 per system in the 1980s, to \$3-\$5 million in the mid 1990s, to \$6-\$10 million for a 1024 pin, 1GHz tester in 2001 [Ref. 7.19].

Although cost per pin is a convenient metric, it is only really appropriate for digital testers. The addition of analog instruments and digital features to support mixed signal tests adds significant fixed cost per system and a small incremental cost per digital pin [Ref. 7.20]. Cost per pin is misleading because it ignores base system costs associated with equipment infrastructure and the beneficial scaling that occurs with increasing pin count. It has been suggested in [Ref. 7.16] that the following expression be used for each tester segment:

$$C_{tester} = b_t + \sum_{i=1}^{n} m_i x_i \tag{7.65}$$

where

b_t = the base cost of a test system with zero pins (scales with capability, performance and features).

m_i = the incremental cost per pin for the ith test segment (depends on memory depth, features, and analog capability).

x_i = the number of pins for the ith test segment.

n = the number of test segments.

Table 7.1. ATE Cost Parameters [Ref. 7.16].

Tester Segment	b_t (K\$)	m (\$)	x (\$)
High-performance ASIC/MPU	250-400	2700-6000	512
Mixed signal	250-350	3000-18000	128-192
DFT tester	100-350	150-650	512-2500
Low-end microcontroller/ASIC	200-350	1200-2500	256-1024
Commodity memory	200+	800-1000	1024
RF	200+	~50000	32

The summation in Equation (7.65) addresses mixed configuration test systems that provide different test pin capability (i.e., analog, RF, etc.).

Both b_t and m are expected to decrease over time for equivalent performance points. Table 7.1 provides the range of values for b_t, m and x.

References

7.1 Turino, J. (1990). *Design to Test – A Definitive Guide for Electronic Design, Manufacture, and Service*, (Van Nostrand Rienhold, New York, NY).

7.2 Rhines, W. (2002). Keynote address at the Semico Summit, Phoenix, AZ, March 2002.

7.3 Bushnell. M. L. and Agrawal, V. D. (2000). Chapter 4 - Fault modeling, *Essentials of Electronic Testing for Digital, Memory and Mixed-Signal VLSI Circuits*, (Kluwer Academic Publishers, Boston, MA).

7.4 Dislis, C., Dick, J. H., Dear, I. D. and Ambler, A. P. (1995). *Test Economics and Design for Testability for Electronic Circuits and Systems*, (Ellis-Horwood, Upper Saddle River, NJ).

7.5 Bushnell, M. L. and Agrawal, V. D. (2000). Chapter 5 - Logic and fault simulation, *Essentials of Electronic Testing for Digital, Memory and Mixed-Signal VLSI Circuits*, (Kluwer Academic Publishers, Boston, MA).

7.6 Williams T. W. and Brown, N. C. (1981). Defect level as a function of fault coverage, *IEEE Transactions on Computers*, 30(12), pp. 987-988.

7.7 Agrawal, V., Seth, S. and Agrawal, P. (1982). Fault coverage requirement in production testing of LSI circuits, *IEEE Journal of Solid-State Circuits*, SC-17(1), pp. 57-61.

7.8 de Sousa, J. T. and Agrawal, V. D. (2000). Reducing the complexity of defect level modeling using the clustering effect, *Proceedings of the IEEE Design and Test in Europe Conference*, pp. 640-644.

7.9 Stapper, C. H. (1975). On a composite model to the IC yield problem, *IEEE Journal of Solid State Circuits*, SC-10 (6), pp. 537-539.

7.10 Williams, R. H., Wagner, R. G. and Hawkins, C. F. (1992). Testing errors: Data and calculations in an IC manufacturing process, *Proceedings of the International Test Conference*, pp. 352-361.

7.11 Henderson, C. L., Williams, R. H. and Hawkins, C. F. (1992). Economic impact of type I test errors at system and board levels, *Proceedings of the International Test Conference*, pp. 444-452.

7.12 Williams, R. H. and Hawkins, C. F. (1990). Errors in testing, *Proceedings of the International Test Conference*, pp. 1018-1027.

7.13 Nag, P. K., Gattiker, A., Wei, S., Blanton, R. D. and Maly, W. (2002). Modeling the economics of testing: A DFT Perspective, *IEEE Design & Test of Computers*, 19(1), pp. 29-41.

7.14 Volkerink, E. H., Khoche, A., Kamas, L. A., Revoir, J. and Kerkhoff, H. G. (2001). Tackling test trade-offs from design, manufacturing to market using economic modeling, *Proceedings of the International Test Conference*, pp. 1098-1107.

7.15 Williams, T. W. (1985). Test length in a self-testing environment, *IEEE Design and Test of Computers*, 2(2), pp. 59-63.

7.16 Test and Test Equipment, *The International Technology Roadmap for Semiconductors, Semiconductor Industries Association*, 2001.

7.17 Bardell, P., McAnney, W. and Savir, J. (1987). *Built-in Test for VLSI, Pseudorandom Techniques*, (John Wiley & Sons, New York).

7.18 Ungar, L. Y. and Ambler, T. (2001). Economics of built-in self-test, *IEEE Design & Test of Computers*, 18(5), pp. 70-79.

7.19 LaPedus, M. (2001). Intel shifts test strategy to battle exploding costs of big ATE systems, *EETimes*, June 19.

7.20 Ortner, W. R. (1998). How real is the new SIA roadmap for mixed-signal test equipment? *Proceedings of the International Test Conference*, p. 1153.

7.21 Landman, B. S. and Russo, R. L. (1971). On a pin versus block relationship for partitions of logic graphs, *IEEE Trans on Computers*, C-20(12), pp. 1469-1479.

Bibliography

There are several basic sources of information on test economics. Good sources of information include the following:

Davis, B. (1994). *The Economics of Automatic Testing*, 2nd Edition, (McGraw-Hill, New York, NY).

IEEE Design & Test of Computers, special issue on test economics, September 1998.

Bushnell, M. L. and Agrawal, V. D. (2000). *Essentials of Electronic Testing for Digital, Memory and Mixed-Signal VLSI Circuits*. (Kluwer Academic Publishers, Boston, MA).

Steininger, A. (2000). Testing and built-in self test – A survey, *Journal of Systems Architecture*, 46, pp. 721-747.

Journal of Electronic Testing Theory and Applications (JETTA), (Kluwer Academic Publishers).

International Test Conference (ITC), IEEE Computer Society.

IEEE Design & Test of Computers, Institute of Electrical and Electronics Engineers, Inc.

Problems

7.1 Assume that you have a process that forms solder balls (for flip chip bonding) on the inner-lead bond pads on bare die. The process produces 220 ppm defects per solder ball. If each die has 484 I/Os (solder balls), what is the number of defects of defect type "defective solder ball" in the die?

7.2 What is the yield of individual die with respect to just the solder-ball forming process in Problem 7.1?

7.3 A defect spectrum is given by $\begin{pmatrix} 0.2 \\ 0.1 \\ 0.130 \end{pmatrix}$, what is the overall board yield?

7.4 Given the following conversion matrix,

$$C = \begin{pmatrix} 0.2 & 0.8 & 0.1 \\ 0.7 & 0 & 0.75 \\ 0.1 & 0.2 & 0.15 \end{pmatrix}$$

Using the data provided in Problem 7.3, determine the fault spectrum. From the fault spectrum, verify the board yield determined in Problem 7.3.

7.5 Assuming fault coverages of $f_{c1} = 0.9$, $f_{c2} = 0.98$, and $f_{c3} = 0.76$, and the data in Problem 7.3, calculate the overall defect coverage from each type of defect.

7.6 Derive Equation (7.21) from Equation (7.20).

7.7 In the limit as Y_{in} approaches zero, what happens to the Y_{out} from Equation (7.25)? Note that this is not a trivial problem. Is the equation even applicable under this condition?

7.8 Derive the Agrawal *et al.* result (Equation (7.26) and Y_{bg}) for outgoing yield, assuming a negative binomial distribution defect density distribution. Note, Y_{bg} is the same as P_{bad}.

7.9 Using the notation in Figure 7.2, and assuming that the test step neither introduces new defects nor repairs existing defects, prove that the net yield out (passed and scrapped) is the same as the yield in.

7.10 Assume that a test step has to be added to the following process flow:

Step	Time (sec/board)	Op Util	Capacity (boards)	Material Cost (per unit of material)	Units of Material (per board)	Tooling Cost	Tooling Life (number of boards)	Equip Cost	Equip Operational Time (fraction)	Defect Density (defects/sq
A	10	1	1	0	0	0	100000	150000	0.6	0.1
B	60	2	1	3.2	1	0	100000	20000	0.6	0.7
C	30	0.5	12	0.1	4	1000	20000	1000000	0.6	0.06
D	110	0.25	1	0	0	0	100000	75000	0.6	0.13
E	100	1	1	0	0	0	100000	25000	0.6	0.3
F	45	0.5	10	2	1	10000	100000	10000	0.6	0.11
G	14	1	2	0	0	5000	100000	15000	0.6	0.02
H	60	1	2	1	3	500	50000	5000	0.6	0.01
I	25	1.5	5	0.5	4	0	100000	200000	0.6	0.5
J	120	1	1	0.2	2	0	100000	0	0.6	0.1
K	90	1	1	0.1	2	0	100000	10000	0.6	0
L	26	0.5	30	50	0.1	0	100000	5000	0.9	0.1
M	200	2	1	0	0	10000	1000	5000000	0.5	0.23

The test step to be added has the following characteristics: $f_c = 0.95$, time = 20 sec/board, operator utilization = 1, no materials are consumed, tooling cost = \$50,000 (only charged once), equipment cost = \$1,000,000 (0.6 equipment operational time), equipment capacity = 1 board, labor rate = \$22/hour, labor burden (b) = 0.8, 100,000 boards will be processed, years to depreciate = 5, there are 8760 hours/year, the board area is 2.1 cm^2, and assume that the Poisson yield equation applies.[14]

If the target is to minimize yielded cost, where should the test step be inserted: a) between steps C and D, b) between steps H and I, c) after step M, or d) don't insert a test step anywhere? Assuming there is only one fault type present. Assume that there is no diagnosis or rework. Assume that the test step does not introduce any new defects and does not generate any false positives.

7.11 Suppose that the test step is defined C_{in} = \$4, and $Y_{in} = 0.91$, is the last step in a process (and there is no rework) and that C_{test} and f_c have the following functional dependency:

$$C_{test} = 5e^{3f_c}, \text{for } 0 \le f_c \le 1$$

Marketing indicates that they expect on average each defective instance of the product shipped to cost the company \$1000 (warranty costs, liability, lost future business, etc.). What is the best f_c to buy if you want to minimize the effective cost of the product, i.e., minimize total cost.

7.12 Compute C_{out}, Y_{out} and S for the following case: C_{in} = \$20, $Y_{in} = 0.82$, $f_c = 0.8$, C_{test} = \$6 (on average, finding false positive production costs about 10% less than the full test cost). Assume that the false positives are incurred prior to the fault coverage and apply to all units ($f_p = 0.2$).

7.13 Rework Problem 7.12 in the case where false positives are applied to only bad units.

7.14 Rework Problem 7.13 assuming that the test step has a yield of 93.5%.

7.15 Derive the outgoing yield and cost and the total scrap when false positives are included and assumed to be incurred after the fault coverage. Under what conditions does the solution for this assumption give the same answer as the example provided in Section 7.5 (Equations (7.47) through (7.49))?

7.16 Can the effects of false positives be rolled into a "false positive coverage" parameter that functionally operates the same way as the fault coverage (i.e., for which the scrap produced in Figure 7.8 has the form $1 - Y_{in}^{f_{p-coverage}}$)? How can you check the validity of the derivation?

7.17 What is the bonepile yield corresponding to the test step with false positives example provided in Section 7.5?

7.18 Determine the outgoing cost and outgoing yield for the case shown in Figure 7.10. Given $C_{test1} = C_{test2}$ and $f_{c1} = f_{c2}$, what do the outgoing cost and yield reduce to? For $f_{c1} = f_{c2}$ and $C_{test1} = C_{test2}$, check the simple cases of $f_c = 0$, $f_c = 1$ and $Y_{in} = 1$; show that your answers reduce to the correct form in these cases.

7.19 Prove Equation (7.51) by following the argument in Section 7.4 for the wafer probe situation.

---

[14] Note, the tooling cost has to be modified after a test step because Q in Equation (2.10) changes due to boards being scrapped by the test step.

7.20 Show that the Williams and Brown derivation reduces to f_c = fraction of defective units when the maximum number of defects per unit is 1.

7.21 Use Rent's Rule,[15] Moore's Law and the cost-per-pin data presented in Table 7.1 to justify (generate) the data in Figure 7.14.

---

[15] Rent's Rule [Ref. 7.21] relates the number of signal and control I/Os on a chip to the number of gates.

Chapter 8

Diagnosis and Rework

When a test or inspection activity is performed, a product that does not pass the test can be either scrapped (disposed of), salvaged (all or part of the product is recovered for reuse in the same or another product), recycled (broken down to its constituent materials), or reworked. The first activity that takes place after a product fails a test is to determine why it failed; this activity is called diagnosis. Once the diagnosis is completed, a decision can be made as to whether a particular unit should be reworked (repaired and sent back into the test) or scrapped. A simple view of diagnosis and rework is shown in Figure 8.1.

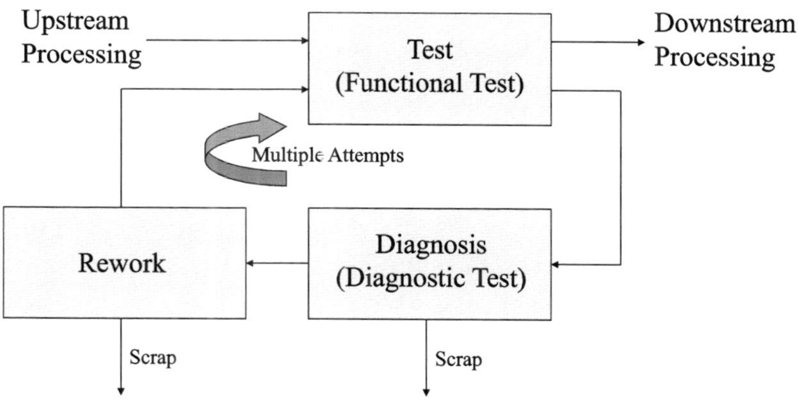

Fig. 8.1. A simple test/diagnosis/rework process.

In the example test/diagnosis/rework process shown in Figure 8.1, all of the products coming from production are tested. A more detailed diagnostic test is applied to all the products that are identified as defective during the test. After diagnosis some products may be reworked and all reworked products are retested. In some cases diagnosis or the rework process may decide to scrap product instances (units). Note that diagnosis and rework are not perfect — they introduce defects, make misdiagnoses, and fail to correctly rework defective units — therefore, a unit may go through testing, diagnosis and rework repeatedly in multiple "attempts".

The goal of analyzing the diagnosis and rework process (coupled with the test) is to determine which units should be reworked (rather than scrapped), and to determine the optimum number of times to attempt to rework a unit before giving up and scrapping it. At a broader level, the challenge is to determine where in the manufacturing process to test and when to diagnose and rework test rejects. In some cases it may be more economical to simply scrap products that do not pass tests than to pay to diagnose and rework them.

8.1 Diagnosis

Diagnosis, also known as fault isolation, refers to determining the type of defect that caused a specific fault and the location of that defect within the faulty unit. Before any decisions are made regarding the disposition of a product deemed faulty by the test step, a diagnosis must be performed. The outcome of the diagnosis will be one of the following:

- No fault found (the test identified a false positive) — If no fault is found, the unit is sent back for retesting without any rework. Note that even if no fault is found, the unit still incurs the cost of the diagnosis and is subject to any defects that may be inserted into the unit by the test and diagnosis processes.
- Defect type and location successfully identified — In this case a decision is made as to whether the defect is repairable or not, and whether it is worth repairing or not. If the defect is not worth repairing, then the unit will be scrapped.

Tests are performed on a product are often categorized as either functional or diagnostic tests. Functional tests are usually relatively quick pass/fail tests with limited diagnostic capability. If rework of a faulty unit is impractical or non-economical, then only functional tests are run. If rework is an option, then a diagnostic test will follow or replace functional testing. A diagnostic test (labeled "Diagnosis" in Figure 8.1) is characterized by a diagnostic resolution. The diagnostic resolution is a measure of the ability of a test to exactly identify the lowest replaceable unit that is faulty [Ref. 8.1]. An ideal diagnostic test would have a diagnostic resolution of 1; a test that could only narrow the defect down to one of two lowest replaceable units would have a diagnostic resolution of less than 1. The diagnostic resolution of a diagnostic activity (or diagnostic test) is related to how well the activity characterizes the faults that can appear in the product. This understanding is often captured in the form of a fault dictionary or diagnostic tree.

A fault dictionary correlates test symptoms and known faults [Ref. 8.2]. Groups of faults that share the same symptoms are referred to as "equivalent faults." By definition, equivalent faults cannot be distinguished from each other using only a fault dictionary. Dictionaries are often augmented with entries corresponding to actual faults found during manufacturing tests, so that the fault dictionary "learns" during the manufacturing process.

Fault dictionaries cannot be used until all tests are applied. In addition, the efficiency of fault dictionaries may be poor for large circuits. An alternative approach uses a diagnostic tree or fault tree. In this approach, tests are applied one at a time and a partial diagnosis is performed using the result of each test. The diagnosis obtained is then used to make a decision about the next test to be performed. For diagnostic trees the average diagnostic length of the diagnosis tree (i.e., the depth) is given by [Ref. 8.3]:

$$D_{avg} = \sum_{i=0}^{N_f} d_i p_i \tag{8.1}$$

where

N_f = the number of distinguishable fault sets.

d_i = the number of tests on the branch from the root to the ith leaf node.

p_i = the probability of occurrence of the fault (or fault set) represented by the ith leaf node.

The average diagnostic length is the average number of test applications before termination of the diagnosis. If, for example, the length of time required for a test application is known, D_{avg} from Equation (8.1) could be used to estimate the cost of diagnostic testing. Bushnell and Agrawal [Ref. 8.3] present several excellent tutorial examples of diagnosis for simple systems.

Several cost impacts are associated with diagnosis. First, the creation of fault dictionaries or trees and correlating them to a product is a significant and very resource-consuming activity. Existing fault dictionaries and trees are rarely directly applicable to a specific application and require considerable resources to be made useful in the diagnosis process. Simply performing the diagnosis process itself consumes resources (labor, tooling, capital, etc.). Diagnostic testing impacts the throughput of the entire test/diagnosis/rework process.

8.2 Rework

Rework is the process of correcting defects in a product during the manufacturing process. Rework is differentiated from repair, which is the process of correcting defects in a product that has failed at some point in time after manufacturing was completed. In the case of repair, the defect could be due to undetected manufacturing defects or damage accumulated during field use. Rework generally plays a more important role when large costs have been invested in products prior to testing. While rework is common for board assembly, it is also performed during some types of integrated circuit fabrication.

Rework is one of the most unpredictable and variable parts of the board assembly process. In fact, no other single activity in the assembly process negatively affects profitability more than rework [Ref. 8.4]. Unfortunately, most electronic assemblers treat rework as an afterthought, clinging to the notion that they can perfect their process to eliminate rework.

In the past, costs of doing rework were not accurately tracked since labor, equipment and work in progress were not overly expensive. With today's complex electronic systems, rework has taken on a whole new meaning. The equipment, training, and engineering support required costs electronics assemblers millions, not to mention the damage/scrap that is being generated. Additionally, the time-to-market factor costs assemblers billions daily by keeping large quantities of boards in work-in-progress to be reworked, unable to be completed and sold. This is especially true for high-volume commercial products whose life cycles are short.

The impacts of rework appear in many forms, such as engineering change orders, product upgrades or revisions, and general process errors. Persons who are responsible for rework most likely ask themselves the following questions on a monthly, if not weekly, basis in an effort to address their rework challenges [Ref. 8.4]:

- How many people should I have performing rework tasks?
- What kind of equipment should I buy?
- How much training is appropriate?
- How can I reduce damage/scrap?
- Why do I spend so much time dealing with rework issues?
- How many times should rework be attempted on the same unit before giving up?

The remainder of this chapter develops rework and diagnosis models that can be coupled with testing and used within process-flow modeling. The models can be used to answer many of the questions posed above for specific applications and manufacturing environments.

8.3 Test/Diagnosis/Rework Modeling

Several existing test/rework models are applicable to process-flow-based cost modeling. The basic test/rework models currently in use are shown in Figure 8.2. In the following description we use the word "unit" to refer to the item being tested (e.g., a board assembly). In the example test/diagnosis/rework models shown in Figure 8.2, all units coming from production are tested; the diagnosis and repair are applied to all the units that are identified as defective during the test, and all reworkable units are retested. Many versions of these models have been developed to support some subset of the variables shown, including single-rework and multiple-rework attempt models [Ref. 8.5] through [Ref. 8.13].

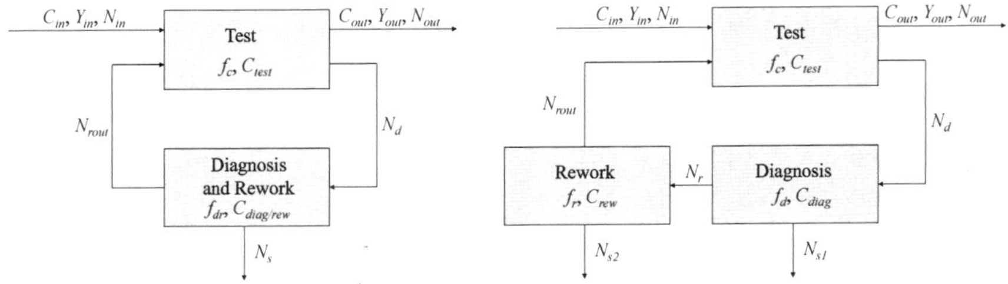

Fig. 8.2. Example test/diagnosis/rework models currently in use for process-flow cost modeling. C = cost, Y = yield, N = number of units, f_c = fault coverage, f_{dr} = fraction of units that are diagnosible and reworkable, f_r = fraction of units that are reworkable, f_d = fraction of units that are diagnosible, and N_s = number of units scrapped.

8.3.1 *Single-Pass Rework Example*

General models of the test/diagnosis/rework process become cumbersome and it becomes difficult to trace units through the process. Therefore, it is helpful to begin our analysis with a simplified scenario in which the following assumptions are imposed:

- Whatever rework claims is repaired is in fact repaired (single-pass rework).
- Rework, diagnosis and test do not introduce any new defects.
- The test step does not have any false positives.

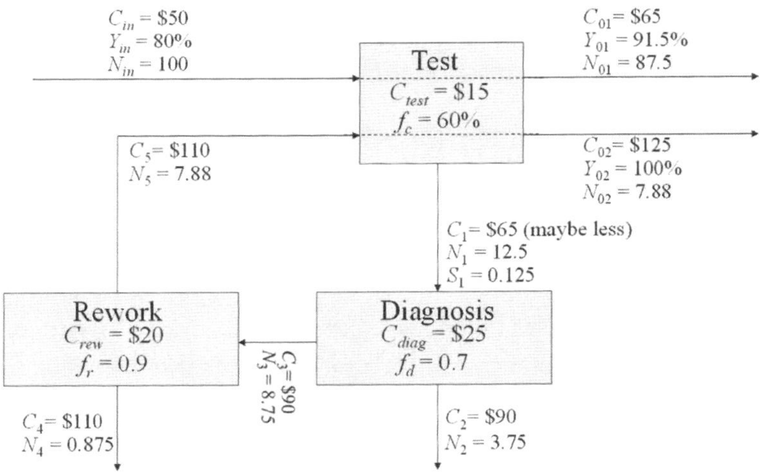

Fig. 8.3. Single-pass rework numerical example.

Figure 8.3 shows an example test/diagnosis/rework combination. Given the inputs C_{in}, Y_{in}, and N_{in}, and the characteristics of each step in the process (shown inside the boxes), the number of units, their cost, and the yield can be computed on each branch (arrow), subject to the three assumptions above. Using the relations developed in Chapter 7 in Equations (7.25) and (7.33), the values of the costs, yields and quantities traced through the process are given by

$$\left.\begin{aligned} C_{01} &= C_{in} + C_{test} = 50 + 15 = 65 \\ Y_{01} &= Y_{in}^{(1-f_c)} = 0.8^{(1-0.6)} = 0.915 \\ N_{01} &= PN_{in} = Y_{in}^{f_c} N = 0.8^{0.6} 100 = 87.5 \end{aligned}\right\} \text{Units passed by the test, ignoring rework}$$

$$\left.\begin{aligned} C_1 &= C_{in} + C_{test} = 50 + 15 = 65 \\ N_1 &= N_{in} - N_{01} = 100 - 87.5 = 12.5 \\ S_1 &= 1 - P = 1 - Y_{in}^{f_c} = 1 - 0.8^{0.6} = 0.125 \end{aligned}\right\} \text{Units rejected by the test}$$

$$\left.\begin{aligned} C_2 &= C_1 + C_{diag} = 65 + 25 = 90 \\ N_2 &= (1 - f_d)N_1 = (1 - 0.7)12.5 = 3.75 \end{aligned}\right\} \text{Units scrapped by the diagnosis}$$

$$\left.\begin{aligned} C_3 &= C_1 + C_{diag} = 65 + 25 = 90 \\ N_3 &= f_d N_1 = (0.7)12.5 = 8.75 \end{aligned}\right\} \text{Units passed by the diagnosis}$$

$$\left.\begin{aligned} C_4 &= C_3 + C_{rew} = 90 + 20 = 110 \\ N_4 &= (1 - f_r)N_3 = (1 - 0.9)8.75 = 0.875 \end{aligned}\right\} \text{Units scrapped by the rework}$$

$$\left.\begin{aligned} C_5 &= C_3 + C_{rew} = 90 + 20 = 110 \\ N_5 &= f_r N_3 = (0.9)8.75 = 7.88 \end{aligned}\right\} \text{Units successfully repaired by the rework}$$

$$\left.\begin{aligned} C_{02} &= C_5 + C_{test} = 110 + 15 = 125 \\ Y_{02} &= 1.0 \\ N_{02} &= N_5 = 7.88 \end{aligned}\right\} \text{Repaired units passed by the test}$$

So the total number of units continuing through the process (ultimately passed by the test) is given by

$$N_{out} = N_{01} + N_{02} = 87.5 + 7.88 = 95.38$$

The yield of the units passed by the test step is

$$Y_{out} = \frac{\text{good units passed by the test}}{\text{all units passed by the test}} = \frac{Y_{01}N_{01} + N_5}{N_{out}} = \frac{87.88}{95.38} = 0.9214$$

The total money spent on all the units in this process is

$$C_{01}N_{01} + C_2N_2 + C_4N_4 + C_{02}N_{02} = \$7106$$

Thus, the effective cost per passed unit and the effective cost per good passed unit (yielded cost) are given by

$$C_{out} = \frac{7106}{87.5 + 7.88} = \$74.50,\ C_Y = \frac{74.50}{0.9214} = \$80.86$$

The total fraction of the original units scrapped by the process is given by

$$S_{total} = \frac{N_2 + N_4}{N_{in}} = 0.046$$

If we consider the process shown in Figure 8.3 without any rework (just scrapping the units that the test step considers bad on the first pass), the output would have been

$$N_{out} = N_{01} = 87.5$$

$$Y_{out} = Y_{01} = 0.915$$

$$C_{out} = \frac{C_{01}N_{01} + C_1N_1}{N_{out}} = \$74.29, C_Y = \frac{74.29}{0.915} = \$81.19$$

$$S_{total} = \frac{N_1}{N_{in}} = 0.125$$

Comparing these results to the results of the diagnosis and rework process, we see that although the cost per passed unit increased when rework was done (obviously), the yielded cost per passed unit decreased. In fact, if the yielded cost per passed unit does not decrease when rework is used, then very possibly units should be scrapped rather than reworked.

The result above for the test step without rework can be generalized as follows. The cost out is,

$$C_{out} = \frac{C_{01}N_{01} + C_1 N_1}{N_{out}} = \frac{C_{01}\dfrac{N_{01}}{N_{in}} + C_1\dfrac{N_1}{N_{in}}}{\dfrac{N_{out}}{N_{in}}} = \frac{C_{01}P + C_1 S}{P}$$

where we have divided the numerator and the denominator by N_{in}. When there is no rework $N_{01}/N_{in} = P$ and $N_1/N_{in} = S$, the pass and scrap fractions respectively. Substituting for C_{01} and C_1 (for the case with no rework), we get (remembering that $S + P = 1$),

$$C_{out} = \frac{(C_{in} + C_{test})P + (C_{in} + C_{test})S}{P} = (C_{in} + C_{test})\left(\frac{P+S}{P}\right)$$

$$= \frac{C_{in} + C_{test}}{P}$$

This result is the same as Equation (7.35) for a test step.

In real processes, rework would not be 100% successful in repairing defects and diagnosis and rework would both potentially insert new defects into the unit. These effects could be included in the simple model and the process of tracing units and their properties could be continued. The next section derives a general model for an arbitrary number of rework attempts.

8.3.2 *A General Multi-Pass Rework Model [Ref. 8.13]*

The objective of this section is to develop a general model for test/diagnosis/rework that accommodates the effects relevant to printed circuit board fabrication and electronic system assembly processes. In these processes, defect insertion during test and rework operations (e.g., from handling and/or probes making physical contact with the board) is not uncommon. False positives can be a significant problem, especially in board fabrication, where multiple rework attempts are made on expensive systems such as multichip modules, and complex rework operations may include reassembly of significant portions of the system.

Figure 8.4 shows the content of a general test/diagnosis/rework model. Inputs to this model are the accumulated cost and yield of upstream processes (C_{in} and Y_{in}). N_{in} is not a required input and is only included for convenience in the formulation of the model.[1] The test portion of the model is the top group of three steps in Figure 8.4. This model can be used to account for defects introduced by the test operation both prior to the actual test (e.g., when loading the unit into the tester or stationing the probes on the unit) and after the test result is recorded (e.g., when unloading the unit from the tester).

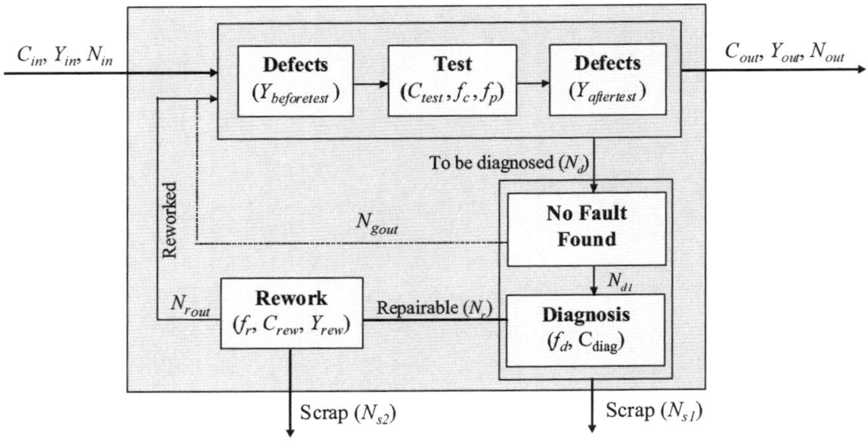

Fig. 8.4. Organization of the general test/diagnosis/rework model. Table 8.1 describes the symbols appearing in this figure. (© 2001 IEEE)

The units that are determined to be faulty go on to the diagnosis step. As mentioned at the beginning of the chapter, three outcomes are possible from diagnosis: (1) no fault is found, in which case the unit goes back for retesting, (2) the unit is determined to be reworkable and is sent on to

---

[1] In general, yield and cost results from this model are independent of N_{in}. However, if equipment, tooling, or other non-recurring costs are included, the results become dependent on N_{in} and can be computed from accumulations of time that specific equipment is occupied or the quantity of tooling used to produce a specific quantity of units (see Equations (8.17) through (8.19) and associated discussion).

rework, or (3) the unit is determined to be non-reworkable (or non-diagnosable) and is sent to scrap. The rework process fixes the reworkable units and scraps units that cannot be successfully reworked. The reworked units are re-tested and if they are found to be faulty again, they are again sent for diagnosis. This rework process can be performed any number of times (attempts). This general model simultaneously considers the effect of fault coverage and false positives on the cost and yield.

Table 8.1. Nomenclature Used in Figure 8.4 and Throughout the Discussion in this Chapter.

C_{in}	Cost of a unit entering the test/diagnosis/rework process	N_{in}	Number of units entering the test/diagnosis/rework process
C_{test}	Cost of test/unit	N_d	Total number of units to be diagnosed
C_{diag}	Cost of diagnosis/ unit	N_{gout}	Number of no fault found units
C_{rew}	Cost of rework/ unit (may be a computed quantity, see Equation (8.20) and Sect. 8.4)	N_{d1}	$N_d - N_{gout}$
C_{out}	Effective cost of a unit exiting the test/diagnosis /rework process	N_r	Number of units to be reworked
f_c	Fault coverage	N_{rout}	Number of units actually reworked
f_p	False positives fraction, or the probability of testing a good unit as bad	N_{s1}	Number of units scrapped by diagnosis process
f_d	Fraction of units that can be diagnosed and are determined to be reworkable	N_{s2}	Number of units scrapped during rework
f_r	Fraction of units actually reworked	N_{out}	Number of a units exiting the test/diagnosis/rework process, including good units and test escapes
Y_{in}	Yield of a unit entering the test/diagnosis/rework process		
$Y_{beforetest}$	Yield of processes that occur entering the test		
$Y_{aftertest}$	Yield of processes that occur exiting the test		
Y_{rew}	Yield of the rework process (may be a computed quantity; see Equation (8.21))		
Y_{out}	Effective yield of a unit exiting the test/diagnosis/ rework process		

Versions of C_{in}, Y_{in} and N_{in} appear both with and without subscripts in the remainder of this chapter. When the variables appear without subscripts they refer to the values entering the process. When they have subscripts, they represent specific rework attempts.

There are several assumptions made in the formulation of this model:

- Defects introduced by the diagnosis step are not explicitly treated.
- False positives (f_p) and fault coverage (f_c) act simultaneously and are independent of each other — that is, the fault coverage acts only on bad units and the false positive acts either only on good units or on all units.

The cost incurred by all the units that eventually pass the test step is given by

$$C_1 = \sum_{i=0}^{n} \left(C_{in_i} + C_{test}\right) N_{out_i} \tag{8.2}$$

where n is the number of rework attempts allowed (the maximum number of attempts to rework an individual unit is n and N_{out_i} is the number of units passed by the test in the ith rework attempt (see Equation (8.7) and its associated discussion). When $i = 0$, C_1 is the total cost of the units that pass the test without ever going through diagnosis or rework. The cost incurred by all the units scrapped by the diagnosis step is given by

$$C_2 = \sum_{i=1}^{n\text{-}1} \left(C_{in_i} + C_{test} + C_{diag}\right) N_{s1_i} \tag{8.3}$$

The cost incurred by all the units scrapped by the rework step is given by

$$C_3 = \sum_{i=1}^{n\text{-}1} \left(C_{in_i} + C_{test} + C_{diag} + C_{rew}\right) N_{s2_i} \tag{8.4}$$

where N_{s1_i} and N_{s2_i} are defined in Equations (8.9) and (8.10).

After the final rework (n^{th} rework attempt), the units that do not pass the test are scrapped. The cost of these final scrapped units is given by

$$C_4 = N_{d_{n+1}}\left(C_{in_n} + C_{test}\right) + N_{in_n} Y_{in_n} Y_{beforetest} f_p \left(C_{in_n} + C_{test}\right) \tag{8.5}$$

The first term in Equation (8.5) accounts for the defective units scrapped by the final test, and the second term accounts for any false positives on good units that are encountered during the final test. Note that this equation is valid for both definitions of f_p (when it applies to only good units and when it applies to all units) because f_p's application to bad units is included in the calculation of N_{in} given in Equation (8.12). N_{in_n}, appearing in Equation (8.5), is defined in Equation (8.12).

The total cost of all the units (including scrapped ones) is the sum of C_1 through C_4. The total effective cost per output unit associated with this model is the total cost divided by the total number of output units (units that are eventually passed by the test):

$$C_{out} = \frac{C_1 + C_2 + C_3 + C_4}{N_{out}} \tag{8.6}$$

Using the results of the false positives discussion in Section 7.5 (Equation (7.41)), where f_p is the probability of testing a good unit as bad, (which should not be confused with the escape fraction, which is the probability of testing bad units as good), the number of units moving through the process is given in Equations (8.7) through (8.12):

$$N_{out_i} = N_{in_i}\left(1\text{-}f_p Y_{in_i} Y_{beforetest}\right)\left(\frac{\left(1\text{-}f_p\right)Y_{in_i} Y_{beforetest}}{1\text{-}f_p Y_{in_i} Y_{beforetest}}\right)^{f_c} \tag{8.7a}$$

$$N_{d1_i} = N_{in_i}\left(1\text{-}f_p Y_{in_i} Y_{beforetest}\right) - N_{out_i} \tag{8.8a}$$

when f_p applies to only good units. Then

$$N_{out_i} = N_{in_i}\left(1 - f_p\right)\left(Y_{in_i} Y_{beforetest}\right)^{f_c} \tag{8.7b}$$

$$N_{d1_i} = N_{in_i}\left(1\text{-}f_p\right) - N_{out_i} + f_p N_{in_i}\left(1\text{-}Y_{in_i} Y_{beforetest}\right) \tag{8.8b}$$

When f_p applies to all units:

$$N_{s1_i} = \left(1\text{-}f_d\right)N_{d1_i} \tag{8.9}$$

$$N_{s2_i} = \left(1\text{-}f_r\right)N_{r_i} \tag{8.10}$$

$$N_{r_i} = f_d N_{d1_i} \tag{8.11}$$

$$N_{in_i} = \begin{cases} N_{in} \text{ when } i = 0 \\ f_r N_{r_{i-1}} + f_p N_{in_{i-1}} Y_{in_{i-1}} Y_{beforetest} \text{ when } i > 0 \end{cases} \tag{8.12}$$

where parameters without subscripts (N_{in}, C_{in}, and Y_{in}) indicate values entering the process (Figure 8.4) and the form of Equation (8.7a) follows from Equation (7.33). The total number of units that successfully pass the test process is given by

$$N_{out} = \sum_{i=0}^{n} N_{out_i} \tag{8.13}$$

The unit counting in Equations (8.7) through (8.12) assumes that all false positives on good units go through diagnosis and back into test without scrapping units in diagnosis or rework. The formulation is also only valid when $f_p < 1$, $Y_{in} > 0$ and $Y_{beforetest} > 0$. The input cost, C_{in_i}, that appears in Equations (8.2) through (8.5) is given by C_{in} when $i = 0$, and by Equation (8.14) when $i > 0$:

$$C_{in_i} = \frac{\left(C_{in_{i-1}} + C_{test} + C_{diag}\right) f_p Y_{in_{i-1}} Y_{beforetest} N_{in_{i-1}}}{N_{in_i}} + \frac{\left(C_{in_{i-1}} + C_{test} + C_{diag} + C_{rew}\right) f_r N_{r_{i-1}}}{N_{in_i}} \tag{8.14}$$

The input yield, Y_{in_i}, that appears in Equations (8.5) and (8.7) through (8.14) is given by Y_{in} when $i = 0$ and by Equation (8.15) when $i > 0$.

$$Y_{in_i} = \frac{f_p Y_{in_{i-1}} Y_{beforetest} N_{in_{i-1}} + Y_{rew} f_r N_{r_{i-1}}}{N_{in_i}} \tag{8.15}$$

The final yield of units that successfully pass the process is given using the general result of Equation (7.25), by

$$Y_{out} = \frac{\sum_{i=0}^{n} Y_{aftertest} N_{out_i} \left(\frac{\left(1-f_p\right) Y_{in_i} Y_{beforetest}}{1-f_p Y_{in_i} Y_{beforetest}} \right)^{1-f_c}}{N_{out}} \tag{8.16a}$$

when f_p applies to only good units, and

$$Y_{out} = \frac{\sum_{i=0}^{n} Y_{aftertest} N_{out_i} \left(Y_{in_i} Y_{beforetest}\right)^{1-f_c}}{N_{out}} \tag{8.16b}$$

when f_p applies to all units. Note that N_{in} cancels out of Equations (8.6) and (8.16), making the total cost per unit and final yield independent of the number of units that start the process. This is intuitively correct, since no volume-sensitive effects (such as material or equipment costs) are included in the model.

In order to support the calculation of equipment costs associated with the test, diagnosis, and rework activities, the total time spent in each activity can be accumulated. The effective tester, diagnosis, and rework time per unit can be formulated using Equations (8.7) through (8.12):

$$T_{total\ test} = \frac{T_{test}}{N_{out}} \sum_{i=0}^{n} N_{in_i} \tag{8.17}$$

$$T_{total\ diag} = \frac{T_{diag}}{N_{out}} \sum_{i=1}^{n} \left(N_{d1_i} + B\right) \tag{8.18}$$

where

$$B = \begin{cases} f_p N_{in_i} Y_{in_i} Y_{beforetest}, \text{ when } f_p \text{ applies to only good units} \\ f_p N_{in_i}, \text{ when } f_p \text{ applies to all units} \end{cases}$$

$$T_{total\ rew} = \frac{T_{rew}}{N_{out}} \sum_{i=1}^{n} N_{r_i} \tag{8.19}$$

where T_{test}, T_{diag}, and T_{rew} represent the times for individual units in the test, diagnosis and rework equipment.

8.3.3 *Variable Rework Cost and Yield Models*

In general, the costs of performing rework and the yield of items that result from it will be dependent on the type and quantity of rework that must be performed. In a variable rework model, C_{rew} and Y_{rew} are not treated as constants (as in the previous section), but are variables based on whatever the dominant defect is.

For electronic module assembly, defects are often associated with defective devices (chips). For example, if the rework of a printed circuit board assembly process is dominated by the replacement of defective devices, C_{rew} and Y_{rew} (the average rework cost and yield per board) for the ith rework attempt could be determined using

$$C_{rew_i} = \sum_{j=1}^{N_{device}} \left(C_{rework\,fixed_j} + C_{device_j}\right)\left(1 - Y_{device_j}\right)^i \tag{8.20}$$

$$Y_{rew_i} = \prod_{j=1}^{N_{device}} Y_{rework\,process_j} Y_{device_j}^{\left(1-Y_{device_j}\right)^i} \tag{8.21}$$

where

C_{device_j}, Y_{device_j} = the cost and yield of the jth device when it enters the board assembly process.

$C_{rework\ fixed_j}$ = the fixed cost per device instance to perform a replacement — that is, the cost of removing the defective device, cleaning the site, and attaching a new device (see Section 8.4). $C_{rework\ fixed_j}$ may be a function of the area of the chip or die being replaced (see Section 8.4 for an example of the computation of $C_{rework\ fixed_j}$).

N_{device} = the total number of devices on the board.

$Y_{rework\ process_j}$ = the yield of a single device replacement action for the jth device.

This is a simple model that assumes that the only type of fault possible is defective devices and that each device reworked is an independent operation. Another form of the rework cost model that is effectively equivalent to Equation (8.20) appears in [Ref. 8.14].

In this model, the rework time for the ith rework attempt is given by

$$T_{rew_i} = \sum_{j=1}^{N_{device}} T_{device_j}\left(1 - Y_{device_j}\right)^i \tag{8.22}$$

where T_{device_j} is the time to rework the jth device (this time depends on many things, but may range from minutes, for high-volume commercial applications, to hours for multichip modules).

8.3.4 *Example Test/Diagnosis/Rework Analysis*

This section presents example results generated using the model discussed in Section 8.3.2, and the application of the model to an electronic power module.

The data used for the first example in this section is given in Table 8.2. The results are presented in terms of yielded cost. Yielded cost is defined as cost divided by yield (see Section 3.4). In electronic assembly, yielded cost represents the effective cost per good (non-defective) assembly for a manufacturing process.

Table 8.2. Baseline Data for Example Results.

C_{in}	\$100	f_c	70%	Y_{in}	90%
C_{test}	\$20	f_r	81%	$Y_{beforetest}$	97%
C_{diag}	\$10	f_d	100%	$Y_{aftertest}$	97%
C_{rew}	\$25	f_p	10%	Y_{rew}	90%
Rework attempts	2	False positives are created on good parts only			

Figure 8.5 shows that when false positives are created and rework yield is low, there is an optimum number of rework attempts per part (two attempts for Y_{rew} = 30%, one for Y_{rew} = 10% or less). If no false positives are created, depending on the rework yield, the cost of performing the rework, and the rework success rate, rework may not be economically viable.

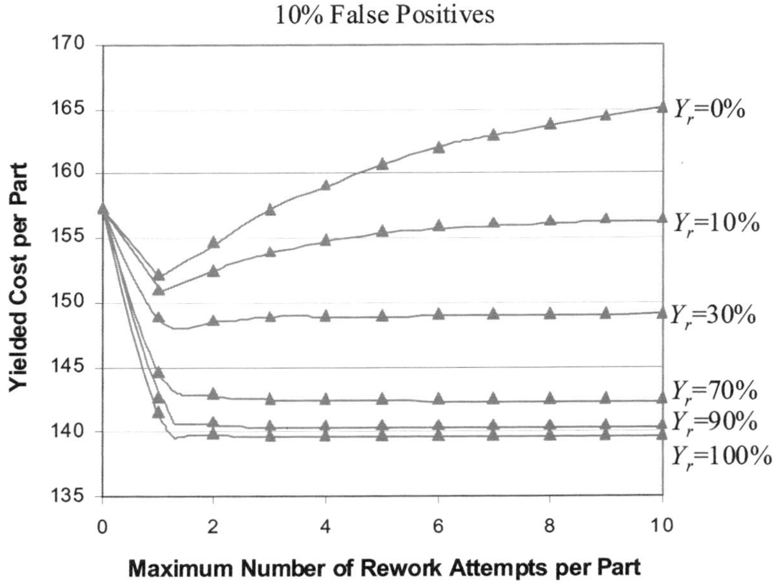

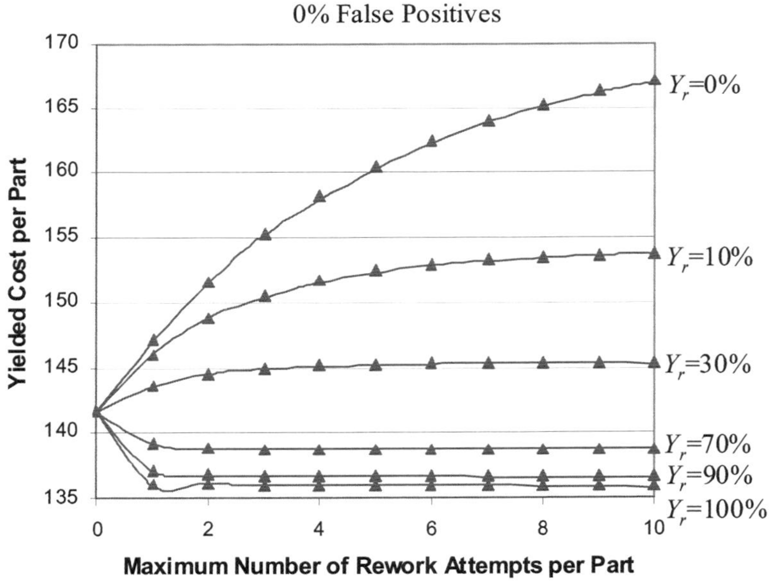

Fig. 8.5. Variation of final yielded cost (cost divided by yield) of parts that pass the test/diagnosis/rework process with the number of allowed rework attempts per part. In this example, false positives are only created on good parts. (© 2001 IEEE)

Figure 8.6 shows the effect of whether the false positives are created on only the good parts or all the parts. With no rework (in the zero rework-attempts case, parts that are identified as defective are scrapped without diagnosis), if a fixed false positive fraction only affects good parts, the resulting per part yielded cost is higher than if the false positives affect all parts. While the same number of parts are scrapped in both cases, when the false positive fraction affects all parts, some defective parts are removed, resulting in a low yielded cost. When many rework attempts are allowed, false positive creation on only good parts results in an overall lower yield part (because the false positive creation didn't remove any defective parts), and also a lower overall cost per part (because fewer parts were reworked). The net effect in this case is that the overall yielded cost per part is lower.

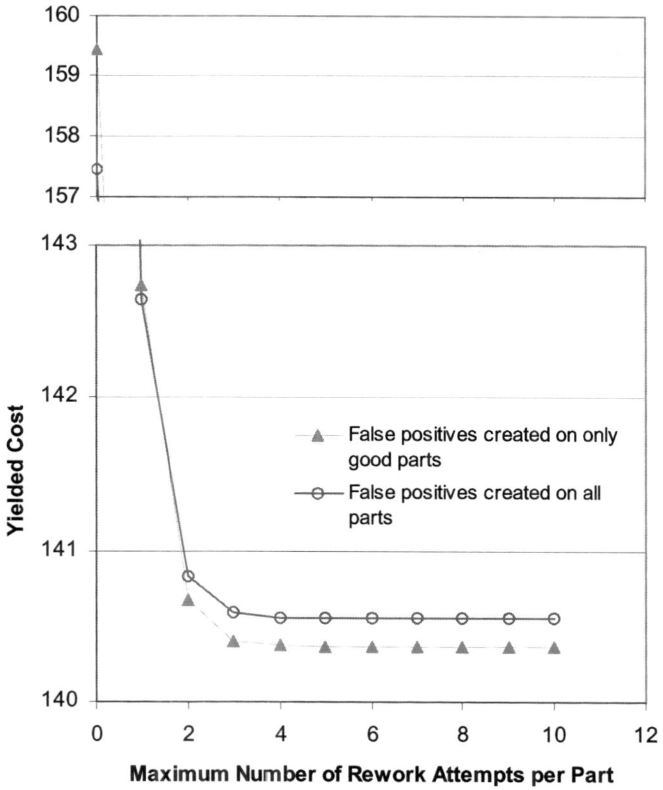

Fig. 8.6. Effect of the false positives definition on the part population. (© 2001 IEEE)

The model developed in this section has been used to plan the location of test/diagnosis/rework operations in the manufacturing process for an advanced electronic power systems (AEPS) module. AEPS refers to a system built around a packaging concept that replaces complex power electronics circuits with a single multi-function device that is intelligent and/or programmable. For example, depending on the application, an AEPS might be configured to act as an AC-to-DC rectifier, DC-to-AC inverter, motor controller, actuator, frequency changer, circuit breaker, and so on. The AEPS module considered here consists of sixteen ThinPak™ devices [Ref. 8.15] as shown in Figure 8.7. A ThinPak™ is a ceramic chip scale package for discrete three-terminal high-power devices. A simplified process flow for the AEPS module is shown in Figure 8.8.[2] The test economics challenge with the AEPS module is to determine where to perform test and rework operations: at the die level, device level, and/or module level.

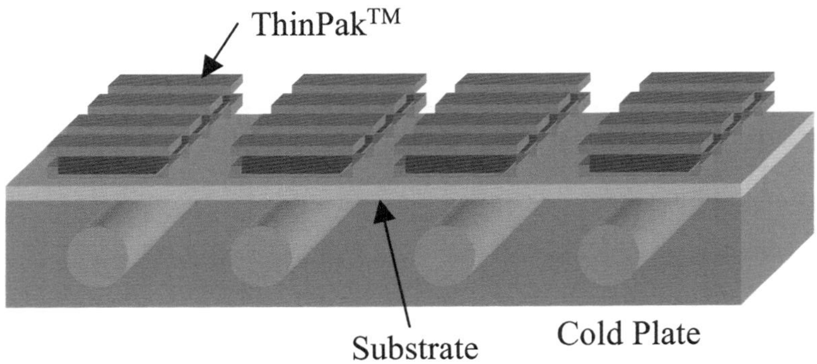

Fig. 8.7. AEPS module (600V half bridge) with 16 ThinPak™ devices mounted on it. (© 2001 IEEE)

[2] The multiplier step, denoted by "M", appears twice in the AEPS module process flow. The "M=2" process step denotes the assembly of two copper straps with the die-alumina lid assembly to complete the ThinPak™ device level assembly. Similarly, the "M=16" process step denotes the assembly of sixteen ThinPak™ devices on the substrate during the module-level assembly.

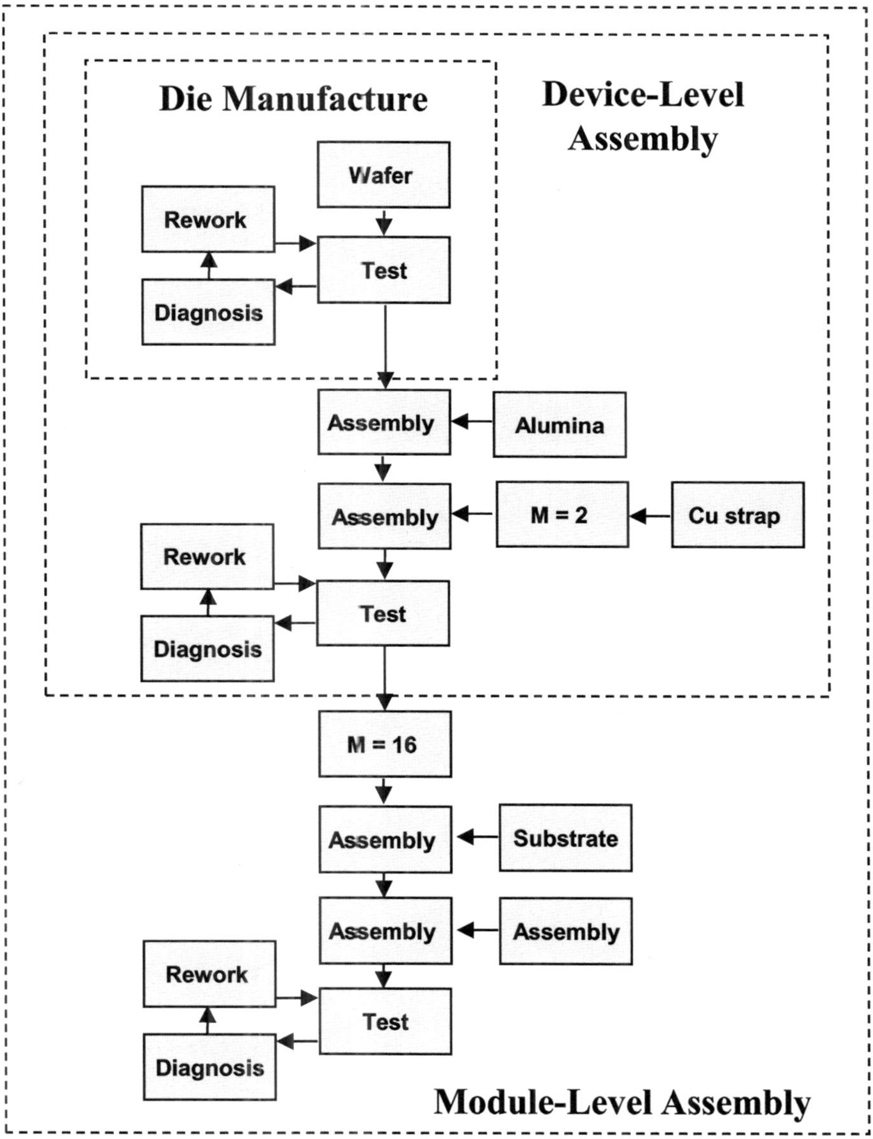

Fig. 8.8. Simplified process flow for the AEPS module, including candidate test/diagnosis/rework operations. (© 2001 IEEE)

Not all possible permutations of test and rework were analyzed. Die-level rework was omitted, because the die used in the ThinPak™ devices are relatively inexpensive and no practical methods of reworking defective die are available. We also did not consider device-level testing or rework in the present analysis.

Figure 8.9 shows the results of an analysis of the AEPS module. When the yield of the die is 100%, the most economical solution is to conduct no testing or rework (this result is intuitive). Module testing is relatively inexpensive and scraps defective modules prior to shipping; however, it has little overall effect on the yielded cost (the ratio of cost to yield). When die testing is introduced, the cost shifts upward by an amount equal to the test cost per die multiplied by 16. Again, performing module testing along with die testing improves the yield of modules exiting the process, but has little effect on the overall yielded cost. When module-level rework is performed, some of the scrapped modules are recovered, thus reducing the cost. For die with yields between 0.998 and 0.952, module testing and rework is the most economical. For 0.952 > yield > 0.942, die and module testing and rework is best. For yield < 0.942, die testing only is the best solution.

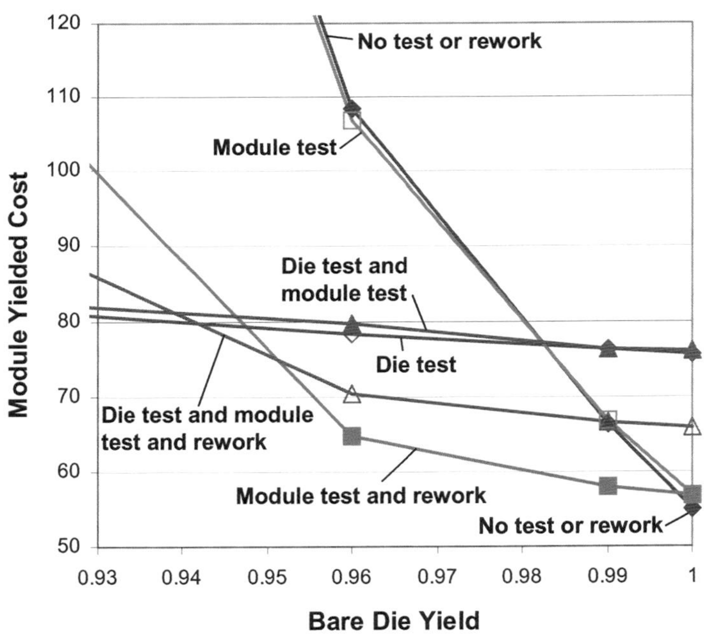

Fig. 8.9. Test/diagnosis/rework placement for an AEPS module containing 16 devices. (© 2001 IEEE)

8.4 Rework Cost ($C_{rework\ fixed}$)

The models for rework developed in this chapter deal with the impact of rework (and diagnosis) on the manufacturing process. We have not, however, addressed how the actual cost of performing the rework is computed, or $C_{rework\ fixed}$ in Equation (8.20).

The so-called fixed rework cost is the cost of reworking a single instance of a component on a board a single time, less the purchase price of the replacement component. An example data set for determining this fixed rework cost was provided in Table 8.3 [Ref. 8.16].

The dataset in Table 8.3 and the associated model results include training, supervision, equipment, floor space, and labor. Using the assumptions in Table 8.3, the following summary of rework costs can be generated (reproducing the specific calculations to obtain the following results is left to the student as exercises, Problems 8.13 and 8.14):

<u>Training Costs</u>	
Generic training	\$83,270/year
Specific training	\$118,670/year
Supervisor	\$2,708/year
Total training costs	\$204,648/year
<u>Equipment and Materials Costs</u>	
Soldering stations (1)	\$600/year
Rework equipment and support (1)	\$23,000/year
Soldering tips	\$2,570/year
Workbenches (1) and consumables	\$2,250/year
Total equipment and materials	\$28,420/year
<u>Work Space Costs</u>	\$275/year
Hours per week doing rework	75
Labor costs of performing rework	\$83,276
Number of components reworked	22,500/year
<u>Total Rework Costs</u>	\$316,619/year

<u>Effective cost per component reworked</u> ($C_{rework\ fixed}$) = \$14

Table 8.3. Data Set for Considering Component Replacement Rework [Ref. 8.16].

Property	Value
LABOR	
Labor rate for rework personnel ($/hour)	15.00
Overhead rate (burden) (%)	33
TRAINING	
Rework trainer's salary and benefits ($/year)	40,000
Number of employees trained per year by an individual trainer	15
Number of training hours per year per trained employee	40
Employers' expected rate of return on an employee's labor rate	2.5
Training floor space used (square feet)	800
Cost of demonstration equipment for training ($)	12,000
Cost of student equipment for training ($)	50,000
Cost of student workbenches for training ($)	15,000
Depreciation for training equipment (years)	5
Cost of training supplies ($/year)	20,000
SUPERVISION	
Salary and benefits of supervisor ($/year)	52,000
Number of personnel supervised	12
REWORK EQUIPMENT AND SUPPLIES	
Cost of one soldering station ($)	3,000
Depreciation for rework equipment (years)	5
Cost of top four soldering tips replaced ($):	
#1	20
#2	35
#3	48
#4	18.50
Average tip life expectancy (hours)	200
Soldering station maintenance (all stations) ($/year)	2,000
Other rework equipment ($)	65,000
Number of engineers supporting rework	1
Salary and benefits of engineer ($/year)	50,000
Utilization of the engineer (%)	20
Workbench cost ($)	1,500
Workbench ESD cost ($/year)	600
Life expectancy of workbench (years)	10
Cost of consumables (assumes 2 inches of solder wick per component reworked and 6 components reworked per hour) ($/hour)	0.40
Floor space (square feet)	25
Rework throughput rate per operator (components reworked/hour)	6
COMMON DATA	
Number of units reworked per week	450
Floor space cost ($/square foot/year)	11
Hours per year (3 shifts)	5760
Weeks per year	50
Equipment depreciation (years)	5

Note that the cost of replacement components is <u>not</u> included in the model above. The example model presented in this section is simple, but provides a good feel for the scope of the rework costs. One glance at the magnitude of the cost of performing rework should make it evident to the reader why, for many types of products, it is more economical to scrap assemblies that do not pass tests than to attempt rework. If the investment in the assembly is less than the effective cost per component reworked, you are better off spending your money to build another board than to rework a defective one.

Obviously this simple model's detail level could be improved by performing an actual cost-of-ownership analysis on the rework process (see Chapter 4).

References

8.1 Kime, C. R. (1970). An analysis model for digital system diagnosis, *IEEE Transactions on Computers*, C-19(11), pp. 1063-1073.

8.2 Richman, J. and Bowden, K. R. (1985). The modern fault dictionary, *Proceedings of the International Test Conference*, pp. 696-702.

8.3 Bushnell, M. L. and Agrawal, V. D. (2000). Chapter 18 - System Test and Core-Based Design, *Essentials of Electronic Testing for Digital, Memory and Mixed-Signal VLSI Circuits*, (Kluwer Academic Publishers, Boston, MA).

8.4 Cudmore, J. (1998). Rework management and optimization, *SMT Magazine*, October.

8.5 Dislis, C., Dick, J. H., Dear, I. D., Azu, I. N. and Ambler, A. P. (1993). Economics modeling for the determination of test strategies for complex VLSI boards, *Proceedings of the International Test Conference*, pp. 210-217.

8.6 Abadir, M., Parikh, A., Bal, L., Sandborn, P. and Murphy, C. (1994). High level test economics advisor, *Journal of Electronic Testing: Theory and Applications*, 5(2/3), pp. 195-206.

8.7 Sandborn, P. A. and Moreno, H. (1994). *Conceptual Design of Multichip Modules and Systems*, (Kluwer Academic Publishers, Boston, MA), pp. 152-169.

8.8 Tegethoff, M. and Chen, T. (1994). Defects, fault coverage, yield and cost, in board manufacturing, *Proceedings of the International Test Conference*, pp. 539-547.

8.9 Scheffler, M., Ammann, D., Thiel, A., Habiger, C. and Troster, G. (1998). Modeling and optimizing the costs of electronic systems, *IEEE Design & Test of Computers*, 15(3), pp. 20-26.

8.10 Dislis, C., Dick, J. H., Dear, I. D. and Ambler, A. P. (1995). *Test Economics and Design for Testability*, (Ellis Horwood, Upper Saddle River, NJ).

8.11 Garg, V., Stogner, D. J., Ulmer, C., Schimmel, D., Dislis, C., Yalamanchili, S. and Wills, D. S. (1997). Early analysis of cost/performance trade-offs in MCM systems, *IEEE Transactions on Component, Packaging and Manufacturing Technology, Part B*, 20(3), pp. 308-319.

8.12 Driels, M. and Klegka, J. S. (1991). Analysis of alternative rework strategies for printed wiring assembly manufacturing systems, *IEEE Transactions on Components, Hybrids, and Manufacturing Technology*, 14(3), pp. 637-644.

8.13 Trichy, T., Sandborn, P., Raghavan, R. and Sahasrabudhe, S. (2001). A new test/diagnosis/rework model for use in technical cost modeling of electronic systems assembly, *Proceedings of the International Test Conference*, pp. 1108-1117.

8.14 Petek, J. M. and Charles, H. K. (1998). Known good die, die replacement (rework), and their influence on multichip module costs, *Proceedings of the Electronic Components and Technology Conference (ECTC)*, pp. 909-915.

8.15 McCluskey, P., Iyengar, R., Azarm, S., Joshi, Y., Sandborn, P., Srinivasan, P., Reynolds, B., Gopinath, D., Trichy, T. K. and Temple, V. (1999). Rapid reliability optimization of competing power module topologies using semi-analytical fatigue models, *Proceedings of the PowerSystems World HFPC'99 Conference*, pp. 184-194.

8.16 http://www.solder.net/main/Rework_Calc.xls, November 2002. Accessed August 2013.

Problems

8.1 Repeat the single-pass rework example in Section 8.3.1 using C_{test} = \$25 and f_c = 70%. Is this a better or worse option than the example provided in the text?

8.2 In the single-pass rework example in the text, what if the rework operation introduces new defects into 6% of the modules it reworks? Assuming that the process remains a single-pass process, i.e., the modules not passed by the test step after rework are scrapped (not diagnosed and reworked again). What is the final effective cost and yield of parts passed by the test step?

8.3 Assuming the test/diagnosis/rework process shown in Figure 8.3 is used, what is the maximum you can afford to pay for diagnosis?

8.4 If all you are concerned with is yielded cost, assuming one rework attempt and given the data used for the single-pass rework example in Section 8.3.1, should the test be done at all? Why or why not?

8.5 If C_{test} = \$10, f_c = 0.87, C_{in} = \$4, Y_{in} = 0.91, and C_{rew} = \$8, calculate C_{out}, Y_{out} for the process shown below. Assume that the rework step does not add any new defects and has a 100% success rate (it fixes everything and the yield of the fixed parts is 100%).

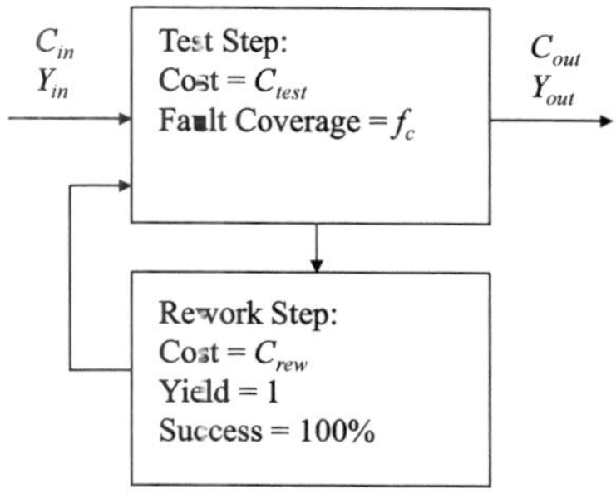

8.6 In Problem 8.5, is the rework worth doing? Why or why not?

8.7 Repeat Problems 8.1-8.3 using the general multi-pass rework model (assuming only a single rework attempt is allowed).

8.8 Reduce the general multi-pass rework model to treat the single-pass case, i.e., generate general equations for the single-pass case.

8.9 Derive Equation (8.7).

8.10 Derive Equation (8.16).

8.11 Determine the effective cost, yield and total scrap fraction under the conditions given in Table 8.2.

8.12 Determine an equation for the number of devices reworked on the ith rework attempt (companion equation to Equations (8.20) through (8.22)).

8.13 Reproduce the model used in Section 8.4 and verify the results given in the text.

8.14 Using the model in Section 8.4 (and Problem 8.13), what happens to the effective cost per component reworked if you add a fourth shift? Note that a fourth shift corresponds to the weekend, and we will assume this represents 16 additional hours per week of production.

Chapter 9

Uncertainty Modeling — Monte Carlo Analysis

Uncertainty is defined as the state of having limited knowledge, which makes it impossible to exactly describe the existing state or the future outcome of a system. Accounting for uncertainties is very important in all types of modeling. Models of costs (or any other property estimated from a model) rarely predict exact answers. If your boss asks you to predict the recurring manufacturing cost of a new electronic system during its design process and your answer is $1345.54 per unit, there is one thing that your boss knows with a 100% certainty, and that is that you are wrong. Chances are excellent that prior to the actual manufacturing of any units, there are some unknowns, and not every unit is going to cost the same (e.g., some may need to be reworked to replace a faulty component, and some may not). After a population of the product you costed has been manufactured, the recurring manufacturing cost per unit is probably best represented by a distribution.

From a modeling standpoint, the sources of error (uncertainty) in the values predicted by models include the following:[1]

- The description of the system may not be fully known — that is, the data going into the models may be unavailable or inaccurate (data or parameter uncertainty).
- The knowledge of the environment in which the system will operate may be incomplete; boundary conditions may be inaccurate or poorly understood, operational requirements may not be clear.

[1] Other taxonomies and types of uncertainty, in addition to those mentioned here, may be relevant depending on the activities being considered, including measurement uncertainties and subjective uncertainties.

- The formulation of the model may be inaccurate, the understanding of the behavior of the system may be incomplete, or the model may represent a simplification of a real world process (model uncertainty) .
- Computational inaccuracies or approximations may occur. Even if the formulation of the model is accurate, numerical fitting techniques may be necessary to execute the model and the solution may only represent an approximation to the actual solution.

The uncertainty in a model can be represented as shown in Figure 9.1. *Epistemic* is defined as, relating to, or involving knowledge. Epistemic uncertainties are due to a lack of knowledge. Collecting more data or knowledge can shrink epistemic uncertainties. For example, the time it takes to perform a process step is an epistemic uncertainty that can be decreased if additional data collection and process observation can establish the duration of the step, thus increasing the body of knowledge.

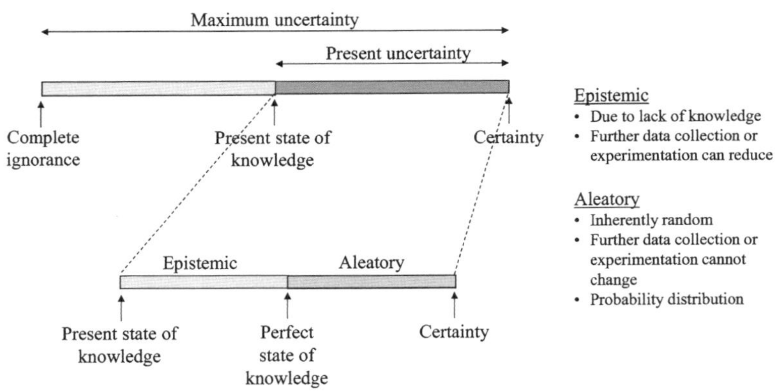

Fig. 9.1. Representation of various types of uncertainty [Ref. 9.1].

Aleatory (or aleatoric) means "pertaining to luck," and derives from the Latin word *alea*, referring to throwing dice. Aleatoric art exploits the principle of randomness. Aleatory uncertainties cannot be reduced through further observation, data collection or experimentation. Aleatory uncertainties have an inherently random nature attributable to true heterogeneity or diversity in a population or an exposure parameter. An example of an aleatory uncertainty in a process step could be the yield associated with a particular random fault in the step.

It is often just as important to understand the size and nature of errors in a predicted value as it is to obtain the prediction. When proposals are made, business cases constructed, and quotations prepared for manufacturing new products, management needs to understand the uncertainties that are present in the prediction. Without a statement of uncertainties, a prediction is incomplete.

Uncertainty Modeling

Methods for sensitivity analysis and uncertainty propagation can be classified into the following four categories [Ref. 9.2]: (a) sensitivity testing, (b) analytical methods, (c) sampling-based methods, and (d) computer algebra-based methods.

Sensitivity testing involves studying a model response for a set of changes in model formulation, and for selected model parameter combinations. In this approach, the model is run for a set of sample points for the parameters of concern or with straightforward changes in model structure (e.g., in model resolution). This approach is often used to evaluate the robustness of the model, by testing whether the model response changes significantly in relation to changes in model parameters and the structural formulation of the model. The application of this approach is straightforward, and it has been widely employed. Its primary advantage is that it accommodates both qualitative and quantitative information regarding variation in the model. However, its main disadvantage is that detailed information about the uncertainties is difficult to obtain. Further, the sensitivity information depends to a great extent on the choice of the sample points, especially when only a small number of simulations can be performed.

Analytical methods involve either differentiating the model equations and subsequently solving of a set of auxiliary sensitivity equations, or reformulating the original model using stochastic algebraic/differential equations. Some of the widely used analytical methods for sensitivity/uncertainty are: (a) differential analysis methods, (b) Green's function method, (c) the spectral-based stochastic finite element method, and (d) coupled and decoupled direct methods. The analytical methods require the original model equations and may require that additional computer code be written for the solution of the auxiliary sensitivity equations--this often proves to be impractical or impossible.

Sampling-based methods involve running a set of models at a set of sample points, and establishing a relationship between inputs and outputs using the model results at the sample points. Widely used sampling-based sensitivity/uncertainty analysis methods are include: (a) Monte Carlo and Latin hypercube sampling methods (the remainder of this chapter focuses on these methods), (b) the Fourier Amplitude Sensitivity Test (FAST), (c) reliability-based methods, and (d) response-surface methods.

Computer algebra-based methods involve the direct manipulation of the computer code, typically available in the form of a high-level language code (such as C or FORTRAN), and estimation of the sensitivity and uncertainty of model outputs with respect to model inputs. These methods do not require information about the model structure or the model equations, and use mechanical, pattern-matching algorithms to generate a "derivative code"" based on the model code. One of the main computer algebra-based methods is automatic (or automated) differentiation.

Many methods have been proposed for characterizing uncertainty in cost estimation [Ref. 9.3]. Most methods are based on probability theory. If sufficient historical data exists, probability distributions can be determined for various parameters (see Section 9.1) and Monte Carlo analysis can be performed. However, other approaches can also be used.

9.1 Representing the Uncertainty in Parameters

In cost modeling, nearly every parameter that appears in the models has both an epistemic and aleatory component. As an example, consider the process time for a step. Observation and data collection for 1000 units results in 1000 step times. When the step times are plotted as a histogram, Figure 9.2 is obtained.

For example, Figure 9.2 indicates that if 1000 products go through the process step, 0.369 or 36.9% of the units will have a step time between 55 and 65 seconds.

The histogram of measured results shown Figure 9.2 can be fit with a known distribution type — in this case represented as a normal distribution with a mean of 67 seconds and a standard deviation of 10 seconds.

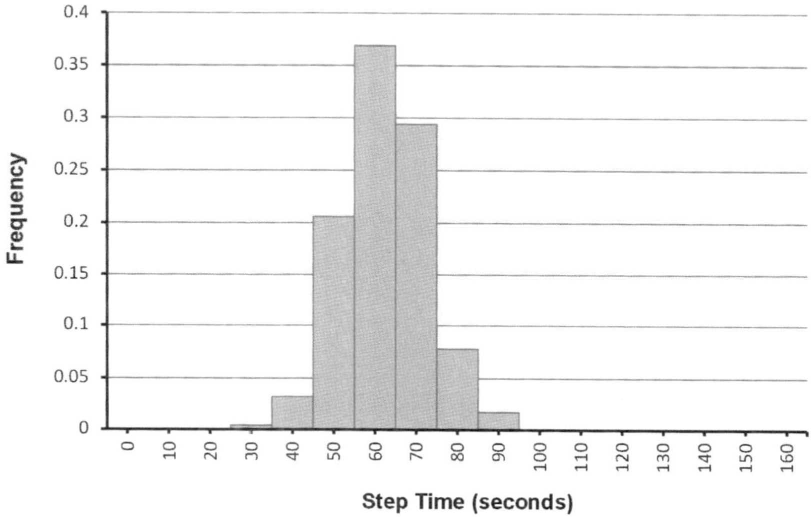

Fig. 9.2. Histogram of measured process step times.

9.2 Monte Carlo Analysis

Monte Carlo refers to a class of algorithms that rely on repeated sampling of probability distributions representing input parameters to develop a histogram of results. Stanislaw Ulam, a mathematician who worked for John von Neumann on the Manhattan Project in the United States during World War II, is reputed to have invented the Monte Carlo method in 1946 by pondering the probabilities of winning a card game of solitaire while convalescing from an illness [Ref. 9.4]. In the 1940s, scientists at Los Alamos Scientific Laboratory (today known as Los Alamos National Laboratory) were studying the distance that neutrons would travel through various materials. Analytical calculations could not be used to solve the problem because the distances depended on how the neutrons scattered during their transit through the material, an inherently random process. von Neumann and Ulam suggested that the problem be solved by modeling the system on a computer.[2] Although von Neumann and Ulam coined the term "Monte Carlo," such methods can be traced as far back as Buffon's needle in the 18th century.

9.2.1 *How Does Monte Carlo Work?*

Suppose we have the following equation to solve:

$$G = B + C \tag{9.1}$$

If we know the values of B and C (say $B = 2$ and $C = 3$) then G is easy to solve for. But what if we don't know exactly what B or C are—that is, there is some uncertainty associated with them. Then what is G? If we knew the range of values that B and C could take (their minimum and maximum values), we could easily establish the largest value and smallest value that G could have. Alternatively, the average values of B and C could be used to find the average value of G from Equation (9.1) (however, this only works if the relationship between G, B and C is linear and B and C are represented by symmetric distributions). These would all be useful results.

Let's generalize the problem a bit. Suppose that B and C were represented as probability distributions like the ones described in Figure 9.3. It is intuitive that the resulting G (from Equation (9.1)) will also be a probability distribution, but how do we find it?

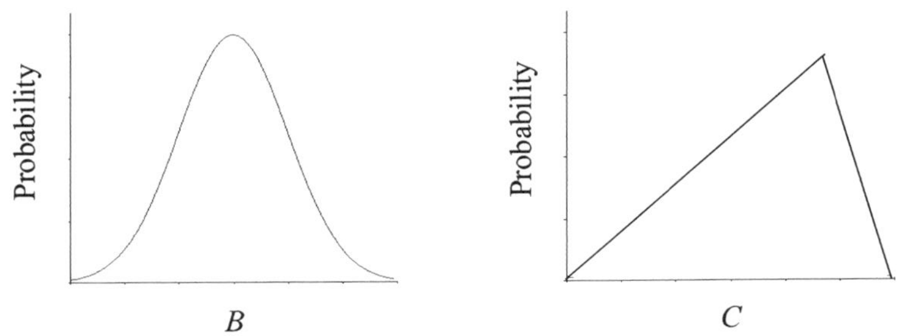

Fig. 9.3. Probability distributions representing B and C.

[2] Since the Manhattan Project was highly secret, the work required a code name. "Monte Carlo" was chosen as a reference to the Monte Carlo Casino in Monaco.

The Monte Carlo method of solving this problem is to sample the *B* and *C* distributions, combine the samples as prescribed in Equation (9.1) to obtain a sample of *G*, and then repeat the process many times to generate a histogram of *G* values. This process is shown in Figure 9.4.

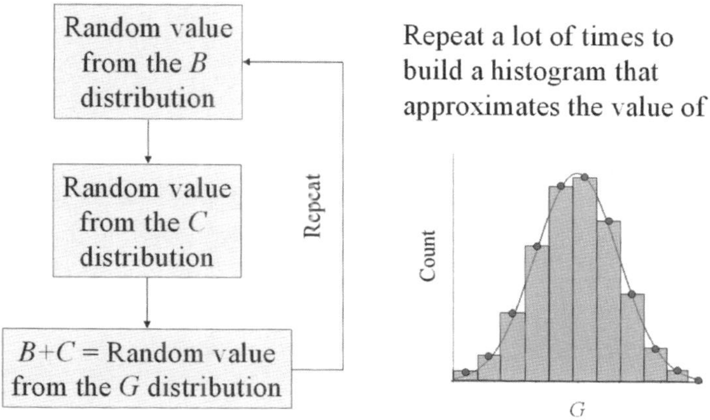

Fig. 9.4. Monte Carlo solution process.

For this process to work, two key questions must be addressed. How do we sample from a distribution in a valid way? And how many times must the process in Figure 9.4 be repeated in order to build a valid distribution for *G*?

It is worthwhile at this point to clarify some terminology. A *sample* is a specific set of observed random variables; one value sampled from the distribution for *B* and one value sampled from the distribution for *C* together are referred to as a single sample. Each sample can be used to independently generate one final value (one value of *G*). The end result of applying one sample to the Monte Carlo process is referred to as an *experiment*. The total number of samples (which corresponds to the total number of computed values of *G*) is referred to as the *sample size* and all the experiments together create summary statistics and a *solution*.

Monte Carlo is not iterative — that is, the results of the previous experiment are not used as input to the next experiment. Each individual experiment has the same accuracy as every other experiment. The overall solution is composed of the combination of all the individual experiments. Each individual experiment in a Monte Carlo analysis can be thought of as the complete and accurate solution for one member of a large population. The end result of using many samples (each sample representing one member of the population) is a statistical representation of the population. The population could represent, for example, many instances of a product or many applications of a process step.

9.2.2 *Random Sampling Values from Known Distributions*

For Monte Carlo to work effectively, the samples obtained from the B and C distributions need to be distributed the same way that B and C are distributed. The question boils down to determining how to obtain random numbers that are distributed according to a specified distribution. For example, the value shown in Figure 9.5 is not a uniformly distributed number, i.e., all values between 0 and 1 are not equally likely.

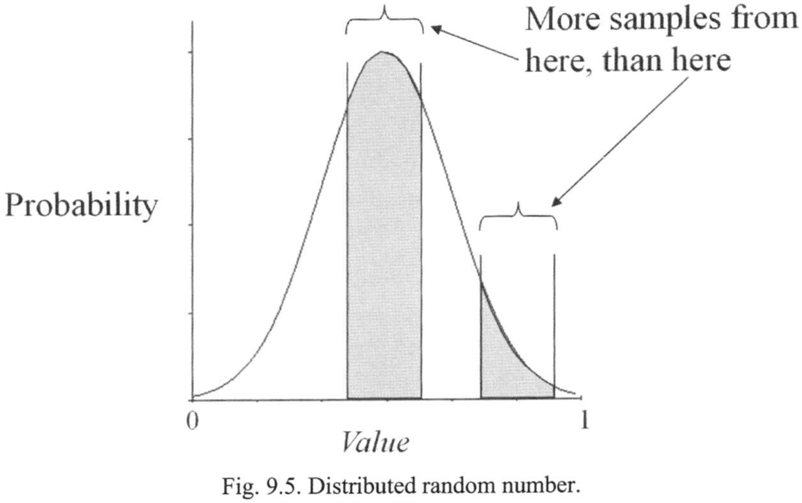

Fig. 9.5. Distributed random number.

In order to obtain samples distributed in a specified way, we need to generate the cumulative distribution function (CDF) that corresponds to a probability distribution (PDF) like that shown in Figure 9.5. In general CDFs are found from the PDF using

$$F(x) = \int_{-\infty}^{x} f(t)dt \tag{9.2}$$

where $f(t)$ is the probability density function (PDF) and x is the point at which the value of the CDF is desired, as shown in Figure 9.6.

To obtain a sample from the distribution (the sample is called a *random variate* or *random deviate*), a uniformly distributed random number between 0 and 1 (inclusive) is generated. This uniform random number (U) corresponds to the fraction of the area under the PDF ($f(t)$) and is the value of the CDF ($F(x)$) that corresponds to the sampled value (x_1). This works because the total area under $f(t)$ is 1.

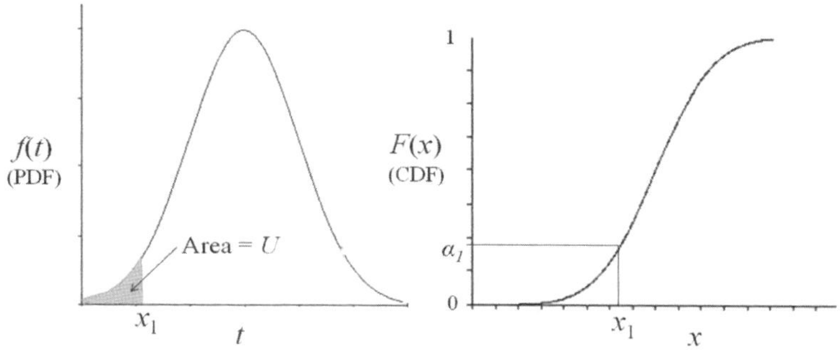

Fig. 9.6. Example PDF and the corresponding CDF.

If a variable is represented by a probability distribution that has a closed-form mathematical expression for its CDF, then sampling the distribution is easy. Simply choose a uniformly distributed random number between 0 and 1 inclusive and set $F(x)$ equal to it, then find the corresponding x. However, not all PDFs have closed-form CDFs. Most notably, there is no closed-form solution to Equation (9.2) for the normal distribution.[3]

The sampling strategies discussed in this chapter are referred to as transformation methods (specifically, inverse transform sampling). An alternative is called the rejection method [Ref. 9.6], which does not require a CDF (it only requires that the PDF be computable up to an arbitrary scaling constant). The rejection method has the advantage of being straightforwardly applicable to multivariate probability distributions. However, rejection methods are much more computationally intensive than transformation methods.

---

[3] Extremely efficient numerical approximations to the CDF for normal distributions do exist; see, for example, [Ref. 9.5].

9.2.3 *Triangular Distribution Derivation*

As an example of a useful distribution for Monte Carlo analysis, consider a non-symmetric triangular distribution. The distribution we wish to develop a sampling process for is shown in Figure 9.7 and is defined by a minimum (α), most likely or mode (β), and maximum (γ) — referred to as a three-point estimator. Triangular distributions are useful because they have controllable minimum and maximum values (α and γ).

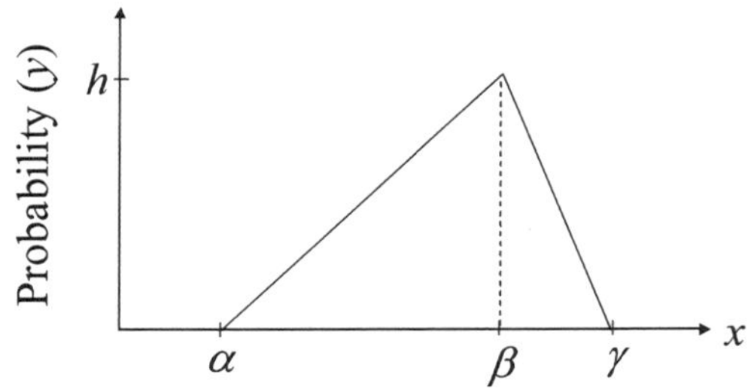

Fig. 9.7. Example triangular distribution PDF.

To be a valid probability distribution, the area under the triangle must equal 1. Based on this constraint, we can solve the following equation for h:

$$\frac{1}{2}(\beta-\alpha)h+\frac{1}{2}(\gamma-\beta)h=1 \tag{9.3}$$

which becomes

$$h=\frac{2}{(\gamma-\alpha)} \tag{9.4}$$

Now solve for y as a function of x for the left and right triangles in Figure 9.7. Considering the left side first,

$$y=\frac{h}{(\beta-\alpha)}x-\frac{h\alpha}{(\beta-\alpha)}=\frac{h}{(\beta-\alpha)}(x-\alpha) \tag{9.5}$$

which is valid when $\alpha \le x \le \beta$. Similarly, for the right side,

$$y=\frac{-h}{(\gamma-\beta)}x+\frac{h\gamma}{(\gamma-\beta)}=\frac{h}{(\gamma-\beta)}(-x+\gamma) \tag{9.6}$$

which is valid when $\beta \le x \le \gamma$. Lastly, $y = 0$ when $\alpha \ge x$ *and* $x \ge \gamma$.

Next we need to determine the area (U) enclosed by the triangle as a function of x. For $x \le \alpha$, $U = 0$. For $\alpha \le x \le \beta$, the area enclosed is

$$U = \frac{1}{2}(x-\alpha)\frac{h}{(\beta-\alpha)}(x-\alpha) \tag{9.7}$$

For $\beta \le x \le \gamma$ the total area enclosed is

$$U = \frac{1}{2}(\beta-\alpha)\frac{h}{(\beta-\alpha)}(\beta-\alpha)+\frac{1}{2}(\gamma-\beta)h-\frac{1}{2}(\gamma-x)\frac{h}{(\gamma-\beta)}(\gamma-x) \tag{9.8}$$

where the first term in Equation (9.8) is Equation (9.7), with $x = \beta$. Finally, for $x \ge \gamma$, $U = 1$.

Now, solving Equation (9.7) for x we get

$$x = \sqrt{\frac{2U(\beta-\alpha)}{h}}+\alpha \tag{9.9}$$

which should be used if $\frac{1}{2}(\beta-\alpha)h \ge U \ge 0$. Solving Equation (9.8) for x,

$$x = \gamma - \sqrt{\frac{-2\left(U-\frac{1}{2}(\beta-\alpha)h-\frac{1}{2}(\gamma-\beta)h\right)(\gamma-\beta)}{h}} \tag{9.10}$$

which should be used if $1 \ge U \ge \frac{1}{2}(\beta-\alpha)h$, where h is given by Equation (9.4).

The value of x in Equations (9.9) and (9.10) is a sample from the triangular distribution defined by α, β and γ, generated using the uniformly distributed random number U between 0 and 1 inclusive.

9.2.4 *Random Sampling from a Data Set*

Sometimes you have a data set that represents observations or possibly the result of an analysis that determines one of the variables in your model. You could create a histogram from the data (like Figure 9.2), fit the histogram with a known distribution form, determine the CDF of the distribution (either in closed form or numerically), and sample it as described in Section 9.2.2. However, why go to the trouble of approximating a data set with a distribution when you already have the data set? A better solution if you have a sufficiently large data set is to directly use the data set for sampling. If the data set has N data points in it,

(1) Sort the date set in ascending order (smallest to largest) — (x_1, x_2, ..., x_N).
(2) Choose a uniformly distributed random number between 0 and 1 inclusive (U).
(3) The sampled value lies between the data point $\lfloor NU \rfloor$ and the data point $\lceil NU \rceil$.

The above algorithm works if you have a large data set, or if you have a small data set and do not have any other information. If you have just a few data points and you know what the distribution shape should be, then you are better off finding the best fit to the known distribution, then proceeding as previously described.

9.2.5 *Implementation Challenges with Monte Carlo Analysis*

There are several common issues that arise when Monte Carlo analyses are implemented.

Because of Monte Carlo's reliance on repeated use of uniformly distributed random or pseudo-random numbers, it is important that an appropriate random number generator is used. Since computers are deterministic, computer-generated numbers aren't really random. But, various mathematical operations can be performed on a provided random number seed to generate unrelated (pseudo-random) numbers. Be careful; if you use a random number generator that requires a seed provided by you, you may get an identical sequence of random numbers if you use the same seed. Thus, for multiple experiments, different random number seeds may have to be used. Many commercial applications use a random number seed from somewhere within the computer system, commonly the time on the system clock, therefore, the seed is unlikely to be the same for two different experiments.

In general you should not use an unknown random number generator; random number generators should be checked (see [Ref. 9.7]). While it is impossible to prove definitively whether a given sequence of numbers (and the generator that produced it) is random, various tests can be run. The most commonly used test of random number generators is the chi-square test;[4] however, there are other tests — for example, the Kolmogorov-Smirnov test, the serial-correlation test, two-level tests, k-distributivity, the serial test, or the spectral test. Lastly, it is generally inadvisable to use *ad hoc* methods to improve existing random number generators.

In general, you do not want to restart your random number generator for each experiment. A common implementation mistake is to choose a single uniform random number and use it to sample the distributions associated with all the variables in the experiment. This is a grave error if all the variables are supposed to be independent. Using the same random number to sample all the distributions effectively couples all the variables together so they are no longer independent. Doing this effectively makes the correlation coefficient between all the variables equal to one. Independent variables need to be sampled using independent random numbers.

Some distributions can produce non-physical values — that is, the tails of the distributions matter. A prime culprit is the normal distribution. Normal distributions may be problematic for parameters that cannot take on negative values since the left tail of a normal distribution goes to $-\infty$.

---

[4] To run a chi-square test, prepare a histogram of the observed data. Count the number of observations in each "bin" (O_j for the jth bin). Then compute the following:

$$D = \sum_{j=1}^{k} \frac{(O_j - E_j)^2}{E_j}, \quad E_j = \frac{\sum_{j=1}^{k} O_j}{k}$$

Since we are interested in the goodness-of-fit to a distribution made up of perfectly random results, the expected frequencies (E_j for the jth bin) are the same for every bin (j) and are equal to the total number of observations divided by the number of bins. D asymptotically approaches a chi-square distribution with k-1 degrees of freedom, and if $D < \chi^2_{\alpha,\nu}$, then the observations are random with a 1-a confidence ($\nu = k$-1, the degrees of freedom).

Normal distributions can also be problematic for parameters that cannot be greater than 1 (e.g., a yield), since the right tail goes to $+\infty$. You may think that if the mean is large enough and/or the standard deviation is small enough, unrealistic numbers won't be generated; however, a few bad samples can skew the results of the analysis. It is tempting to simply screen the samples taken from the distributions and, if they are negative (for example), simply sample again; however, this practice does not produce valid distributions. Don't do it![5] Other distributions may be preferred that have controllable minimum and/or maximum values, such as triangular distributions.

Many simple tests are possible to verify the implementation of a Monte Carlo analysis model. A histogram of the values sampled can be plotted from the input distributions to verify that the sampled values result in the same distribution as the input. If the problem is linear (like Equation (9.1)) and symmetric input distributions (e.g., for B and C) are used, then the mean value of the resulting G distribution should be equal to the G calculated using the mean values of B and C. A distribution of the mean output from each Monte Carlo solution should always be normal (if the sample size is large enough — see Section 9.3).

9.3 Sample Size

A fundamental question with Monte Carlo analysis is how many samples must be produced (or experiments must be performed) to generate an acceptable solution? The sample size (n) is the quantity of data points or observations that need to be collected from a single Monte Carlo analysis to form a solution. Because Monte Carlo is a stochastic method, we will get a different set of summary statistics every time we perform the analysis. As the sample size increases, the difference between repeated solutions decreases.

There are two ways to approach answering the sample size question. The practical answer is that you need to run experiments until the quantity you want from your analysis — that is, the precision of the estimate of the

---

[5] Note that there are mathematically valid truncated normal distributions that are bounded below and/or above. For an example, see [Ref. 9.8].

mean or precision of the estimate of the cumulative distribution — stops changing. As long as the uniform random number generator is not reset or does not otherwise begin repeating random numbers, more experiments can be run and added to the experiments you already have. For example, when you run 100 more experiments and there is no change in the summary statistics you are interested in, you are done.

The sampling problem can also be treated in a mathematically rigorous way as well. The sample mean is an estimation of the mean of the true population. So how accurate is this estimation? It is obvious that the mean is not the same when the analysis is repeated.

If you repeat the Monte Carlo simulation and record the sample mean μ each time, based on the Central Limit Theorem, the distribution of the sample mean will follow a normal distribution. The Central Limit Theorem states that if random samples are selected from a population with mean μ and a finite standard deviation σ, as the sample size n increases, the mean of the sample set (sample mean) approaches a normal distribution with a mean of μ and a standard deviation equal to the standard error, $\sigma/\sqrt{n}$ (referred to as the standard error of the mean). If the population is sufficiently large, this is independent of the shape of the sampled population.

The standard error is a useful indicator of how close the estimate from the Monte Carlo solution is to the unknown estimand (the parameter being estimated). A common practical stopping criterion for Monte Carlo analysis is to stop when the standard error of the mean is less than 1%:[6]

$$\frac{\sigma}{\sqrt{n}} \leq 0.01\mu \tag{9.11}$$

Using the standard error we can calculate confidence intervals for the true population mean. For a two-sided confidence interval, the upper confidence limit (*UCL*) and lower confidence limit (*LCL*) on the true population mean are calculated as

$$UCL = \text{true population mean} + z\frac{\sigma}{\sqrt{n}} \tag{9.12a}$$

[6] Equation (9.11) is used as a stopping criteria, i.e., it is not used to determine the number of samples ahead of time, but rather to figure out if you have done enough samples.

$$LCL = \text{true population mean} - z\frac{\sigma}{\sqrt{n}} \tag{9.12b}$$

where z is the z-score (standard normal statistic — the distance from the sample mean to the population mean in units of standard error). The value of z used depends on the desired confidence level. The area under the normal distribution of the sample set means (μ) between $-z$ and $+z$ is the desired confidence level. Since the distribution of the sample set means is a normal distribution, the values of z are tabulated in statistics textbooks, as in Table 9.1.

Table 9.1. Values of z Corresponding to Various Two-Sided Confidence Levels.

Confidence Level Desired	z
90%	1.645
95%	1.960
99%	2.576

Equation (9.12) means that we have a given confidence that the true population mean is between the *LCL* and the *UCL*.

9.4 Example Monte Carlo Analysis

In this section we present a simple analysis performed using the Monte Carlo method. Suppose that a particular process produces printed circuit boards that cost \$25 each. The individual printed circuit boards have an area of 3 square inches and are fabricated on a larger panel. The process that makes the panel is somewhat erratic, producing panels with defect densities that are constant across a panel but that vary from panel-to-panel. The cost of performing recurring functional testing with a fault coverage of 0.85 on the boards also varies from board to board. You wish to determine the confidence that the cost per board (after test for the boards that pass the test) is less than \$44.

The input data for this example is:

- C_{in} = \$25.

- C_{test} = triangular distribution with α = \$4, β = \$5 and γ = \$7 (h = 0.667).
- f_c (fault coverage) = 0.85.
- A (area of the board) = 3 in^2.
- D_0 (defect density, defects/in^2) = triangular distribution with α = 0.1, β = 0.15 and γ = 0.16 (h = 33.333).
- Assume that the Poisson yield model holds and that there is no rework of the boards that do not pass the test (they are scrapped).
- Assume that the test cost and defect density are independent (in reality, they may not be).

The applicable equations for calculating the cost of boards that pass the test are (7.35) and (3.20), which, when combined, give

$$C_{out} = \frac{C_{in} + C_{test}}{e^{-AD_0 f_c}} \tag{9.13}$$

If we solve Equation (9.13) using the most likely values of the C_{test} and D_0 (the values of β) we obtain C_{out} = \$43.98/board.

To solve Equation (9.13) using a Monte Carlo analysis requires that we sample the distributions for C_{test} and D_0. As an example, one sample could be[7]

$$C_{test}\text{: } U = 0.927,\ \frac{1}{2}(\beta - \alpha)h = 0.333,$$

which is less than U, so using Equation (9.10), $x = 6.338$

$$D_0\text{: } U = 0.138,\ \frac{1}{2}(\beta - \alpha)h = 0.833,$$

which is greater than U, so using Equation (9.9), $x = 0.120$.

The combination of C_{test} = \$6.338 and D_0 = 0.120 represent one sample. Note that different uniform random numbers (U) were used for C_{test} and D_0 because we are assuming that they are independent. Using this sample in Equation (9.13), we calculate the final value of C_{out} = \$42.59 corresponding to the sample. This process represents one experiment.

---

[7] You can easily check your implementation of the sampling process by forcing the random number, U, to be 0, in which case x should equal α; and if you force $U = 1$, x should be γ.

Taking n = 1000 samples (each with a new pair of uniform random numbers), we obtain the histogram of 1000 values of C_{out} shown in Figure 9.8. The mean value of C_{out} obtained is \$43.01 (standard deviation = \$1.67). To find the confidence that the final C_{out} is less than \$44, we simply count the number of experiments that produced C_{out} values that were below \$44 (717) and divide it by the number of experiments done (1000) to obtain 0.717, or 71.7% confidence.

Using Equation (9.11) to solve for the number of samples needed to obtain a standard error on the mean of less than 1%, we get $n > 15$ samples. Does this make sense? 1% of the mean is 0.43. Looking at the bottom plot in Figure 9.8, it takes very few experiments for the mean to approach its final value within 0.43.

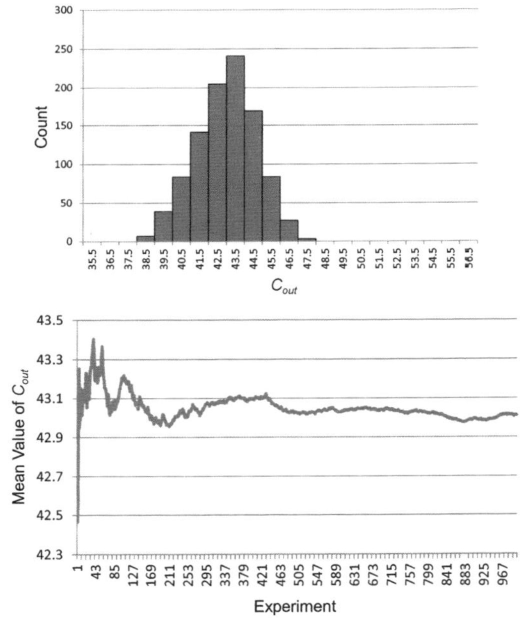

Fig. 9.8. Top – histogram of C_{out} values, Bottom – variation of the mean C_{out} as a function of the number of experiments.

9.5 Stratified Sampling (Latin Hypercube)

The methodology considered so far in this chapter assumes random sampling from the prescribed distributions — that is, we are using uniformly distributed random numbers between 0 and 1 inclusive to extract distributed random numbers.

Stratified sampling can characterize the population equally as well as simple random sampling, but with a smaller sample size. In stratified sampling, the data is collected to occupy prearranged categories or strata. The form of stratified sampling we are going to consider in this section is called Latin Hypercube.

9.5.1 *Building a Latin Hypercube Sample (LHS)*

To building a Latin hypercube sample, four steps are required [Ref. 9.9]:

(1) The range of each variable is divided into n_I non-overlapping intervals each representing equal probability.
(2) One value from each interval for each variable is selected using random sampling.
(3) The n_I values obtained for each variable are paired in a random manner to form n_I k-tuplets (the LHS).
(4) The LHS is used as the data to determine the overall solution.

First the range of each variable is divided into n_I non-overlapping intervals, each representing equal probability, as shown in Figure 9.9. In this example, the range of the variable V is divided into n_I = 5 equal probability (0.2) intervals.

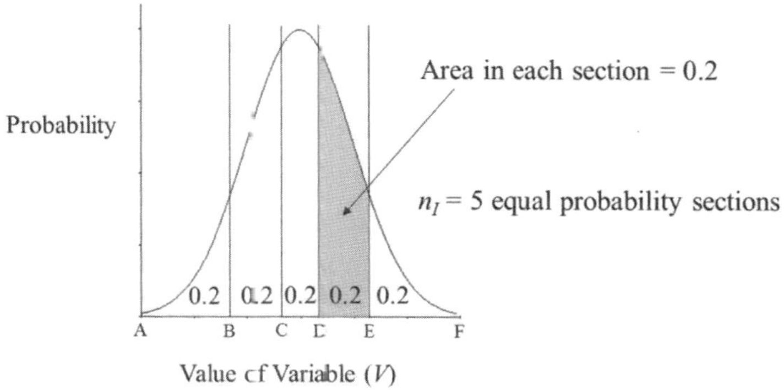

Fig. 9.9. Division of the PDF into n_I equal probability intervals.

Next, one value from each interval for each variable is selected using random sampling, as shown in Figure 9.10. The sampling from each interval is performed essentially identically to the random sampling discussed in Section 9.2.

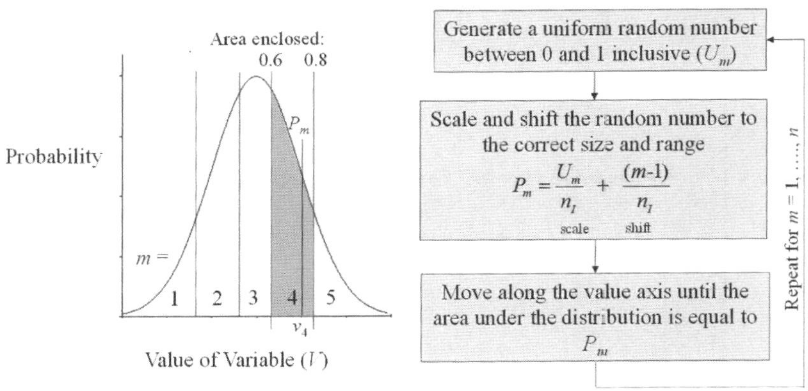

Fig. 9.10. Selecting one value from each interval via random sampling.

In the third step, the n_I values $(v_1, \ldots, v_{n_I})$ obtained for each variable are paired in a random manner (equally likely combinations) forming n_I k-tuplets (k is the number of variables considered), this is called the Latin hypercube sample (LHS). For $k = 2$ (two variables, V and Z with distributions) and $n_I = 5$ intervals, we pair two random permutations of (1, 2, 3, 4, 5): Permutation Set 1: (3, 1, 5, 2, 4) and Permutation Set 2: (2, 4, 1, 3, 5), as shown in Table 9.2.

Table 9.2. Two 5-Tuplets That Define the LHS for a Problem with Two Random Variables (V and Z).

Computer Run Number	Interval used for V	Interval used for Z
1	3	2
2	1	4
3	5	1
4	2	3
5	4	5

Figure 9.11 shows a representation of the LHS of size 5 for V and Z. Note that only the generation of the V values was shown in Figure 9.9, Z is another variable with a similar generation process. In Figure 9.11 v_4 is

the $m = 4$ interval sample from the variable V and z_5 is the $m = 5$ interval sample from the variable Z. In general, Figure 9.11 would be k dimensional and have n_I^k cells in it and produce n_I k-tuplets of data.

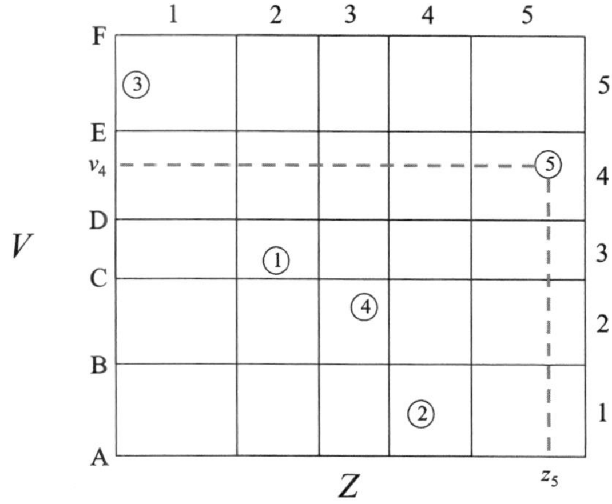

Fig. 9.11. Two-dimensional representation of one possible LHS of size 5 with two variables.

Finally, we use the LHS as the data to determine the overall solution. The data pairs specified by Table 9.2 are used: (v_3,z_2), (v_1,z_4), (v_5,z_1), (v_2,z_3), (v_4,z_5). These five data pairs are used to produce five possible solutions.

9.5.2 *Comments on LHS*

LHS forms a random sample of size n_I that appropriately covers the entire probability space. LHS results in a smoother sampling of the probability distributions — that is, it produces more evenly distributed (in probability) random values and reduces the occurrence of less likely combinations (e.g., combinations where all the input variables come from the tails of their respective distributions). Random sampling required n samples (n is the sample size from Section 9.3) of k variables = kn total samples. LHS requires n_I samples (intervals) of k variables = kn_I total samples. It is not unusual for LHS to require only a fifth as many trials as Monte Carlo with simple random sampling.

To determine n_I, apply the standard error on the mean criteria (e.g., Equation (9.11)) to each interval.

Even though variables are sampled independently and paired randomly, the sample correlation coefficient of the n_I k-tuplets of variables, in general, is not zero (due to sampling fluctuations). Restricting the way in which variables can be paired can be used to induce a user-specified correlation among selected input variables. See [Ref. 9.10] for more discussion.

9.6 Discussion

Monte Carlo simulation methods are particularly useful for studying systems that have a large number of coupled degrees of freedom. Monte Carlo methods are also useful for modeling systems with highly uncertain inputs. Monte Carlo methods are not deterministic (i.e., there is no set of closed-form equations to solve for an answer).

Monte Carlo is independent of the formulation of the model — for example, the model does not have to be linear. Monte Carlo also does not constrain what form the distributions take, and the distributions need not necessarily even have a mathematical representation. Monte Carlo also has the advantage that even though it is computationally intensive, it will always work.

The main argument against Monte Carlo is that it is a "brute force" computationally intensive solution. Another potential drawback is that Monte Carlo implicitly assumes that all the parameters are independent. Correlation of the parameters in Monte Carlo analyses can be done. In general, the parameters are uncorrelated because independent random numbers are used to generate the samples. The degree to which the parameters are correlated depends on the how correlated the random numbers used to sample them are (see, e.g., [Ref. 9.11]).

There are many software packages for performing Monte Carlo analysis today — Palisade, @Risk®, Minitab, and Crystal Ball® are available for Excel. A treatment of Monte Carlo implementation within Excel is provided in [Ref. 9.12].

References

9.1 Aughenbaugh, J. M. and Paredis, C. J. J. (2005). The value of using imprecise probabilities in engineering design, *Proceedings of the ASME Design Engineering Technical Conference (DETC)*.

9.2 Isukapalli, S. S. (1999). *Uncertainty Analysis of Transport-Transformation Models*, Ph.D. Dissertation, The State University of New Jersey at Rutgers. Available at: http://www.ccl.rutgers.edu/ccl-files/theses/Isukapalli_1999.pdf. Accessed April 22, 2016.

9.3 Goh, Y. M., Newnes, L. B., Mileham, A. R., McMahon, C. A. and Saravi, M. E. (2010). Uncertainty in through-life costing – Review and perspectives, *IEEE Transactions on Engineering Management*, 57(4), pp. 689-701.

9.4 Eckhardt, R. (1987). Stan Ulam, John von Neumann, and the Monte Carlo method, *Los Alamos Science, Special Issue*, 15, pp. 131-137.

9.5 West, G. (2005). Better approximations to cumulative normal functions, Wilmott Magazine, 9, pp. 70–76. https://lyle.smu.edu/~aleskovs/emis/sqc2/accuratecumnorm.pdf. Accessed May 8, 2016.

9.6 von Neumann, J. (1951). Various techniques used in connection with random digits, *National Bureau of Standards Applied Mathematics Series*, No. 12, pp. 36-38.

9.7 Park, S. K. and Miller, K. W. (1988). Random number generators: Good ones are hard to find, *Communications of the ACM*, 31(10), pp. 1192-1201.

9.8 Greene, W. H. (2003). *Econometric Analysis*, 5th Edition (Prentice Hall, Upper Saddle River, NJ).

9.9 McKay, M. D., Conover, W. J. and Beckman, R. J. (1979). A comparison of three methods for selecting values of input variables in the analysis of output from a computer code, *Technometrics*, 21(2), pp. 239-245.

9.10 Iman, R. L. and Conover, W. J. (1982). A distribution-free approach to inducing rank correlation among input variables, *Communications in Statistics*, B11(3), pp. 311-334.

9.11 Touran, A. (1992). Monte Carlo technique with correlated random variables, *Journal of Construction Engineering and Management*, 118(2), pp. 258-272.

9.12 O'Connor, P. and Kleyner, A. (2012). Chapter 4 – Monte Carlo simulation, *Practical Reliability Engineering*, 5th Edition (John Wiley & Sons, West Sussex, England).

Bibliography

In addition to the sources referenced in this chapter, there are many books and other good sources of information on Monte Carlo modeling including:

Hazelrigg, G. A. (1996). *Systems Engineering: An Approach to Information-Based Design*, (Prentice Hall, Upper Saddle River, NJ).

Kalos, M. H. and Whitlock, P. A. (1986). *Monte Carlo Methods, Vol. 1: Basics*, (John Wiley & Sons, New York, NY).

Ross, S. (1998). *A First Course in Probability*, 5th Edition, (Prentice-Hall International Inc., Upper Saddle River, NJ).

Hammersley, J. M. and Handscomb, D. C. (1964). *Monte Carlo Methods*, (John Wiley & Sons, Inc., New York, NY).

Metropolis N. and Ulam, S. (1949). The Monte Carlo method, *J. American Statistical Association*, 44(247), pp. 335-341.

Problems

Monte Carlo problems appear in other places in this book. See Problems 12.10 and 15.9.

9.1 Given a random variable, x, with a non-symmetric triangular distribution defined by $\alpha = 2$, $\beta = 4$ and $\gamma = 6$, construct the CDF of x. Sample the CDF of x and show that you can rebuild the original distribution function.

9.2 Derive the PDF and CDF for a uniform distribution (also called a rectangular distribution) with a minimum value of α and a maximum value of γ. Show how you would set up a scheme to sample from this distribution using a uniform random number between 0 and 1 (U), i.e., derive the analog of Equations (9.9) and (9.10).

9.3 Write an algorithm that appropriately interpolates between two sorted data set points, $\lfloor NU \rfloor$ and $\lceil NU \rceil$. See Section 9.2.4 for the relevance of this problem.

9.4 Assume that you have generated 2000 uniformly distributed random numbers between 0 and 1 inclusive. When you sort them you obtain the following number of observations in ten equal size bins: 208, 200, 201, 189, 210, 178, 198, 201, 220, 195. By applying the chi-square test, determine if this is an acceptable random number generator.

9.5 Suppose that you have run a Monte Carlo analysis (sample size of n) and you wish to cut the standard deviation in half. What is the required sample size?

9.6 An current in an electric circuit was modeled with 1000 experiments. The output has a mean value of 20 amps with the standard deviation of 10 amps. Estimate the sample size (number of experiments) required to obtain 1% accuracy (standard error on the mean) with 95% two-sided confidence.

9.7 Use Equation (9.12) to determine what the stopping criterion in Equation (9.11) implies about the combination of confidence level and error size.

9.8 Given the following probability distribution,

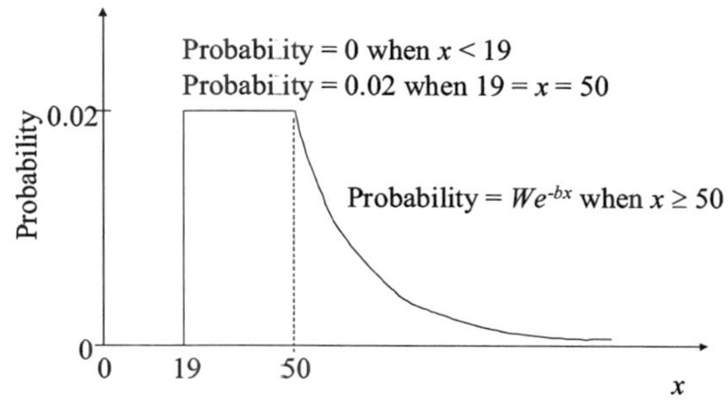

a) What is the value of the parameter W?

b) If the uniform random number is 0.62, what value of x is returned after sampling the above distribution? Hint: you do not need to solve part a) to work this part.

c) If the uniform random number is 0.7, what value of x is returned after sampling the above distribution?

d) If you sampled the above distribution and obtained x = 39.0, what was the uniform random number? Hint: you do not need to solve part a) to work this part.

9.9 Starting with the example in Section 9.4, model the cost of test (C_{test}) using a uniform distribution ranging from \$4 to \$7. Find the new C_{out} distribution.

9.10 A process is characterized by the following data:

Unit	Unit Time
1	1500
2	1300
3	950
5	850
23	712
51	598
100	510
275	500
500	400
1000	330
1100	320
2540	310
3000	300
3200	298
3780	298
3900	290
4000	287

Unit	Unit Time
4150	288
4600	285
5000	284

a) Write an expression of the unit learning curve (see Chapter 10) and predict the time required to build unit number 6120.
b) Assume that each of the parameters in your learning curve expression (first unit time[8] and s; see Equation (10.6)) can be represented by an asymmetric triangular distribution with a mode equal to the value found in part a), a low limit equal to 92% of the mode, and a high limit equal to 110% of the magnitude of the mode. Plot a histogram of the predicted time required to build unit number 6120 for 10,000 samples.
c) Using your result from part b), for an 80% confidence level, what is the build time for unit 6120? There are several ways to interpret an 80% confidence level. Explain what 80% confidence means for the solution you provide. Hint: you do not have to "fit" the result from part b) to any known distribution form to determine the answer to this question.

9.11 Use Latin hypercube sampling to solve part b) of Problem 9.10.

9.12 A random variable X used in a Monte Carlo analysis has a distribution defined by,

$$f(x)=\begin{cases} 0 & \text{for } x \leq 0 \\ 2wx & \text{for } 0 \leq x \leq 3 \\ 3w(5-x) & \text{for } 3 \leq x \leq 5 \\ 0 & \text{for } 5 \leq x \end{cases}$$

a) What does the value of w have to be?
b) If a random number between 0 and 1 equal to 0.68 is selected to sample this distribution, what value of X is produced by this sampling?

9.13 If a variable time is represented as a Weibull distribution ($\beta = 4$, $\eta = 10^5$ hours and $\gamma = 20{,}000$ hours) and the modeling program chooses the value of a random number (between 0 and 1, inclusive) equal to 0.27, what is the sample value that a Monte Carlo analysis will returned from the distribution? The Weibull distribution is described in Section 11.2.3.

[8] Not the intercept! (first unit time = $10^{intercept}$).

Chapter 10

Learning Curves

When forecasting or estimating production costs, engineers are always looking for relationships between production variables and the resulting product cost. One of the most widely applied cases is the relationship between cumulative production volume and the cost of production. Even before World War II, product manufacturers knew that production costs decrease with cumulative output.

One factor that increases output while lowering cost is the *learning curve* of production personnel. When a person performs a repetitive activity, learning takes place. This learning, when it is actively practiced, results in a decrease in the time needed to perform the activity. It also often results in an increase in quality of the resulting output. Learning curves were observed empirically as early as 1925 in aircraft production. The earliest quantitative treatments involved airframes [Ref. 10.1] and machine tools [Ref. 10.2], but subsequently, relationships between production costs and the number of units produced have been identified for a wide variety of industries, including automobile manufacturing [Ref. 10.3], construction [Ref. 10.4], chemical processing [Ref. 10.5], software development [Ref. 10.6], and integrated circuits [Ref. 10.7]. Learning curves have even been used to model writing books [Ref. 10.8].

Learning is not confined to manual production activities, even fully automated production "learns." For example, a pick and place operation in an electronics assembly facility is programmed by an engineer, based on experience with other products. After production of a specific board begins and experience assembling the board is accumulated, engineers can apply that knowledge and edit the programming of the machine to optimize the speed and quality of the operation.

The concept of learning curves — also called improvement curves, progress curves, progress functions, or experience curves — grew from the basic idea that the more of a product you build, the less time it takes to build each one. It takes fewer hours because the skill input into the production operation increases. Increased skill may be due to any or all of the following:

- Operator learning – Individuals or groups of employees become increasingly familiar with the process.
- Improvements in methods, processes, tooling, machines, software, and so on.
- Management learning – improvements in scheduling and work planning.
- Incentives.
- Debugging – decreases required engineering time.

Quantitatively, learning curves denote the relationship between unit cost and unit defect rates and cumulative output in a stable process. Learning-curve modeling makes sense for the production of high-volume, labor-intensive products, when production is uninterrupted, there are no major technological changes, and there is continuous pressure to improve.

10.1 Mathematical Models for Learning Curves

The rate of learning improvement is not arbitrary; it is a function of the process itself. A rate of improvement for a process cannot simply be chosen. To improve, the process itself must be changed to remove limitations to improvement. This often requires a capital investment to improve tools and skills and the removal of the limitations inherent in the process. Such an investment must genuinely improve the process and not just reshuffle the work or reflect wishful thinking.

Many mathematical models for learning curves have been proposed. The four most common relations are

$$\text{Log-linear [Ref. 10.1]: } y = Hx^s \qquad (10.1)$$

$$\text{Stanford-B [Ref. 10.9]: } y = H(x+B)^s \quad (10.2)$$

$$\text{De Jong [Ref. 10.10]: } y = C + Hx^s \quad (10.3)$$

$$\text{S-Curve [Ref. 10.11]: } y = C + H(x+B)^s \quad (10.4)$$

In Equations (10.1) through (10.4), the dependent variable y represents the individual unit learned quantity, the cumulative average of the learned quantity or the marginal quantity,[1] and x is the unit number. The log-linear equation (Equation (10.1)) is the simplest and most common equation and it applies to a wide variety of processes. Figure 10.1 shows a simple log-linear learning curve.

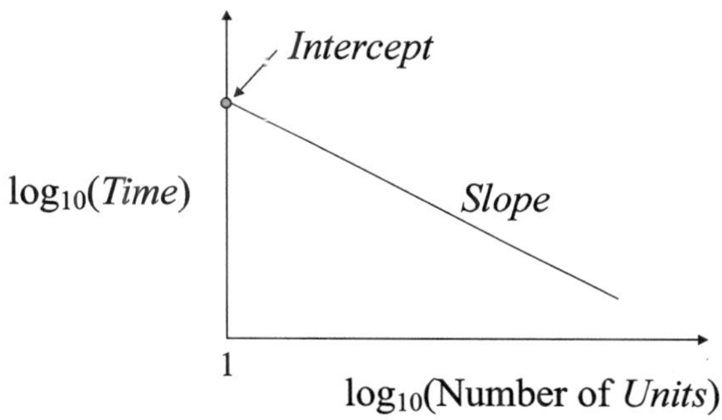

Fig. 10.1. Example of a log-linear learning curve.

The equation for the straight line shown in Figure 10.1 is

$$\log_{10}(Time) = (Intercept) + (Slope)\log_{10}(Unit) \quad (10.5)$$

which reduces to

$$Time = 10^{Intercept}\ Unit^{Slope} = H\ Unit^{s} \quad (10.6)$$

where $H = 10^{Intercept}$ is the time for the first unit to be manufactured, and s is the learning index (*Slope*).

The "Stanford-B" model assumes that prior learning can be captured and utilized on new designs if the new design is consistent with the old

[1] Sections 10.1 – 10.6 are presented in terms of "time" as the learned quantity; however, everything developed in these sections is applicable to other learned quantities, e.g., cost.

design and has as similar degree of complexity. The factor "B" in Equation (10.2) represents the number of units theoretically produced prior to the first unit acceptance, or the equivalent units of experience available at the start of a manufacturing process; H is the cost of the first unit when $B = 0$, as shown in Figure 10.2. The Stanford-B model has been used to model airframe production and mining.

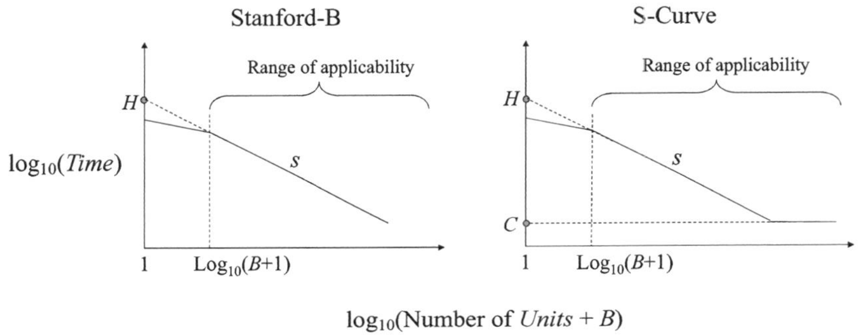

Fig. 10.2. Stanford-B and S-Curve learning curve models.

The De Jong model is used to characterize processes where a portion of the process cannot improve. In Equation (10.3), C represents the fixed component of the learning curve. The De Jong equation is often used in factories where the nature of the assembly line ultimately limits improvement. The S-Curve model combines the Stanford-B and De Jong models to model processes when the experience carries over from one production run to the next and a portion of the process cannot improve. Figure 10.2 shows examples of Stanford-B and S-Curve learning curve models.

The log-linear model has been shown to model future productivity very effectively. In some cases, the De Jong and Stanford-B models work better. The S-Curve model often models past productivity more accurately, and usually models future productivity less accurately, than the other models. The remainder of this chapter will focus on modeling learning with log-linear relations.

The next three sections provide examples and discuss the unit, cumulative average, and marginal forms of the learning curve in the context of the log-linear model. Casting the examples in the other basic learning curve model forms is straightforward.

10.2 Unit Learning Curve Model

The simplest learning curve model is the unit learning curve, also known as the Crawford or Boeing model [Ref. 10.12]. This model has the form shown in Equation (10.6), where the left-hand side of Equation (10.6) or Equation (10.1) is interpreted as the unit time or cost. In the unit learning curve model, an 80% unit learning curve means that each doubling of production brings the unit time (or cost) required to 80% of its former value. Figure 10.3 shows an example of the unit learning curve with a learning rate of 0.8.

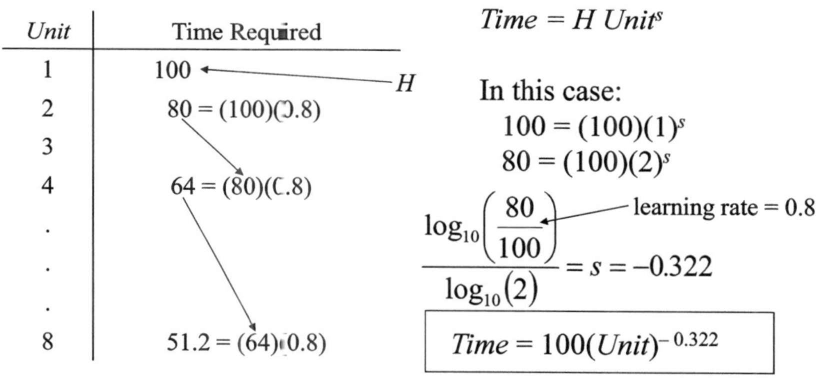

Fig. 10.3. Unit learning curve example for an 80% learning curve.

10.3 Cumulative Average Learning Curve Model

Wright's original work on learning curves generated a cumulative average, Wright, or Northrop model [Ref. 10.1]. This model has the form shown in Equation (10.6) where the left-hand side of Equation (10.6) or Equation (10.1) is interpreted as the cumulative average time (or cost). In the cumulative average learning curve model, an 80% unit learning curve means that each doubling of production brings the cumulative average time (or cost) required to 80% of its former value. Figure 10.4 shows an example of the unit learning curve with a learning rate of 0.8.

Average time over all units up to and including this one

Total time for 2 units

Time for the first unit

Unit	Average Time Required	Unit Time
1	100	100
2	$80 = (100)(0.8)$	$60 = (2)(80)\text{-}(100)$
3	$70.2 = (100)(3)^{-0.322}$	$50.6 = (3)(70.2)\text{-}(100+60)$
4	$64 = (80)(0.8)$	45.4

$$\begin{array}{c}\text{Average Cost or Time}\\ \text{for Units 1 through } X\end{array} = H(X)^s$$

Same as other model:

$$100 = (100)(1)^s$$
$$80 = (100)(2)^s$$
$$s = -0.322$$

$$\boxed{\text{Cumulative Average Time} = 100(\textit{Unit})^{-0.322}}$$

Fig. 10.4. Cumulative average learning curve example for an 80% learning curve.

Note that in both the unit and cumulative average learning curve examples, for a learning rate of 0.8, the learning index (s) is the same (it only depends on the learning rate). Also the learning curve equations are the same. The only difference is in the interpretation of the left-hand side of the equation.

Unit information can be extracted from the cumulative average learning curve (see Section 10.5.1).

10.4 Marginal Learning Curve Model

For the marginal learning curve, the left-hand side of Equation (10.6) or Equation (10.1) is interpreted as the marginal time or cost. In the marginal learning curve model, an 80% unit learning curve means that each doubling of production brings the marginal time or cost required to 80% of its former value.

The marginal time or cost is the change in time or cost when changing the unit by one — that is, instead of a learning curve on the unit time or cost, this is a learning curve on the difference in time or cost between adjacent units. Figure 10.5 shows an 80% marginal learning curve example.

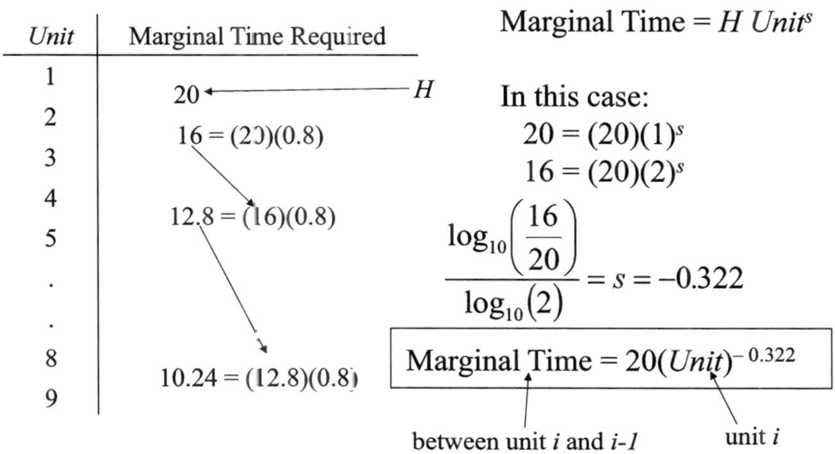

Fig. 10.5. Marginal learning curve example for an 80% learning curve.

10.5 Learning Curve Mathematics

Armed with the basic definitions of a learning curve in Equation (10.1), we can develop the mathematics necessary to facilitate useful work with learning curve data. In this section we will confine the discussion to the log-linear form of the learning curve; however, the formulations developed can be extended to treat the other learning curve model forms.

10.5.1 *Unit Learning Data from Cumulative Average Learning Curves*

Consider the cumulative average hours (or cost) for N units described by

$$\overline{T}_N = T_1 N^s \tag{10.7}$$

Following from Equation (10.7), the total number of hours for all N units would be

$$T_N = N\overline{T}_N \tag{10.8}$$

Substituting Equation (10.7) into Equation (10.8) and solving for T_N and T_{N-1} we obtain

$$T_N = NT_1N^s = T_1N^{s+1} \tag{10.9a}$$

$$T_{N-1} = T_1\left(N-1\right)^{s+1} \tag{10.9b}$$

The time (or cost) of the Nth unit is therefore given by

$$U_N = T_N - T_{N-1} = T_1N^{s+1} - T_1\left(N-1\right)^{s+1} = T_1\left(N^{s+1} - \left(N-1\right)^{s+1}\right) \tag{10.10}$$

Equation (10.10) allows the unit time or cost to be computed, assuming you have the cumulative average learning curve.

As an example application of the derivation above, consider the following simple problem. Assume that the total number of hours to produce 100 units is 1500, and the total number of hours for 200 units is 2850. How long does it take to build unit number 150? From Equation (10.9a), the total times to produce 100 and 200 units are given by

$$T_{100} = T_1\left(100\right)^{s+1} \text{ and } T_{200} = T_1\left(200\right)^{s+1}$$

The first step is to find the value of the learning index (s). By taking the ratio of the relations for T_{100} and T_{200}, we obtain

$$\frac{T_{100}}{T_{200}} = \frac{T_1\left(100\right)^{s+1}}{T_1\left(200\right)^{s+1}} = \left(\frac{100}{200}\right)^{s+1} = \frac{1500}{2850}$$

$$\ln\left(\frac{1500}{2850}\right) = \left(s+1\right)\ln\left(\frac{100}{200}\right)$$

When solved for s this gives $s = -0.074$. Next we need to find the value of the first unit's time (T_1) from either of the original two given data points:

$$T_{100} = 1500 = T_1\left(100\right)^{-0.074+1}$$

which gives $T_1 = 21.09$ hours. Now the time for the 150th unit is given by Equation (10.10) as,

$$U_{150} = 21.09\left(150^{-0.074+1} - 149^{-0.074+1}\right) = 13.48 \text{ hours}$$

10.5.2 *The Slide Property of Learning Curves*

The example at the end of Section 10.5.1 demonstrates the use of a property of the power law called the "slide" property. Generalizing the example,

$$T_i = T_1 X_i^s \text{ and } T_j = T_1 X_j^s \tag{10.11}$$

$$\frac{T_i}{T_j} = \frac{T_1 X_i^s}{T_1 X_j^s} = \left(\frac{X_i}{X_j}\right)^s \tag{10.12}$$

$$T_i = T_j \left(\frac{X_i}{X_j}\right)^s \tag{10.13}$$

Equation (10.13) is the "slide" formula; it allows any point to be found on a learning curve if s and one other point on the curve are known. It is valid independent of the interpretation of T — that is, T could be the unit cost, cumulative average cost, or marginal cost.

10.5.3 *The Relationship between the Learning Index and the Learning Rate*

The learning rate is the fraction (or percentage) by which the time or cost decreases due to a doubling in production. Starting from the general relation

$$T_i = T_1 X_i^s \tag{10.14}$$

the learning rate (r_l) is defined by,

$$r_l T_i = T_1 \left(2X_i\right)^s \tag{10.15}$$

Substituting Equation (10.14) for T_i in Equation (10.15) and canceling, we obtain

$$r_l = 2^s \text{ or } s = \frac{\log\left(r_l\right)}{\log\left(2\right)} \tag{10.16}$$

10.5.4 *The Midpoint Formula*

The midpoint formula allows the accumulation of total hours when a unit learning curve is used. The midpoint formula was developed prior to the advent of digital computing and was useful because it allowed the accumulation of a large number of terms that would have otherwise been extremely tedious to work with. Starting with the formulation for a unit learning curve,

$$U_N = U_1 N^s \tag{10.17}$$

the total hours or cost for units 1 through N is given by

$$T_N = \sum_{n=1}^{N} U_n = U_1 \sum_{n=1}^{N} n^s \tag{10.18}$$

The sum in Equation (10.18) is tedious for large N. Alternatively, it can be shown (see Problem 10.9) that for large N there is a unit, k, between the first and last units in the run such that

$$T_{F,L} = U_k N \tag{10.19}$$

where

$T_{F,L}$ = time to manufacture units F through L inclusive.
F = the first unit.
L = the last unit.
N = the number of units in the run = L-F+1.
k = the "midpoint" unit, $F < k < L$.

The midpoint unit, k, is given by

$$k = \left[\frac{\left(L + \frac{1}{2}\right)^{1+s} - \left(F - \frac{1}{2}\right)^{1+s}}{N(1+s)}\right]^{\frac{1}{s}} \tag{10.20}$$

The determination of the midpoint unit (k) can be used to compute the total time or cost associated with a range of units manufactured.

The learning index (s) in Equation (10.20) is from the unit (not the cumulative average) learning curve. There is no analog to k for the cumulative average learning curve. The difficulty with Equation (10.20) is that it cannot be used if the learning index (s) is unknown. Alternatively, one can use the algebraic midpoint of the units. The algebraic midpoint is given by [Ref. 20.13],

First Lot:
$$k = \frac{N+1}{3} + \frac{1}{2} \tag{10.21a}$$

Subsequent Lots:
$$k = \frac{N}{2} + F - 1 \tag{10.21b}$$

where "lot" refers to a block of units and the first lot is the block that starts with the first unit. Equations (10.21a) and (10.21b) are an approximation to the midpoint that works when the lot sizes are small.

An example of the use of midpoint formula follows. Assume that the first unit takes 45 hours to manufacture. If an 80% unit learning curve is applied, what is the total time for the first 5 units? First solve for the learning index (s) using Equation (10.16):

$$s = \frac{\log(0.8)}{\log(2)} = -0.322$$

The exact total time could be computed using Equation (10.18) as

$$T_5 = \sum_{n=1}^{5} U_n = U_1 \sum_{n=1}^{5} n^s = 45\left(1^s + 2^s + 3^s + 4^s + 5^s\right) = 168.2 \text{ hours}$$

The approximate solution using the midpoint formula is found using

$$k = \left[\frac{\left(5+\frac{1}{2}\right)^{1+(-0.322)} - \left(1-\frac{1}{2}\right)^{1+(-0.322)}}{5\left(1+(-0.322)\right)}\right]^{\frac{1}{(-0.322)}} = 2.4166$$

The total time for the first 5 units is found, using Equation (10.19), to be 169.4 hours. The time for the midpoint unit calculated using $U_k = U_1 k^s$

is 33.87 hours. Note, the cumulative average time for unit number 5 (by definition) would be 168.2/5 = 33.6 hours, the unit time for the kth unit is an approximation of this.

For this example, the algebraic midpoint given by Equation (20.21a) is

$$k = \frac{5+1}{3} + \frac{1}{2} = 2.5$$

10.5.5 *Comparing Learning Curves*

In order to gain insight into the formulation of learning curves, let's compare the unit, cumulative average and total times predicted by the models. Assume that we have fit our data to a cumulative average learning curve for time and obtained the following relation:

$$\overline{T}_N = 50\, N^{-0.25}$$

From Equation (10.8), the total time is given by

$$T_N = N\overline{T}_N$$

From Equation (10.10) the unit time is given by

$$U_N = 50\left(N^{0.75} - \left(N - 1\right)^{0.75}\right)$$

The above three relations are plotted versus the number of units (N) in Figure 10.6. All the curves in Figure 10.6 begin at time 50 and the plot of $\overline{T}_N$ is a straight line (T_N is also a straight line), but the plot of U_N is not a straight line. You can choose to fit your data to either a cumulative average curve or a unit curve; usually one model will represent your data better than the other. The learning index that results from the fit you choose will differ depending on your choice of curve. You can determine the unit result from the cumulative average curve or vice versa, but the result will never be a straight line in both cases, and in general, the learning index will not be the same for unit and cumulative average learning curves fit to the same data.

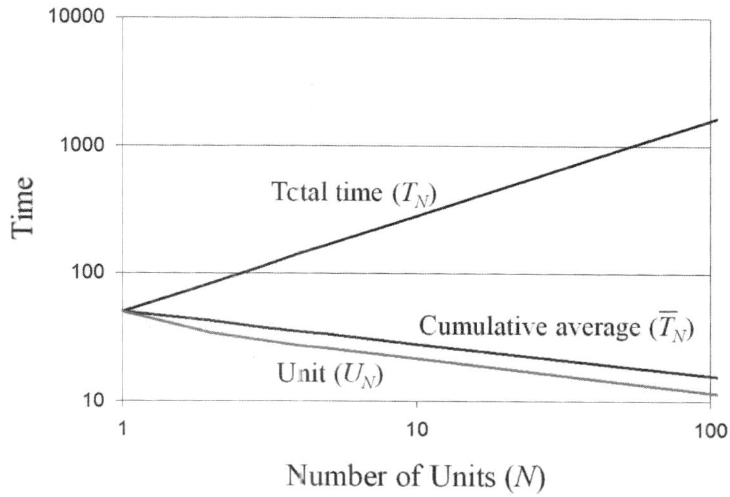

Fig. 10.6. Comparison of cumulative learning curve and derived unit learning curve and total time.

Now let's assume that we are starting with a unit learning curve:

$$U_N = 50N^{-0.25}$$

From Equation (10.19) and Equation (10.20), the total time is given by ($F = 1$, $L = N$, $s = -0.25$, $U_1 = 50$):

$$T_N = T_{1,N} = \frac{50}{0.75}\left[\left(N + \frac{1}{2}\right)^{0.75} - \left(\frac{1}{2}\right)^{0.75}\right]$$

By definition the cumulative average time is given by

$$\overline{T}_N = \frac{T_N}{N}$$

The above three relations are plotted versus the number of units (N) in Figure 10.7. In this case, U_N is the only straight line. Also note that we used the midpoint formula to determine the total time.

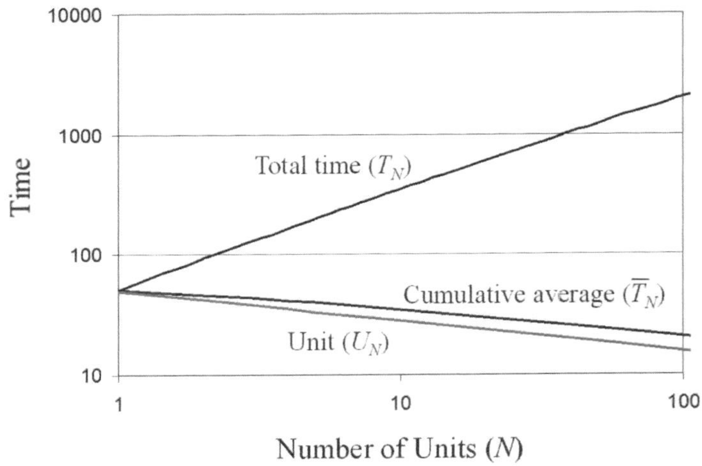

Fig. 10.7. Comparison of unit curve and derived cumulative average learning curve and total time.

10.6 Determining Learning Curves from Actual Data

The best source for learning curves is actual data from production processes; however, there are several problems that make obtaining good data sets difficult, including

- production interruptions
- changes to the product
- inflation
- overhead charges
- changes in personnel.

The actual process being modeled determines whether the unit, cumulative average, or marginal quantity is used. The available data may determine the form used, or if multiple types of data are available, the data that is best fit by a straight line on a log-log plot should be used.[2]

[2] The best fit is determined by performing loglinear regression and obtaining the correlation coefficient (R^2). The data with the highest correlation coefficient is the preferred data set.

The learning curves defined in Equation (10.1) through (10.4) all have simple linear transformations (they come from straight line fits to data on log-log graphs).

$$U_N = U_1 N^s \rightarrow y = sx + b \tag{10.22}$$

where

$y = \log(U_N)$.
$x = \log(N)$.
$b = \log(U_1)$.

10.6.1 *Simple Data*

Consider the simple data shown in Figure 10.8. In this case, unit number versus unit hours is available. We wish to generate a unit time learning curve from the data. The values of s and b are determined using a simple least squares fit where

$$b = \frac{\sum y \sum x^2 - \sum x \sum xy}{M\sum x^2 - \left(\sum x\right)^2} \tag{10.23}$$

$$s = \frac{M\sum xy - \sum x \sum y}{M\sum x^2 - \left(\sum x\right)^2} \tag{10.24}$$

Unit (N)	Hours (U_N)
1	100
2	91
3	85
4	80

Fit $U_N = U_1 N^s$ to this data

N	$x = \log N$	U_N	$y = \log U_N$	x^2	xy
1	0	100	2	0	0
2	0.301	91	1.959	0.0906	0.5897
3	0.4771	85	1.929	0.2276	0.9203
4	0.6021	80	1.903	0.3625	1.146
	$\Sigma x = 1.3802$		$\Sigma y = 7.791$	$\Sigma x^2 = 0.6807$	$\Sigma xy = 2.656$

Fig. 10.8. Simple learning curve data.

For the data in Figure 10.8, $b = 2.00$ and $s = -0.157$. Substituting this data into Equation (10.22), we obtain

$$\underbrace{\log U_N}_{y} = \underbrace{-0.157}_{s}\underbrace{\log N}_{x} + \underbrace{2.00}_{b}$$

Raising both sides to the base of the log we obtain the resulting unit learning curve equation:

$$U_N = 100\,N^{-0.157}$$

10.6.2 *Block Data*

Data does not usually appear as simple unit data. More often the data exists in block form, as in Table 10.1.

Table 10.1. Example Block Data.

Unit	Total Cost
1 – 50	\$2,290,000
51 – 200	\$4,640,000
201 – 225	\$690,000

Using the data in Table 10.1 we determine the cumulative average learning curve for the production cost in Figure 10.9. The last two columns in Figure 10.9 are the only places on the curve that we have actual cumulative average data (we can use this data to check our curve when we are done). As in the case with simple data, we will write the linear transformation corresponding to the data we have and fit the data using a least squares method. The relation needed for this case is given in Equation (10.9a) where we are using C for cost instead of T for time; its linear transformation is

$$C_N = C_1 N^{s+1} \rightarrow y = hx + b \tag{10.25}$$

where C_1 is the cost of the first unit, C_N is the total cost of N units, and

$$y = \log(C_N) \qquad x = \log(N)$$
$$b = \log(C_1) \qquad h = s+1$$

(not cumulative) (not $\overline{C}_N$) $\left(\frac{2290}{50}\right)$

Unit	Total Cost (K$)	Avg Unit Cost (K$)	Cumulative Cost (K$) C_N	Unit	$\overline{C}_N$
1 - 50	2290	45.8	2290	50	45.8
51 - 200	4640	30.9	6930	200	34.7
201 - 225	690	27.6	7620	225	33.9

$\frac{6930}{200}$ $\frac{7620}{225}$

given block data

only know for three units

$\left(\frac{4640}{150}\right)$ $2290 + 4640$

Fig. 10.9. Data for determining the cumulative average cost learning curve.

The least squares curve fit data is shown in Figure 10.10.

Unit (N)	Total Cost (C_N)
50	2290
200	6930
225	7620

Fit $C_N = C_1 N^{s+1}$ to this data

N	$x = \log N$	C_N	$y = \log C_N$	x^2	xy
50	1.699	2290	3.360	2.887	5.709
200	2.301	6930	3.841	5.295	8.838
225	2.352	7620	3.882	5.532	9.130
	$\Sigma x = 6.352$		$\Sigma y = 11.083$	$\Sigma x^2 = 13.714$	$\Sigma xy = 23.677$

Fig. 10.10. Block data learning curve.

The values of h and b are determined using Equations (10.23) and (10.24), where we find $b = 2.0098$ and $h = 0.7956$. Substituting this data into Equation (10.25), we obtain

$$\underbrace{\log C_N}_{y} = \underbrace{0.7956}_{h}\ \underbrace{\log N}_{x} + \underbrace{2.0098}_{b}$$

Raising both sides to the base of the log we obtain the resulting total cost Equation (10.254) and the resulting learning curve equations:

$$C_N = 102.3N^{0.7956}, \quad \overline{C}_N = 102.3N^{-0.2044}$$

The predicted values of $\overline{C}_N$ derived above can be checked against the actual $\overline{C}_N$ shown in the last column in Figure 10.9. Note, an identical solution could have been found by fitting the unit versus $\overline{C}_N$ data in Figure 10.9.

Our analysis above resulted in functional forms for C_N and $\overline{C}_N$. How do we determine the unit learning curve? From Equation (10.10),

$$U_N = C_N - C_{N-1} = 102.3\left(N^{0.79561} - (N-1)^{0.79561}\right)$$

It is also possible to find the unit learning curve for the block data shown in Table 10.1. Table 10.2 shows the unit calculation. In this case the midpoint of each block (lot) cannot be computed from Equation (10.20) since the learning index corresponding to the unit learning curve is not known. Instead solve the first two block unit learning curves simultaneously (i.e., solve Equation (10.17) at $N = k$ using the values of k calculated from Equation (10.21) shown in Table 10.2); this gives $s = -0.1997$ and $C_1 = 81.11$.[3] A more accurate value of s can be obtained by using this value of s in Equation (10.20) to compute midpoints, then using those midpoints to recalculate the learning index and iterating the process.

Table 10.2. Unit Cost Learning Curve from the Block Data.

Unit	*N*	*F*	*k*	NU_k	U_k	*Unit Learning Curve*
1-50	50	1	17.5	2290	45.8	$45.8=C_1(17.5)^s$
51-200	150	51	125	4640	30.93	$30.93=C_1(125)^s$
201-225	25	201	212.5	690	27.6	$27.6=C_1(212.5)^s$

[3] The s for the cumulative average learning curve in this case is $s = h - 1 = -0.2044$ and $C_1 = 102.3$.

10.7 Learning Curves for Yield

Sections 10.1 through 10.6 of this chapter represent a generic discussion of learning curves, applicable to all types of products and systems from airplanes and automobiles to books. All of the development in these sections can and has been used for electronic systems; however, some additional concepts are needed to complete our discussion for such systems.

The first systematic investigation into learning curves for the semiconductor industry was made by Webbink in 1977 [Ref. 10.14]. Webbink estimated the learning curves for different types of semiconductor devices and products and found evidence that learning curves differed greatly across product types. The best developed work on learning curves in the semiconductor industry is for memory chips.

So far this chapter has focused on learning curves associated with time and cost. In electronic products, an equally important aspect of the manufacturing process is yield. In the manufacturing process, yield is initially low due to the following:

- Parametric processing problems: Mechanical stressing of wafer causes changes in wafer size that exceeds design tolerance.
- Circuit sensitivities: Circuit design may not account for variations in device parameters.
- Point Defects: These can occur from dust or photolithographic effects.

During the production life of the product, yield is improved (learned) as the above problems are mitigated. In this section we need to make a distinction between "yield learning" and learning curves on yield. Yield learning is a learning process by which yield can be improved during manufacturing [Ref. 10.15] and is not treated here. Learning curves for yield are analytical models where yield is derived as a function of time (or number of units). This section is only concerned with learning curves on yield.

A high yield leads to low unit cost and a high marginal profit, both of which are crucial to the competitiveness of semiconductor fabrication businesses. Thus, in the highly competitive semiconductor industry, continuing yield improvement is essential to the survival of the semiconductor fabricator.

10.7.1 *Gruber's Learning Curve for Yield*

The best known learning model for yield is from Gruber [Refs. 10.16 and 10.17]. In Gruber's model, yield is modeled as

$$Y = Y_0(D,A,\theta)L_e(Y) \tag{10.26}$$

where Y_0 is the asymptotic yield,[4] which is a function of the defect density (D), the die area (A), and a set of parameters unique to the specific yield model (θ). The asymptotic relation for Y_0 is the appropriate yield model for the assumed defect distribution corresponding to the die being fabricated. The learning effects, $L_e(Y)$, are often described by exponential functions. Gruber's general learning curve model for yield can be rewritten as

$$Y_t = Y_0 e^{-\frac{\beta}{t}+r(t)} \tag{10.27}$$

where

t = the time that a product has been in production.
Y_t = the instantaneous (average) yield during time period t.
Y_0 = the asymptotic yield.
β = a learning constant.
$r(t)$ = an error term.

The conventional approach to parameterizing Gruber's model is by fitting historical results. The linear transformation of Gruber's model is

$$\ln(Y_t) = \ln(Y_0) - \frac{\beta}{t} + r(t) \tag{10.28}$$

---

[4] The asymptotic yield is the post-learning yield due to the fundamentals of the process and application, and is attained after a long period of time. "Yield learning" addresses improving the asymptotic yield; learning curves on yield address the removal of all other factors over the production history.

Note, in this case, Equation (10.28) is specifically written in terms of natural logs. Previously in this chapter we worked in terms of $\log_{10}$ and really any base would have worked, but here it must be base e. For the simple data shown in Table 10.3 we can perform a least squares fit to Equation (10.28) ignoring $r(t)$.

Table 10.3. Example Yield Data for 10 Months of 16M DRAM Production [10.17].

Time (month)	Y_t (%)
1	37.3
2	58.5
3	54.1
4	74.1
5	61.7
6	80.0
7	71.2
8	71.7
9	59.0
10	72.4

We obtain the following learning curve model:

$$Y_t = 0.769e^{-\frac{0.697}{t}+r(t)}$$

The error term, $r(t)$, that appears in Guber's model, is more accurately described as a homoscedastical,[5] serially noncorrelated error term. The term $r(t)$ is generally assumed to be represented by a normal distribution, with a mean of zero and a variance-covariance matrix. Additional discussion of the error term appears in [Refs. 10.17 and 10.18].

10.7.2 *Hilberg's Learning Curve for Yield*

A different type of learning curve model for yield was developed by Hilberg [Ref. 10.19]. The Hilberg model is based on the use of elementary probability theory to describe the accumulation of knowledge and ability of human workers to improve a process. At the start of production of a

[5] A scatterplot or residual plot shows homoscedasticity if the scatter in vertical slices through the plot does not depend much on where you take the slice.

new device, the new production processes are generally poorly controlled and therefore the yield is very low, but after some period of time, process control is improved and yield increases. The work that needs to be done to create an ideal process with 100% yield can be represented by a volume, V. This volume must be mastered or "learned" by a number of individuals (N) located in different places in a process (research, development, and production). Figure 10.11 shows a geometric illustration in which individuals start work at different places within V and their contributions increase over time. Representing the work performed by an individual as an elementary volume, V_E, V_E increases around the starting point until it collides with the volume associated with another individual. Since the same knowledge or ability can be gained by multiple individuals, the elementary volumes can overlap, as shown in the right side of Figure 10.11. In order to build a model around this concept, assume that the behavior of all the elementary volumes is equal on average, so that at time t the mean individual volume is $V_E(t)$. Let V_L be the total volume inside V that has been mastered or "learned" (the shaded area on the right side of Figure 10.11). An approximation to V_L is given by

$$Y_c = \frac{V_L}{V} = 1\text{-}e^{-\frac{N}{V}V_E(t)} \tag{10.29}$$

where Equation (10.29) assumes that the distribution of N in V is given by the Poisson distribution. Further in Equation (10.29) we postulate that the yield of products produced by the process is given by V_L/V. The rate of growth of V_E is measured in work per unit time and referred to as productivity (P):

$$P = \frac{dV_E}{dt} \tag{10.30}$$

When productivity, the number of individuals, and the learning volume are all constant at P_0, N_0, and V_0, integrating Equation (10.30) and substituting it into Equation (10.29) gives

$$Y_c = 1\text{-}e^{-\frac{N_0 P_0}{V_0}t} = 1\text{-}e^{-\frac{t}{\tau}} \tag{10.31}$$

where τ is a time constant. Often in practice, however, V_E and N rise exponentially and can be approximated by

$$V_E = V_{E0} e^{\alpha t}, \qquad N = N_0 e^{\beta t} \tag{10.32}$$

Substituting Equation (10.32) into Equation (10.29) we obtain,

$$Y_c = 1 - e^{-\frac{N_0 V_{E0}}{V_0} e^{(\alpha+\beta)t}} \tag{10.33}$$

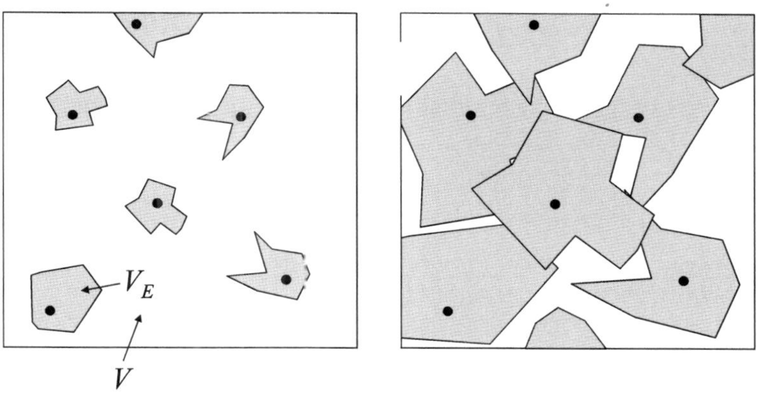

Fig. 10.11. Hilberg learning volume model [10.18]. Left = initial learning, right = learning level at a future time.

10.7.3 *Defect Density Learning*

An alternative to a learning curve for yield is a learning relation for the defect density. Stapper *et al.* [Ref. 10.20] developed the following approach to modeling defect density learning.

(1) Project the defect density from historical defect density learning charts. These are obtained from test sites and chip yields and usually appear as relative defect density versus year, with many different generations of devices displayed on the same graph.
(2) Determine the average number of faults for each circuit type:

$$\lambda_j = \sum_{i=1}^{m} A_{ji} D_i \tag{10.34}$$

where

j = circuit types.
i = defect types.
A_{ji} = the critical areas for each defect type.
D_i = the defect density for defect type i

(3) Determine the yield using

$$Y = Y_0 \left(1 + \frac{\lambda}{\alpha}\right)^{-\alpha} \tag{10.35}$$

where α is a cluster factor and Y_0 is the asymptotic yield.

References

10.1 Wright, T. P. (1936). Factors affecting the cost of airplanes, *Journal of Aeronautical Science*, 3(2), pp. 122-128.

10.2 Hirsch, W. Z. (1952). Manufacturing progress functions, *Review of Economics and Statistics*, 34(2), pp. 143-155.

10.3 De Jong, J. R. (1964). Increasing skill and reduction of work time - concluded, *Time and Motion Study*, October, pp. 20-33.

10.4 Everett, J. G. and Farghal, S. (1994). Learning curve predictors for construction and field operations, *Journal of Construction Engineering and Management*, 120(3), pp. 603-616.

10.5 Lieberman, M. B. (1984). The learning curve and pricing in the chemical processing industries, *Rand Journal of Economics*, 15(2), pp. 213-228.

10.6 Raccoon, L. B. S. (1996). A learning curve primer for software engineers, *Software Engineering Notes*, 21(1), pp. 77-86.

10.7 Dick, A. R. (1991). Learning by doing and dumping in the semi-conductor industry, *Journal of Law Economics*, 34(2), pp. 134-159.

10.8 Ohlsson, S. (1992). The learning curve for writing books: Evidence from professor Asimov, *Psychological Science*, 3(6), pp. 380-382.

10.9 Asher, H. (1956). Cost-quality relationships in the airframe industry, Report No. R-291, The Rand Corporation, Santa Monica, CA, July 1.

10.10 De Jong, J. (1958). The effects of increasing skill on cycle time and its consequences for time standards, *Ergonomics*, 1(1), pp. 51-60.

10.11 Carr, G. W. (1946). Peacetime cost estimating requires new learning curves, *Aviation*, 45(April).

10.12 Crawford, J. R. (1944). Learning curve, ship curve, ratios, related data, Lockheed Aircraft Corporation.

10.13 Liao, S. S. (1988). The learning curve: Wright's model vs. Crawford's model, *Issues in Accounting Education*, (Fall), pp. 302-315.

10.14 Webbink, D. W. (1977). The semiconductor industry: A survey of structure, conduct, and performance, *Staff Report to the FTC*, Washington, DC, US Government Printing Office.

10.15 Nag, P. K., Maly, W. and Jacobs, H. J. (1997). Simulation of yield/cost learning curves with Y4, *IEEE Transactions. on Semiconductor Manufacturing*, 10(2), pp. 256-266.

10.16 Gruber, H. (1994). *Learning and Strategic Product Innovation: Theory and Evidence for the Semiconductor Industry* (North-Holland, Amsterdam).

10.17 Chen, T. and Wang, M. J. (1999). A fuzzy set approach for yield learning modeling in wafer manufacturing, *IEEE Transactions. on Semiconductor Manufacturing*, 12(2), pp. 252-258.

10.18 Joskow, P. L. and Rozansky, G. (1979). The effects of learning by doing on nuclear power plant operating reliability, *Review of Economics and Statistics*, 61(May), pp. 161-168.

10.19 Hilberg, W. (1980). Learning processes and growth curves in the field of integrated circuits, *Microelectronics Reliability*, 20(3), pp. 337-341.

10.20 Stapper, H., Patrick, J. A. and Rosner, R. J. (1993). Yield model for ASIC and process chips, *Proceedings of the IEEE International Workshop on Defect and Fault Tolerance in VLSI*, pp. 136-143.

Bibliography

There are over sixty years' worth of technical publications on learning curves. Many significant papers, as well as several books, have been published on the topic. In addition to the publications referenced in this chapter, the following sources may also be useful.

Abernathy, W. J. and Wayne, K. (1974). Limits of the learning curve, *Harvard Business Review, No. 74501,* pp. 109-113.

Badiru, B. (1992). Computational survey of univariate and multivariate learning curve models, *Transactions on Engineering Management*, 39(2), pp. 176-188.

Belkaoui, A. (1986). *The Learning Curve: A Management Accounting Tool* (Quorum Books, Westport, CN).

Fries, A. (1993). Discrete reliability-growth models based on a learning-curve property, *IEEE Transactions on Reliability*, 42(2), pp. 303-306.

Harvey R. A. and Towill, D. R. (1981). Applications of learning curves and progress functions: Past, present, and future, *Industrial Applications of Learning Curves and*

Progress Functions, (Institution of Electronic and Radio Engineers, London). pp. 1-15.

Jarmin, R. S. (1994). Learning by doing and competition in the early rayon industry, *Rand Journal of Economics*, 25(3), pp. 441-454.

Kemerer, C. F. (1992). How the learning curve affects CASE tool adoption, *IEEE Software*, 9(3), pp. 23-28.

Pierson, G. (1981). Learning curves make productivity gains predictable, *Engineering and Mining Journal*, 182(8), pp. 56-64.

Spence, M. (1981). The learning curve and competition, *Bell Journal of Economics*, 12(1), pp. 49-70.

Stump, E. J. (1988). Parametrics tools of the trade: Learning curve analysis, *International Software Process Association (ISPA) Workshop.*

Learning by new experiences: Revisiting the flying fortress learning curve, in *Learning by Doing: in Markets, Firms, and Countries*, edited by N. R. Lamoreaux, D. M. G. Raff, and P. Temin, The University of Chicago Press (National Bureau of Economic Research), 1999.

Problems

Learning curve problems appear in other places in this book. See Problem 9.10.

10.1 A manufacturing process's cost follows a 72% unit learning curve. The cost of the first unit is $224. What is the cost of the 7th unit?

10.2 A manufacturing process's time follows an 86% cumulative average learning curve; the cumulative average time for the first 15 units is 156 minutes. What was the time to produce the first unit?

10.3 A manufacturing process's cost follows a marginal learning curve. The difference in cost between units 29 and 30 is $1.02 and between 51 and 52 is $0.53. What is the learning index? What is the marginal cost of the first unit?

10.4 In Problem 10.2, assume that the total time to produce the first 15 units is 156 minutes. What was the time to produce the first unit?

10.5 The cumulative average time to produce N units is always less than the time to produce the Nth unit. True or false?

10.6 If there is no learning curve, what is the learning rate?

10.7 Your company needs to obtain a printed circuit board. One of your employees has discovered that you could outsource the board's fabrication out to another company for $39/board. Alternatively, if you choose to make the board in-house you will experience a 75% unit learning curve (unit learning curve model), there will be a $5 million one-time setup fee, and the first board will cost $35.

a) If there was no learning curve, how many boards would you have to make in-house in order to make a business case to your management[6] that the board fabrication should be done in-house rather than outsourced?

b) If you now consider the unit learning curve, how many boards would you have to make in-house in order to make a business case to your management that the board fabrication should be done in-house rather than outsourced? Assume that every outsourced board is $39 (no learning curve for the outsourced boards).

10.8 Unit 12 is the first unit in a range of units being manufactured, and unit 102 is the last. If a 65% unit learning curve is assumed, what is the midpoint unit of this range? If it takes 15 minutes to produce the midpoint unit,

a) how long does it take to produce all the units in the range?

b) how long does it take to produce unit 81?

10.9 Derive the midpoint formula Equation (10.20) used to determine the midpoint unit in a manufacturing process. Explain what the statement, "accurate for large production runs" means.

10.10 What value of the learning index (s) gives k to be exactly half way between F and L?

10.11 In Problem 9.10, what is the cumulative average time for the first 2356 units?

10.12 Two companies (Alpha and Beta) quote the same job, but in different ways:

Alpha: $Part_1 = \$1000$, $Part_{200} = \$900$

Beta: $Part_1 = \$1100$, cumulative average cost at $Part_{300} = \$800$

You must have a total of 2000 parts manufactured. Who should you award the contract to?

10.13 Considering the data given below, use a least squares fit to determine the cumulative average learning curve on the production time.

Unit	Time/unit (hours)
1	3.2
2	3.14
3	3.05
4	3.05
5	3.01
6	2.98
7	2.9

10.14 Considering the data given below, use a least squares fit to determine the cumulative average learning curve on the production time.

[6] A business case is made by showing that it is less expensive to build the board in-house than outsource it.

Unit	Total Time (hours)
1-20	60
21-43	54
44-100	100
101-200	200
201-300	190
301-400	185
401-500	184

10.15 You are contracted by a system integration company to disassembly circuit boards that are returned by their consumers. For the current type of board you are disassembling, you have determined a cumulative average learning curve described by:

$$\overline{C_N} = 34.59N^{-0.2784}$$

where N is the unit number and $\overline{C_N}$ is the cumulative average cost.

a) What is the cumulative average cost of the first 88 disassemblies?

b) What is the total cost of disassembling the first 88 boards?

c) What do you expect the unit disassembly cost of the 88th board to be?

d) The system integration company has come to you and expressed an interest in giving you a contract to disassembly more of the same boards described on the previous page. Your current contract is to do 100 board disassemblies, which you would complete prior to starting the new job. The company has requested a quote for 200 more disassemblies. What total price should you quote the company for the additional 200 disassemblies assuming that you can take advantage of everything you learned disassembling the first 100 boards and that you can follow the learning curve that you did for the first 100. To make thing simple, you can assume 0 profit.

e) The time to disassembly the first unit of the original 100 from the first contract was 1 hour (this is the only time that you know). Assuming that the disassembly time follows the same learning curve (same learning index) as the cost, how much time should you budget for the 200 additional disassemblies you are bidding.

10.16 Your company builds small boats for the Russian Navy. The company has 10 skilled workers. These workers can each provide 2500 labor hours per year (per worker). You are about to sign a new contract to build a new style of boat. The first boat is expected to take 6000 labor hours to complete and you think that you will have a 90% learning curve (0.9 learning rate). How many boats can you make in the first year?

a) If you assume a "cumulative average" learning curve

b) If you assume a "unit" learning curve

10.17 If a mistake was made and the yield figure for month 2 in Table 10.3 was revised to 45%, derive the new learning curve on yield.

10.18 If the area of the DRAM die considered in Table 10.3 was 0.04 cm^2, and a Murphy yield law is used for the asymptotic yield, draw and correctly label (with numbers) the defect distribution for the die.

Part II

Life-Cycle Cost Modeling

The manufacturing cost of a product only represents a portion of the actual costs associated with a product. While most traditional cost analyses focus on manufacturing costs, which are extremely important for inexpensive high-volume products, other costs, for example, operation costs and sustainment costs, can dominate for other types of systems.

Life-cycle cost (sometimes referred to as LCC) is the sum of all recurring and non-recurring costs over the complete life of a product or service. Life-cycle cost includes the design cost, manufacturing cost, installation cost, operating costs, maintenance and upgrade costs, the remaining (residual or salvage) value at the end of ownership or useful life, and disposal costs.

One needn't look too far to find products that have life-cycle costs that are significantly greater than their manufacturing costs. A good example (which will be discussed in more detail in Chapter 20) is an ink jet printer. Making a decision to buy a printer solely based on the purchase price of the printer is somewhat naive since the cost of the printer is quickly eclipsed by the cost of ink. A common example of forecasted operating costs is articulated in the EnergyGuide labels that are attached to many appliances, e.g., Figure II.1.

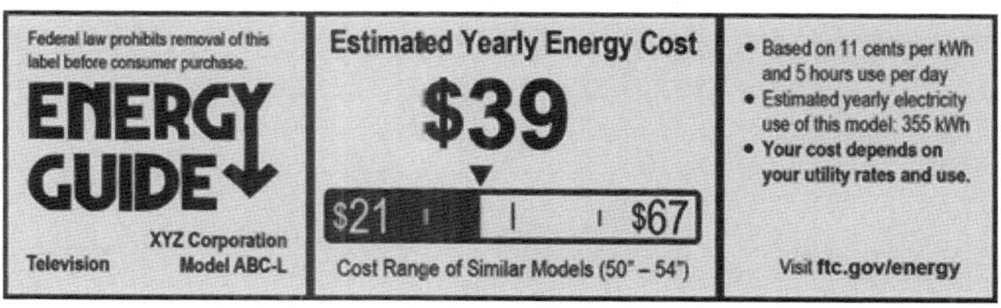

Fig. II.1. An EnergyGuide label for a television that shows the estimated annual energy cost.

The factors that influence the total cost of ownership of a product or system are shown in Figure II.2. Generally the total cost of ownership captures all costs that are passed along to the owner or user through the product, system or service price; plus the infrastructure and business process costs borne by the owner or user.[1] For low-cost, high-volume products, the manufacturer of the product seeks to maximize their profit by minimizing its cost. For higher-cost products, a more important customer requirement for the product may be the minimization of the total cost of ownership of the product. For the majority of products, the contributions to the price of the product are the responsibility of the manufacturer, while operation and support are the responsibility of the customer. However, depending on the product, there may be life-cycle costs that are shared by the customer and manufacturer. In some cases (e.g., large military systems), the design and R&D costs may also be paid for by the customer.

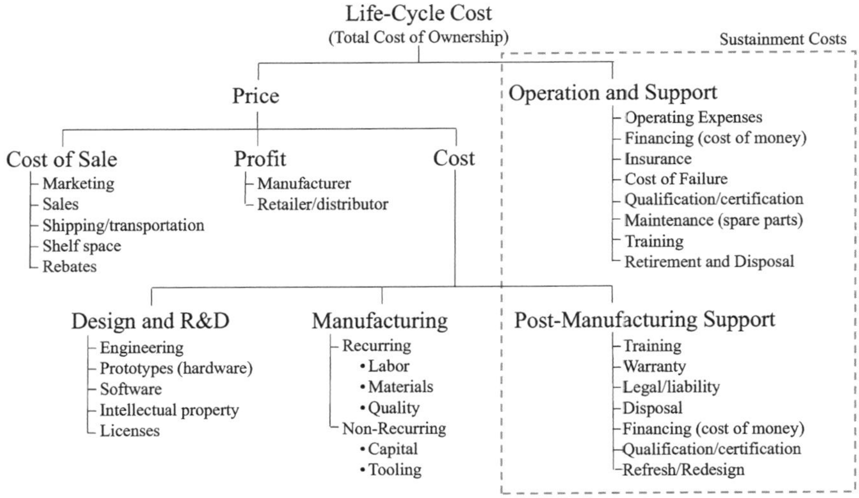

Fig. II.2. The scope of cost analysis (after [Ref. II.1]).

[1] In addition to the costs shown in Figure II.2, total ownership cost of a product could also account for the remaining (residual or salvage) value at the end of ownership or useful life.

For products where the operation and sustainment costs are significant, it is important for the customers to understand all the life-cycle costs when making purchase decisions. In Figure II.3, we have shown the breakdown of costs for an F-16 aircraft. This figure shows that an F-16's sustainment costs represent 78% of the overall cost of the aircraft. However, one need not consider such a complex system to see the dominance of sustainment costs. Consider a network of 25 personal computers managed by one full-time system administrator, also shown in Figure II.3.

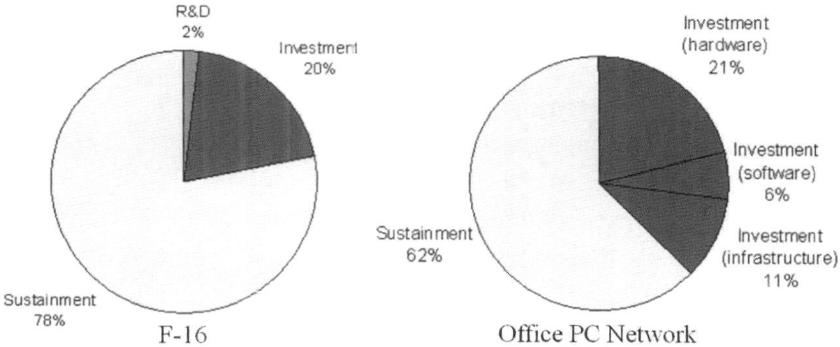

Fig. II.3. Total cost of ownership breakdown for an F-16 fighter jet [Ref. II.2] and an office PC network with a system administrator [Refs. II.3 and II.4].

In Part II of this book we can't cover all the possible contributions to the total cost of ownership, but we have attempted to address some of the key issues that are most relevant to electronic systems. The majority of non-manufacturing costs of ownership are related to either reliability (how and how often a product fails), operation (usage costs), and other sustainment costs associated with keeping a product manufacturable and maintainable. All the topics in this portion of the book are rooted in these three effects.

II.1 System Sustainment

In Figure II.2 we have broadly classified all the post-manufacturing support, and operation and support costs as sustainment costs.

The origin of the word *sustain* is the Latin work *sustenare*, which means "to hold up" or "to support". The modern use of the word *sustain* is to keeping something going or to extend its duration, [Ref. II.5], where the most common synonym for *sustain* is *maintain*. It is not uncommon for *sustain* and *maintain* to be interchangeably used, however, maintenance usually refers to activities that are targeted at correcting problems, while sustainment is a more general term referring to the management of system evolution.

The concept of sustainability is connected to nearly every discipline [Ref. II.6], e.g., environmental sustainability, business or corporate sustainability and technology sustainment, however, in this book we are interested in technology sustainment.[2]

Technology sustainment refers to all activities necessary to [Ref. II.6]: a) keep an existing system operational so that it can successfully complete its intended purpose; b) continue to manufacture and install versions of the system that satisfy the original requirements; and c) manufacture and install revised versions of the system that satisfy evolving requirements.

A good general definition of sustainment is "development, production, operation, management, and end-of-life of systems that maximizes the availability of goods and services while minimizing their footprint". In this case the terms in the definition mean:

- "footprint" could represent any kind of impact that is relevant to the system's stakeholders, e.g., cost (economics), human health, energy required, environmental, and/or other resource consumption (water, materials, labor, expertise, etc.).
- "availability" represents the fraction of time that a good or service is in the right state, supported by the right resources, and in the right place when the customer requires it.
- "customer" could be an individual, a company, a city, a geographic region, a specific segment of the population, etc.

---

[2] Although sustainability and sustainment are closely related in a semantic sense, environmental sustainability organizations almost never refer to what they are doing as sustainment or sustainment engineering. However, organizations that maintain systems (sustainment organizations) use both sustainment and sustainability to describe what they do.

Note that this definition is consistent with both environmental and technology sustainment concerns.

This book will refer to "sustainment-dominated" systems. A sustainment-dominated system is defined as a system for which the lifetime footprint significantly exceeds the footprint associated with making it [Ref. II.6]. Footprint refers to any kind of impact one is interested in (or is relevant to the specific stakeholders associated with the system), e.g., environmental, cost (economics), human health, energy required, and/or other resource consumption (water, materials, labor, expertise, etc.). Defining sustainment-dominated systems provides the opportunity to make a distinction between high-volume, low cost consumer products and more complex, higher-cost systems such as airplanes, infrastructure, and institutions. Figure II.4 demonstrates a fundamental difference between sustainment-dominated products and non-sustainment-dominated products. Non-sustainment-dominated products, which tend to be high-volume products, have relatively little investment in sustainment activities and the total time period associated with the product is short (short manufacturing cycle, short field life). Alternatively, sustainment-dominated products, which tend to be relatively low-volume expensive systems, have large sustainment costs and long manufacturing and/or field lives.

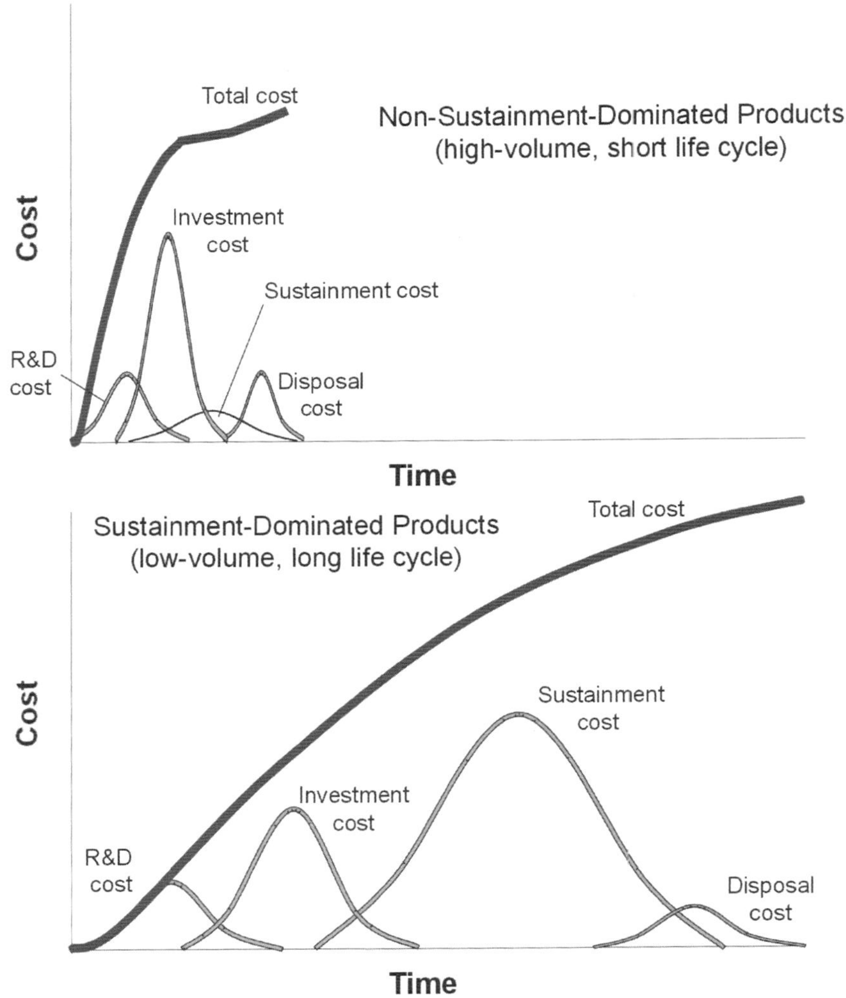

Fig. II.4. Life-cycle spending profile for non-sustainment-dominated and sustainment-dominated products.

II.2 Cost Avoidance

In the case of manufacturing processes, it is reasonable to characterize the value of process, equipment and yield changes as cost *savings*. However, the value of life-cycle management activities (i.e., sustainment) is usually quantified as a cost *avoidance*. "Cost avoidance is a cost reduction that results from a spend that is lower than the spend that would have otherwise been required if the cost avoidance exercise had not been undertaken" [Ref. II.7]. A simpler definition of cost avoidance is a reduction in costs that have to be paid in the future to sustain a system.

The reason that cost avoidance is used rather than cost savings, is that if the value of an action is characterized as a cost savings, then someone wants the saved money back. In the case of sustainment activities there is no money to give back. In cash flow analysis, cost avoidance is never a cash inflow, it is a reduction in future cash outflows.

Unfortunately, making business cases based on a future cost avoidance argument is often more difficult than business cases that are based on cost savings, therefore, there is a greater need to be able to provide detailed quantification of sustainment costs. Requesting resources to create a cost avoidance is not as persuasive as making a cost savings or a return on investment argument.

II.3 Should-Cost

Should-cost is the result of cost modeling done by a customer to determine what it should cost them to purchase a product, service or system support [Ref. II.8]. Should costing is based on the customer's accounting and their understanding of the product or service. When a customer knows a product or service's should-cost, then they know what they should pay for it.

For many common products, market competition protects the customer from significant overpayments (assuming the customer shops around). However, for unique products and services (e.g., military programs, contracted services, etc.), where there is no market competition that sets prices at fair levels, the should-cost is critical when a customer wishes to guard against overpayment. If you know roughly how much you should be paying, then you can assess whether or not there is sufficient savings to justify a sourcing effort.

II.4 Time Value of Money

Money isn't free. The cost of money constitutes the financial costs that are part of the economics of both producing the product and buying it. Whether money has to be borrowed to fund the development, manufacture and sustainment of a product, or money has been saved that could be used to fund other things, there is a cost associated with the use of the money involved. For a customer, there are similar costs. If a buyer has to obtain financing to purchase the product or has to use cash on hand that could otherwise be earning money in another investment, these are costs that are associated with acquiring the product.

The most common way to incorporate the cost of money into calculations is to obtain its present value.[3] The premise for computing the present value is that money available today can be invested and grow, whereas money spent today cannot. The equivalent present value[4] of a future amount of money V_n that is n_t time periods (in discrete time) from the present is given by

$$Present\ value = V_n \left[\frac{1}{(1+r)^{n_t}} \right] \tag{II.1}$$

where r is the discount rate per time period and the term in brackets is commonly called the discount factor. Throughout this book we will refer to r as the "discount rate", since it is used to discount future costs. The terminology used for r varies depending on the domain and the application of Equation (II.1). In engineering economics and engineering design, r may simply be called the interest rate, however this infers an unrealistically narrow view of what r represents. In cost modeling, discount rate is preferred. In the financial world "discount rate" is also used to refer to the interest rate charged to commercial banks and other

---

[3] A significant portion of the discipline of engineering economics is concerned with converting future cash flows to equivalent present worths for comparison purposes.

[4] The present value in Equation (II.1) is at the end of year 0 (the beginning of year 1). This is consistent with the standard end-of-period convention used by engineering economics. In the remainder of this book we will refer to the present values calculated using Equation (II.1) as "year 0 dollars" meaning the dollar value at the end of year 0 (the beginning of year 1).

depository institutions for loans received from the Federal Reserve Bank's discount window. The financial world would more generally call r the weighted average cost of capital (see Appendix B). Depending on the application of Equation (II.1), r could be the yield rate, hurdle rate, discount yield, minimum attractive rate of return on investment, or any of several other application-dependent terms.

Many forms for Equation (II.1) have been developed for different cash flow scenarios (see any engineering economics text for these). Equation (II.1) represents the simplest discounted cash flow case of a single-payment discrete compounding present worth formula. The value of after-tax[5] discount rates (r) vary, depending on the business sector: public, private, non-profit, or government. In 2015, the discount rate for the U.S. government varied between 0.3 and 1.5%,[6] and for moderate-growth public companies the rate can range from 10 to 12% or higher.

The effective discount rate assumed by an organization is not just the interest rate that could be earned if the money was placed in a bank or invested in the stock market; it may be the return on the money that could be obtained if the money was used for another purpose. For a high-growth-rate company with limited resources, if \$1,000,000 can be invested in opportunity A and the investment returned \$1,500,000 at the end of one year, then the discount rate that must be used for the cost of using the \$1,000,000 for a different opportunity is 50% ($V_n = 1{,}500{,}000$, which is the value 1 year in the future; $r = 0.5$, and $n_t = 1$ in Equation (II.1) gives a present value of \$1,000,000). An alternative example is the present value of spending the \$1,500,000 one year in the future if $r = 0.5$ is \$1,000,000.

In general, the discount rate includes a combination of the cost of money from an array of sources (companies don't get all their money from just one source and different sources perceive "risk" in different ways). There is also no single discount rate for a company or organization, different projects carry different risk and are financed in different ways. Additionally, Equation (II.1) implicitly assumes that the discount rate is

---

[5] After all applicable taxes have been deducted.

[6] The effective discount rate for the United States government is often taken as the forecasted interest rate on treasury notes and bonds reported by the Office of Management and Budget (OMB). The numbers shown here are "real" discount rates.

constant over time, though in general it is not. Accurately determining a discount rate can be more art than science — see Appendix B for a detailed treatment of the determination of the discount rate.

II.4.1 *Inflation*

Inflation is the decrease in the purchasing power of money, i.e., one unit of money buys less goods or services in the future. Inflation affects future cash flows in addition to the source of the money. The discount rate may or may not include the effects of inflation. There are two ways that inflation can be accounted for when calculating present value:

- Nominal method: converting real cash flows to nominal cash flows and discounting them using the nominal discount rate.
- Real method: estimating real cash flows and discounting them using the real discount rate.

The present value is same for both methods, but care must be taken not to mix nominal cash flows with real discount rates and vice versa.

The nominal (also called "market") discount rate is the cost of money that includes a market adjustment for inflation; this is the interest rate you get on a bank account or pay on a car loan or mortgage. The real discount rate is the money paid for capital that does not include a market adjustment for anticipated inflation. The real discount rate measures the real growth of money with inflation removed (not ignored, just removed). The nominal and real future cash flows and discount rates are related by,

$$V_n^{nominal} = V_n^{real}(1+f)^n \tag{II.2}$$

$$r^{nominal} = (1+r^{real})(1+f)-1 \tag{II.3}$$

where f is the inflation rate.

Both real and nominal cash flow analyses have a purpose: if one wishes to only analyze the interaction of "real" cost factors in determining future costs then the effects of inflation can be ignored. If, however, one wishes to project the actual funds (actual dollars) that will be needed for different alternatives in a particular budget year, nominal costs should be used.

II.5 Logistics

Logistics has been integral to human survival since mankind began hunting and gathering food. The origin of the word *logistics* is in the French verb *loger* to lodge or to quarter. Its original use was to describe the movement, supply and maintenance of military forces in the field. In the military, logisticians were historically assigned quartermaster duties and were responsible for: feeding soldiers, providing fodder for horses, procuring uniforms, equipment, weapons, ammunition, and supporting equipment.

The common definition of modern logistics is "the management of the flow of goods, information and other resources, including energy and people, between the point of origin and the point of consumption in order to meet the requirements of consumers." The basic functions that comprise logistics include material flow, distribution, and the life-cycle maintenance and support of a system throughout its intended period of utilization [Ref. II.8]. From the systems support perspective, logistics is composed of supply support, test and support equipment, personnel and training, facilities, transportation and material handling, other required resources, and data management.

For many types of systems and in many contexts (e.g., many business schools), the treatment of logistics is confined to a focus on production support: supply-chain management, distribution planning, transportation management, inventory management, and purchasing — these are important factors in manufacturing, but they do not form a complete picture of logistics. In this book we will take a post-manufacturing service view of logistics that is equally limited in its view of logistics and focus on the concepts of sparing, maintainability, availability, mission capability, readiness and obsolescence.

II.6 References

II.1 Magrab, E. B., Gupta, S. K., McCluskey, F. P. and Sandborn, P. A. (2010). *Integrated Product and Process Design and Development - The Product Realization Process*, 2nd Edition (CRC Press, Boca Raton, FL).

II.2 Cost Analysis Improvement Group (CAIG) (1992). *Operating and Support Cost-Estimating Guide*, Office of the Secretary of Defense, http://www.dtic.mil/pae/ Accessed May 1, 2016.

II.3 Gateway Inc. (December 2001). www.gateway.com.

II.4 Shields, P. (2001). Total cost of ownership: Why the price of the computer means so little, http://www.thebusinessmac.com/features/tco_hardware.shtml. Accessed June 2013.

II.5 Sutton, P. (2004). What is sustainability? *Eingana*, 27(1), pp. 4-9.

II.6 Sandborn, P. and Myers, J. (2008). Designing engineering systems for sustainability, in *Handbook of Performability Engineering*, K.B. Misra, Editor, pp. 81-103 (Springer, London).

II.7 Ashenbaum, B. (2006). Defining cost reduction and cost avoidance, CAPS Research.

II.8 Carter, A. B. and Mueller, J. (2011). Should cost management: Why? How?, *Defense AT&L: Better Buying Power*, Sept-Oct, pp. 14-18.

II.9 Blanchard, B. S. (1992). *Logistics Engineering and Management*, 4th Edition (Prentice Hall, Englewood Cliffs, NJ).

Chapter 11

Reliability

Reliability is the most important attribute of many types of products and systems — more important than cost. Reliability is quality measured over time; it is the probability that a product or system will operate successfully for a specific period of time and under specified conditions when used in the manner and for the purpose intended. High reliability may be necessary in order for one to realize value from the product's performance, functionality, or low cost.

The ramifications of reliability on a product or system's life cycle are linked directly to sustainment cost through spare parts requirements and warranty return rates. Indirectly, reliability impacts customer satisfaction, breach of trust, loss of market, and a host of other factors that influence other costs. The combination of how often a system fails and the efficiency of performing maintenance when a system does fail determine the system's availability. The cost of failure avoidance (for example, preventative maintenance) is also linked to reliability.

Reliability is related to safety and quality. Safety can be defined as "freedom from those conditions that can cause death, injury, occupational illness, or damage to or loss of equipment or property, or damage to the environment" [Ref. 11.1]. Safety is not the same as reliability. Reliability is associated with the probability of failure; safety is associated with the probability of a failure resulting in a bad outcome. Highly reliable systems are often assumed to also be safe; however, reliability does not necessarily infer safety or vice versa. The safest car may be the car that is always broken down and never leaves your driveway — a car that we would view as having poor reliability.

Quality is also not the same as reliability. The clearest difference is that quality does not depend on time and reliability does. Quality is a static photograph taken at the end of manufacturing and reliability is a movie of the product over time. Defects in a product at the end of the manufacturing process that escaped detection can negatively affect a product's quality.[1] Defects that develop into problems that negatively affect the product's operation over time are considered reliability issues.

The objective of this chapter is to provide a sufficient introduction to reliability to enable the various cost ramifications of it to be discussed in subsequent chapters. This chapter is by no means a definitive treatment of reliability. There are many fine books on reliability engineering that are much more comprehensive than this chapter.

11.1 Product Failure

Customers, manufacturers and sustainers care about the failure of products or systems in the field. Failure is defined as the inability of a product or system to perform its intended function for a specified period of time under specified environmental conditions.

Field failures of products and systems occur for many different reasons. In some cases there are manufacturing defects that are not detected (or do not become evident) until later in the product's life. There may be fundamental design defects that result in failure, for example, the explosion of the Hindenburg airship is usually considered to be due to a design defect, (although an exact cause could never be pinpointed).

Generally products and systems fail due to one or more of the following:

- Wear-out is deterioration, wear, and/or fatigue over time. For example, car tires, shoes, and carpeting simply wear-out with repeated use. Many electronic products never reach wear-out; electronic components can wear-out, but in many cases the product is either discarded or fails due to some other cause prior to wear-out

[1] The concept of yield (Chapter 3) is a measure of quality. Recurring functional tests (Chapter 7) are part of the manufacturing process and are specifically designed to improve the yield (and thereby the quality) of products that are shipped to customers. However, neither yield nor recurring functional test are necessarily associated with reliability.

occurring. Mechanical systems are more prone to wear out, since moving parts in contact tend to wear and structural elements fatigue. Electronic packaging is more likely to wear out than the actual semiconductor portions of the system — for example, solder joints can suffer from fatigue cracking with repeated thermal cycling.

- Overstress results from unintentionally subjecting a product to environmental stress that is beyond the design specification. An example of overstress would be an electronic system that is struck by lightning.
- Misuse is knowingly subjecting a product or system to environmental stresses that are beyond its design specifications.

Note that products and systems may contain defects or develop defects that are never encountered by their users, either because the users will never use the product or system under certain environmental stresses or because the function of the product or system that is impaired is never exercised by the user. In these cases, the defects, although present, never result in system failure and never incur the associated costs of failure or resolution.

If you kept track of all the failures of a particular population of fielded products over its entire lifetime (until every member of the population eventually failed), you could obtain a graph like the one shown in Figure 11.1. Figure 11.1 assumes, for simplicity, that failed product instances are not repaired. We will work exclusively in terms of time in this chapter, but in general the time axis in Figure 11.1 could be replaced by another usage measure, such as thermal cycles or miles driven.

Three distinct regions of the graph in Figure 11.1 are evident. Early failures due to manufacturing defects (perhaps due to defects induced by shipping and handling, workmanship, process control or contamination) are called infant mortality. The region in the middle of the graph in which the cumulative failures increase slowly is considered the useful life of the product. It is characterized by a nearly constant failure rate. Failures during the useful life are not necessarily due to the way the product was manufactured, but are instead random failures due to overstress and latent defects that don't appear as infant mortality. Finally, the increase in failures on the right side of the graph indicates wear-out of the product due to deterioration (aging or poor or non-existent preventative maintenance). An alternative way to look at the failure characteristics of a product is via the failure rate. Figure 11.2 shows the failure rate that corresponds to the cumulative failures shown in Figure 11.1. Figure 11.2 is known as the "bathtub" curve.

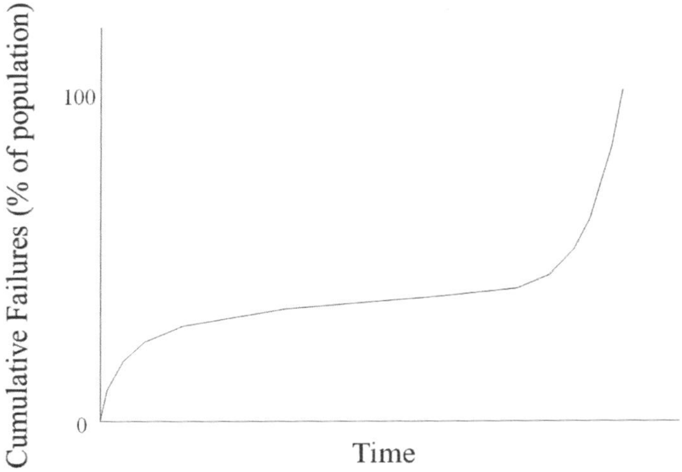

Fig. 11.1. Observed failures versus time for a population of fielded products.

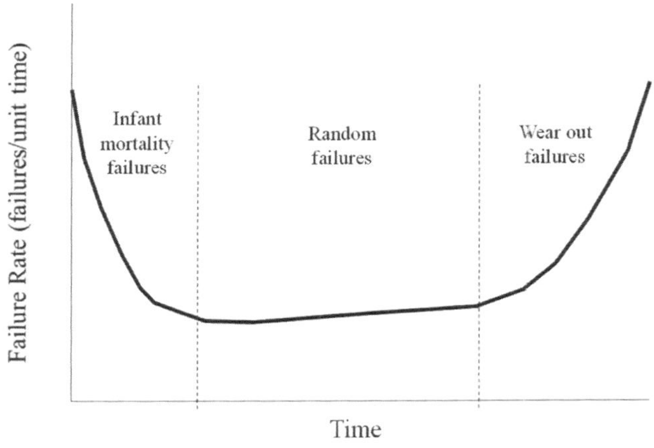

Fig. 11.2. Failure rate versus time observed for a population of fielded products – bathtub curve.

In general, for modeling the life-cycle costs of products, we care more about the cost that represents a population of products than we do about the cost of any one particular instance in the population. While the performance of a particular member of the population is interesting, we have to plan, budget, and characterize based on the whole population. The next section quantitatively describes the failure rate for a population of products in terms of reliability.

11.2 Reliability Basics

If a total of N_0 product instances are tested from time 0 to time t, the following relation must be true at any time t:

$$N_s(t) + N_f(t) = N_0 \tag{11.1}$$

where

$N_s(t)$ = the number of the N_0 product instances that survived to t without failing.

$N_f(t)$ = the number of the N_0 product instances that failed by t.

If none of the product instances were failed at time 0 ($N_f(0) = 0$), the probability of no failures in the population of product instances from time 0 to time t is given by

$$R(t) = \Pr(T > t) = \frac{N_s(t)}{N_s(0)} = \frac{N_s(t)}{N_0} \tag{11.2}$$

where T is the failure time. In Equation (11.2), if $N_s(t) = 0$ at some time t, then the probability of no failures at time t is 0. Alternatively, if $N_s(t) = N_0$ at some time t, then the probability of no failures at time t is 1 (100%). Alternatively, the probability of one or more failures between 0 to t is given by

$$F(t) = \Pr(T \le t) = \frac{N_f(t)}{N_0} \tag{11.3}$$

$R(t)$ is known as the reliability and $F(t)$ is the unreliability of the product at time t. The cumulative failures plotted in Figure 11.1 is $F(t)$. Equations (11.1) through (11.3) imply that for all t,

$$R(t) + F(t) = 1 \tag{11.4}$$

The reliability $R(t)$ can be constructed graphically from Figure 11.1, as shown in Figure 11.3.

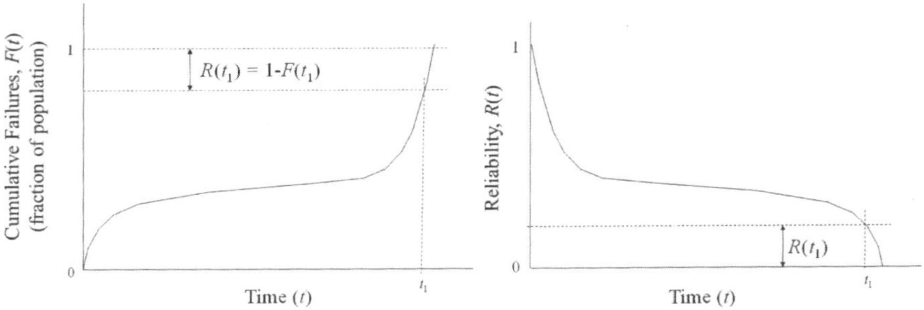

Fig. 11.3. Reliability as a function of time.

11.2.1 *Failure Distributions*

Suppose we perform the following test. Start with 100 instances of a product. All the instances are operational (unfailed) at time 0. If we subject all the instances to exactly the same set of environmental stresses, over time the product instances fail, but they don't all fail at the same time — that is, they are all slightly different (manufacturing and material variations). This gives the example data in Table 11.1.

Plotting the fraction of products failing per time period as a histogram, we obtain Figure 11.4. The fraction of failures at time t, $f(t)$, plotted in Figure 11.4, is known as a failure distribution; it is a probability distribution function (PDF). Assuming that the test was run until all the product instances failed, the total area under the probability distribution in Figure 11.4 is 1, $\Pr(0 \leq t \leq \infty) = 1$. The area under the probability distribution up to time t_1 (to the left of time t_1) is the probability that the part will fail between 0 and t_1, which is the unreliability $F(t_1)$. Therefore, the area under the $f(t)$ curve to the right of t_1 is the reliability. In general,

$$F(t) = \int_0^t f(\tau)d\tau \tag{11.5}$$

Table 11.1. Data Collected From Environmental Testing of $N_0 = 100$ Product Instances, No Repair Assumed.

Time period (hours)	Number of products failing during this time period	Fraction of products failing during this time period (f)	Total number of products failed at the end of this time period (N_f)	Total number of products surviving at the end of this time period (N_s)	Reliability at the end of this time period (R)	Unreliability at the end of this time period (F)	Hazard rate at the end of this time period (h)
0-100	1	0.01	1	99	0.99	0.01	0.010
101-200	3	0.03	4	96	0.96	0.04	0.031
201-300	10	0.1	14	86	0.86	0.14	0.116
301-400	21	0.21	35	65	0.65	0.35	0.323
401-500	31	0.31	66	34	0.34	0.66	0.912
501-600	19	0.19	85	15	0.15	0.85	1.267
601-700	12	0.12	97	3	0.03	0.97	4.000
701-800	2	0.02	99	1	0.01	0.99	2.000
801-900	1	0.01	100	0	0.00	1.00	∞

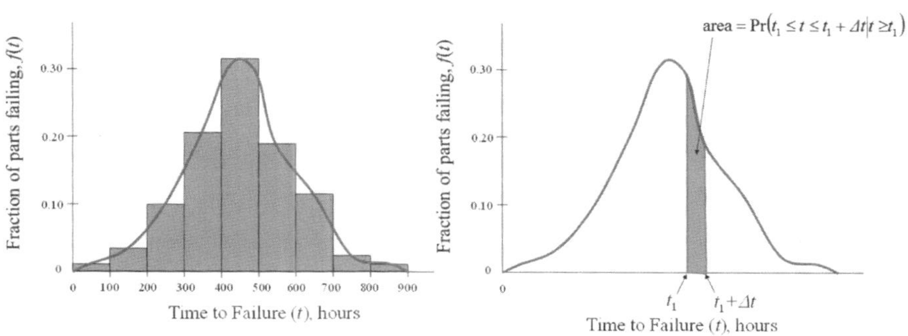

Fig. 11.4 Failure distribution.

and therefore, the area under the $f(t)$ curve to the right of t is the reliability, given by

$$R(t) = 1 - F(t) = 1 - \int_0^t f(\tau)d\tau \tag{11.6}$$

Equation (11.5) is the definition of the cumulative distribution function (CDF). The unreliability is the CDF that corresponds to the probability distribution, $f(t)$. Taking the derivative of Equation (11.6), we obtain

$$\frac{dR(t)}{dt} = -f(t) \tag{11.7}$$

The area within the slice of the distribution between t_1 and $t_1+\Delta t$ in Figure 11.4 is the probability that a part will fail between t_1 and $t_1+\Delta t$ when it has already survived to t_1.

$$\int_{t_1}^{t_1+\Delta t} f(\tau)d\tau = F(t_1+\Delta t) - F(t_1) = R(t_1) - R(t_1+\Delta t) \tag{11.8}$$

The failure rate is defined as the probability that a failure per unit time occurs in the time interval, given that no failure has occurred prior to the start of the time interval:

$$\frac{R(t) - R(t+\Delta t)}{\Delta t R(t)} \tag{11.9}$$

In the limit as Δt goes to 0 and using Equation (11.7), Equation (11.9) gives the hazard rate, or instantaneous failure rate:

$$h(t) = \lim_{\Delta t \to 0} \frac{R(t) - R(t+\Delta t)}{\Delta t R(t)} = -\frac{1}{R(t)}\frac{dR(t)}{dt} = \frac{f(t)}{R(t)} \tag{11.10}$$

The hazard rate is a conditional probability of failure in the interval t to $t+dt$, given that there was no failure up to time t. Restated, hazard rate is the number of failures per unit time per the number of non-failed products left at time t. Figure 11.2 is a plot of the hazard rate.

Once a product has past the infant mortality (or early failure) portion of its life, it enters a period during which the failures are random due to changes in the applied load, overstressing conditions, and variations in the materials and manufacturing of the product.[2] Depending on the type of product or part, different distributions can be used to model the reliability during the random failure (field use) portion of the product's life. The following sections describe two commonly used distributions for electronic systems.[3]

11.2.2 *Exponential Distribution*

The simplest assumption about the field-use (random failures) portion of the life of a product is that the failure rate is constant:

$$h(t) = \lambda \tag{11.11}$$

Using Equations (11.10) and (11.7), we can solve for the PDF:

$$f(t) = h(t)R(t) = \lambda - \lambda \int_0^t f(\tau)d\tau \tag{11.12}$$

Taking the derivative of both sides of Equation (11.12) gives us

$$\frac{df(t)}{dt} = -\lambda f(t) \tag{11.13}$$

Equation (11.13) is satisfied if

$$f(t) = \lambda e^{-\lambda t} \tag{11.14}$$

where $f(t)$ is an exponential distribution. The corresponding CDF and reliability are given by

$$F(t) = \int_0^t \lambda e^{-\lambda\tau} d\tau = 1 - e^{-\lambda t} \tag{11.15}$$

$$R(t) = 1 - F(t) = e^{-\lambda t} \tag{11.16}$$

---

[2] See Chapter 14 for a discussion of burn-in. Burn-in is used to accelerate early failures so that products are already beyond the infant mortality portion of the bathtub curve before they are shipped to customers.

[3] Many other distributions can be used. Readers can consult nearly any reliability engineering text for information on other distributions.

The mean of $f(t)$ is given by the expectation value of $f(t)$:

$$\mathrm{E}[T] = \int_0^{\infty} t f(t) dt = \int_0^{\infty} t \lambda e^{-\lambda t} dt = \frac{1}{\lambda} \tag{11.17}$$

E[*T*] is also known as the mean time to failure (*MTTF*) or, if the failed products are repaired to "good as new" condition after each failure, the E[*T*] is the mean time between failures (*MTBF*). Note that at $t = MTBF = 1/\lambda$, $R(t) = 1/e = 0.37$. This means that $F(t) = 1 - 0.37 = 0.63$ or 63% of the population has failed by $t = MTBF$.

The exponential distribution assumes that products fail at a constant rate, regardless of accumulated age. This is not a good assumption for many real applications. Describing a product using an *MTBF* as a reliability metric usually implies that the exponential distribution was used to analyze the data, in which case the mean completely characterizes the distribution. However, if the data was modeled using any other distribution, the mean is not sufficient to describe the data.[4]

11.2.3 *Weibull Distribution*

The Weibull distribution is much more widely used for electronic devices and systems than exponential distributions because of the flexibility it has in accommodating different forms of the hazard rate. The PDF for a three-parameter Weibull is given by

$$f(t) = \frac{\beta}{\eta}\left(\frac{t-\gamma}{\eta}\right)^{\beta-1} e^{-\left(\frac{t-\gamma}{\eta}\right)^{\beta}} \tag{11.18}$$

where β is the shape parameter, η is the scale parameter, and γ is the location parameter. The corresponding CDF, reliability, and hazard rate are given by

---

[4] In some cases, the use of an exponential distribution for electronics may indicate the use of a reliability prediction model that is not based on actual data, but rather utilizes compiled tables of generic failure rates (exponential failure rates) and multiplication factors (e.g., for electronics, MIL-HDBK-217 [Ref. 11.2]). These analyses provide little insight into the actual reliability of the products in the field [Ref. 11.3].

$$F(t) = 1 - e^{-\left(\frac{t-\gamma}{\eta}\right)^{\beta}} \tag{11.19}$$

$$R(t) = e^{-\left(\frac{t-\gamma}{\eta}\right)^{\beta}} \tag{11.20}$$

$$h(t) = \frac{\beta}{\eta}\left(\frac{t-\gamma}{\eta}\right)^{\beta-1} \tag{11.21}$$

With an appropriate choice of parameter values, the Weibull distribution can be used to approximate many other distributions, e.g., $\beta = 1$, $\gamma = 0$ corresponds to an exponential distribution, $\beta = 3$, $\gamma = 0$ approximates a normal distribution.

Additional properties of the exponential and Weibull distributions will be developed as needed in subsequent chapters.

11.2.4 *Conditional Reliability*

Conditional reliability is the conditional probability that a product will survive for an additional time t given that it has already survived up to time T. The system's conditional reliability function is given by:

$$R(t,T) = \frac{R(t+T)}{R(T)} \tag{11.22}$$

If $R(20) = 0.4$ and $R(10) = 0.6$ then $R(10,10)$, the probability of survival for an additional 10 time units given that the system has already survived 10 time units is 0.67.

The conditional PDF, $f(t,T)$ is given by,

$$f(t,T) = -\frac{d}{dt}R(t,T) = -\frac{\frac{d}{dt}R(t+T)}{R(T)} = \frac{f(t+T)}{R(T)} \tag{11.23}$$

Note, $R(T)$ is not a function of time.

11.3 Qualification and Certification

Many types of products require extensive qualification and/or certification in order to be sold or used. Qualification is the process of determining a product's conformance with specified requirements. The specified requirements may be based on performance, quality, safety, and/or reliability criteria. Certification is the procedure by which a third party provides assurance that a product or service conforms to specific requirements. The terms qualification and certification are sometimes used interchangeably. Figure 11.5 shows the back of a power supply for a laptop computer. Many of the symbols shown on the back of the power supply represent certifications obtained by Dell for the power supply. Examples of certifications required for some products in the United States include:

- The Food and Drug Administration (FDA) requires that certain standards be met for food, cosmetics, medicines, medical devices, and radiation-emitting consumer products, such as microwave ovens and lasers. Products that do not conform to these standards are banned from being sold in the United States and from being imported into the United States.
- The Federal Communications Commission (FCC) requires certification of all products that emit electromagnetic radiation, such as cell phones and personal computers. Devices that intentionally emit radio waves cannot be sold in the United States without FCC certification.
- The Environmental Protection Agency (EPA) certification is required for every product that exhausts into the air or water, including all vehicles (cars, trucks, boats, ATVs), heating, ventilating and air conditioning systems (air conditioners, heat pumps, refrigerators, refrigerant handling and recovery systems), landscaping and home maintenance equipment (chain saws and snow blowers), stoves and fireplaces, and even flea and tick collars for pets.
- Federal Aviation Administration (FAA) certification certifies the airworthiness of all aircraft operating in the United States. The FAA also certifies parts and subsystems used on the aircraft.

Fig. 11.5. Power supply from a Dell Laptop computer showing the wide array of certifications obtained by Dell for the power supply.

Assigning a specific cost to certifications is difficult because in addition to the cost of performing the qualification testing, substantial cost is incurred in designing the product so that it will meet the requirements. The direct cost of certification includes application fees, time to manage the appropriate paperwork, and the cost of legal and other expertise necessary to navigate the certification requirements processes. The indirect costs of certification, which are usually the larger portion of its costs, result from performing required qualification testing prior to seeking certification, product modifications and redesign if qualification requirements are not met and/or certification is not granted, and the time required to gain the certification, which can be years in some cases. Some certifications are relatively inexpensive — for example, the cost for an FCC certification of a new personal computer by an approved third party ranges from $1500 to $10,000 and can be obtained in a few days. However, the average time for FDA approval of a new drug from the start of clinical testing was approximately 90 months in 2003, with estimated costs that can exceed $500 million.

Other certifications, although not required by law, may be required by the retailer or customers of the product. For example, Underwriter Laboratories (UL) provides certification regarding the safety of products, but UL certification is not required by law. The cost of obtaining a UL certification can range from $10,000 to $100,000 for one model of one product. In addition, there are annual fees that are required to maintain the certification. Another example of an optional approval is the EPA's Energy Star program for products that meet energy efficiency guidelines.

General certifications (UL, FDA, FCC, etc.) are usually non-recurring costs borne by the manufacturer. However, qualification of products for specific uses may be borne by either the manufacturer or the customer. For example, the manufacturer of a new electronic part will run a set of qualification tests that correspond to a common standard and then market the part as compliant with that standard. When customers decide to use the part they may perform additional qualification tests to ensure that the part functions appropriately within their usage environment. Manufacturer and customer qualification testing can range from a few thousand dollars to hundreds of thousands of dollars for simple parts. For complex systems, such as aircraft, qualification testing costs millions to tens of millions of dollars. Generally, these are one-time non-recurring expenses; however, they may have to be partially or completely repeated if changes are made to the part or the system using the part.

11.4 Cost of Reliability

Reliability isn't free. The cost of providing reliable products includes costs associated with designing and producing a reliable product, testing the product to demonstrate the reliability it has, and creating and maintaining a reliability organization. The more reliable the product is, the less money will have to be spent after manufacturing on servicing the product. Reliability is, however, a tradeoff and there is an optimum amount of effort that should be expended on making products reliable, as shown in Figure 11.6.

Several of the remaining chapters in this book address estimating the costs directly associated with reliability. Chapters 12 and 13 discuss the calculation of spare requirements and warranty costs, Chapter 14 describes a burn-in cost model, and Chapter 15 describes models for maintainability and availability.

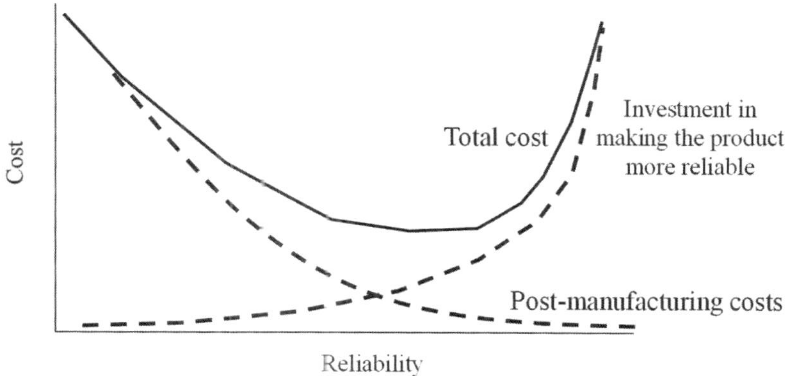

Fig. 11.6. Relationship between reliability and cost.

References

11.1 U.S. Department of Defense, (1993). *Military Standard: System Safety Program Requirements*, MIL-Std-882C.

11.2 U.S. Department of Defense, (1991). Military Handbook: Reliability Prediction of Electronic Equipment, MIL-HDBK-217F(2).

11.3 ReliaSoft (2001). Limitations of the Exponential Distribution for Reliability Analysis, *Reliability Edge*, 2(3).

Bibliography

In addition to the sources referenced in this chapter, there are many good sources of information on reliability and reliability modeling including:

Elsayed, E. A. (1996). *Reliability Engineering* (AddisonWesley Longman, Inc., Reading MA).

O'Connor, P. and Kleyner, A. (2012). *Practical Reliability Engineering, 5th edition* (John Wiley & Sons).

Problems

11.1 Show that the following is true:

$$\lim_{t \to \infty} \int_0^t h(\tau)d\tau = \infty$$

11.2 If the time to failure distribution (PDF) is given by $f(t) = gt^{-4}$ $(t > 2)$ and $f(t) = 0$ for $t \le 2$

a) What is the value of g?

b) What is the mean time to failure?

c) What is the instantaneous failure rate?

11.3 The reliability of a printed circuit board is,

$$R(t) = \begin{matrix} (1 - t/2t_0)^2, & 0 \le t \le 2t_0 \\ 0, & t \ge 2t_0 \end{matrix}$$

a) What is the instantaneous failure rate?

b) What is the mean time to failure (MTTF)?

11.4 Show that Equation (11.17) is equivalent to

$$\mathrm{E}[T] = \int_0^\infty R(t)dt$$

11.5 A manufacturer of capacitors performs testing and finds that the capacitors exhibit a constant failure rate with a value of 4×10^{-8} failures per hour. What is the reliability that can be expected from the capacitors during the first 2 years of their field life?

11.6 A customer performs the test on the capacitors considered in Problem 11.5. A sample size of 1000 capacitors is used and tested for the equivalent of 5000 hours in an accelerated test. How many capacitors should the customer expect to fail during their test?

11.7 An electronic component has an *MTBF* of 7800 operational hours. Assuming an exponential failure distribution, what is the probability of the component operating for at least 5 calendar years? Assume 2000 operational hours per calendar year.

11.8 Your company manufactures a GPS chip for use in marine applications. Through extensive environmental testing, you found that 5% of the chips failed during a 400 hour test. Assuming a constant failure rate and answer the following questions:

a) What is the probability of one of your GPS chips at least 5000 hours?

b) What is the mean life (*MTBF*) for the GPS chips?

11.9 Show that the exponential distribution is a special case of the Weibull distribution.

11.10 The failure of a group of parts follows a Weibull distribution, where $\beta = 4$, $\eta = 10^5$ hours, and $\gamma = 0$. What is the probability that one of these components will have a life of 2×10^4 hours?

11.11 In Problem 11.10, suppose that the user decides to run an accelerated acceptance test on a sample of 2000 parts for an equivalent of 25,000 hours, 12 parts fail during this test, is this consistent with the provided distribution, i.e., are the part better or worse than the provided Weibull distribution implies)?

11.12 If the hazard rate for a part in a system is,

a) 0.001 for $t \leq 9$ hours

b) 0.010 for $t > 9$ hours

What is the reliability of this part at 11 hours?

11.13 Develop expressions for the reliability associated with an *f*(*t*) given by the triangular distribution shown in Figure 9.7.

Chapter 12

Sparing

One of the major elements of logistics is supply support. Supply support for systems includes the spare parts and associated inventories that are necessary to support scheduled and unscheduled maintenance of the system.[1]

When a system fails, one of the following things happens:

- No further action – The system is disposed of and the functionality or role that the system performed is deleted.
- The system is repaired – If your car has a flat tire, you don't dispose of the car, and you may not dispose of the tire either — you get it fixed.
- The system is replaced – If repair is impractical, the failing portion of the system or the entire system is replaced — if a chip fails, you can't repair the chip, you have to replace it.

To expand on these examples, what happens if a tire on your car blows out on the highway and it can't be repaired? You have to replace it. What do you replace your tire with? If you have a spare tire you can change the tire and be on your way. If you don't have a spare you have to have one brought to the car, have the car towed somewhere that has a replacement or, if no one has a replacement, you may have to have one manufactured for you (not a likely scenario for car tire, but for other types of parts in old

[1] Besides spare parts, supply support also includes repair parts, consumables, and other supplies necessary to support equipment; software, test and support equipment; transportation and handling equipment; training equipment; and facilities [Ref. 12.1].

systems this could be the case). A tire that replaces a non-repairable tire is referred to as a permanent spare.

So, why do spares exist? Fundamentally, spares exist because the availability of a system is important to its owner or users. *Availability* is the ability of a service or a system to be functional when it is requested for use or operation. Availability is a function of an item's reliability (how often it fails) and maintainability (how efficiently it can be restored when it does fail). Having your car unavailable to you because no spare tire exists is a problem. If you run an airline, having an airplane unavailable to carry passengers because a spare part does not exist or is in the wrong location is a problem that results in a loss of revenue. (The determination of availability is the topic of Chapter 15.)

Items for which spares exist are generally classified into non-repairable and repairable, which are defined in [Ref. 12.1]. A repairable item is one that, upon removal from operation due to a preventative replacement or failure, is sent to a repair or reconditioning facility, where it is returned to an operational state. Non-repairable items have to be discarded once they have been removed from operation, since it is uneconomical or physically impossible to repair them.

Challenges with Spares

There are numerous issues that arise when managing spares. The most obvious issue is, how many spares do you need to have? There is no need to purchase or manufacture 1000 spares if you will only need 200 to keep the system operational (available) at the required rate for the required time period. The calculation of the quantity of spares is addressed in Section 12.1. The second problem is, when are you going to need the spares? The number of spares I need is a function of time (or miles, or other accumulated environmental stresses); as systems age, the number of spares they need may increase. If possible, spares should be purchased over time rather than all at once at the beginning of the life cycle of the product. The disadvantages of purchasing all the spares up front are the cost of money and shelf life. However, in some cases the procurement life of the spares (see Chapter 16) — may preclude the purchase of spares over time.

The issues with spares extend beyond quantity and time. Spares also have to be stored somewhere. They should be distributed to the places where the systems will be when they fail or, more specifically, where the failed system can be repaired. (Is a spare tire more useful in your garage or in the trunk of your car?) On the other hand, does it make sense to carry a spare transmission in the trunk of the car? Probably not — transmissions fail more rarely than tires and a transmission cannot be installed into the car on the side of the road.

12.1 Calculating the Number of Spares

There are many models for spare part inventory optimization. In general in inventory control problems, infinite populations are assumed. Alternatively, considering the problem from a reliability engineering perspective assumes that the spare demand rate depends on the number of units fielded. From a maintenance perspective, the goal of the inventory model is to ensure that the support of a population of fielded systems meets operational (availability) requirements.

The tradeoff with spares is that too much inventory (too many spares) may maximize availability, but is costly — large amounts of capital will be tied up in spares and inventory costs will be high. On the other hand, having too few spares results in reduced availability because customers must wait while their systems are being repaired, which may also be costly. The situation when the inventory of spares runs out is referred to as "stock-out."

Spare part quantities are a function of demand rates and are determined by how the spares will actually be used. Generally, spares can be used to:

1. Cover actual item replacements occurring as a result of corrective and preventative maintenance actions.
2. Compensate for repairable items that are in the process of undergoing maintenance.
3. Compensate for the procurement lead times required for replacement item acquisition.
4. Compensate for the condemnation or scrapage of repairable items.

Basic sparing calculations can be developed from reliability analysis. From Equation (11.6), the reliability of a system at time t is given by

$$R(t) = 1 - \int_0^t f(\tau)d\tau \tag{12.1}$$

Most models assume that the demand for spares follows a Poisson process. If the time to failure is represented by an exponential distribution,

$$f(t) = \lambda e^{-\lambda t} \tag{12.2}$$

where λ is the failure rate,[2] then the demand for spares is exactly a Poisson process for any number of parts.[3] Substituting Equation (12.2) into Equation (12.1), the probability of no defects occurring in time t assuming that the system was not failed at time 0, is

$$\Pr(0) = R(t) = 1 - \int_0^t \lambda e^{-\lambda\tau} d\tau = 1 + e^{-\lambda\tau}\Big|_0^t = e^{-\lambda t} \tag{12.3}$$

which is the same result given by Equation (11.16). For a unique system with no spares, the probability of surviving to time t is Pr(0). Similarly, the probability of exactly one failure in time t (assuming that the system was not failed at time 0) is given by

$$\Pr(1) = \lambda t e^{-\lambda t} \tag{12.4}$$

Generalizing (similarly to the generalization of Equation (3.15)), we obtain the Poisson equation:

$$\Pr(x) = \frac{(\lambda t)^x e^{-\lambda t}}{x!} \tag{12.5}$$

---

[2] If maintenance activities were confined to only failed items, then λ is the failure rate. However, in reality, non-failed items also appear in the repair process requiring time and resources to resolve that needs to be accounted for as well, so in this context λ is more generally the replacement or removal rate.

[3] If the number of identical units in operation is large, the superposed demand process for all the units rapidly converges to a Poisson process independent of the underlying time to failure distribution [Ref. 12.2].

So, the probability of surviving to time t with exactly one spare is

$$\Pr(0) + \Pr(1) = e^{-\lambda t} + \lambda t e^{-\lambda t} \tag{12.6}$$

and in general,

$$\Pr(x \le k) = \sum_{x=0}^{k} \frac{(\lambda t)^x e^{-\lambda t}}{x!} \tag{12.7}$$

Equation (12.7) is the probability of k or fewer failures in time t, or the probability of surviving to time t with k spares. $\Pr(x \le k)$ is the confidence that your system can survive to time t (assuming it was functional at time 0) with k spares. The derivation in Equations (12.1) through (12.7) is relatively simple; however, it can be interpreted in several different ways.

Our first interpretation is that spares are used to permanently replace failed items (this is the non-repairable item assumption). In this case we assume that (a) no repair of the original failed item is possible (it is disposed of when it fails); (b) λ is the failure rate of the original item; (c) the failed item is replaced instantaneously; and (d) the spare item has the same reliability as the original item it replaces. Under these assumptions, t is the total time the original unit has to be supported. In this interpretation, for a constant failure rate, calculating the number of spares from Equation (12.7) is the same as using a renewal function to compute the number of renewals for warranty analysis (see Section 13.2).[4]

Our second interpretation is that spares are only used to temporarily replace failed items while they undergo repair (the repairable item assumption). If the spares are intended to just cover the repair time for the original items, then we are really modeling the probability of failure of the spares in time t (where t is the repair time for the failed original units) — that is, we are figuring out how many spares we need to cover t, assuming that (a) the spares can't be restored (repaired) if they fail during t; (b) the spares can be restored if necessary between failures of the original unit, and (c) the spares are always good as new. In this case, λ is the failure rate of the spare items (the original item could have a different failure rate). In this case, the original item can be supported forever, assuming that the

[4] Equation (12.7) produces the same result as the renewal function (see Section 13.2) for the constant failure rate assumption when $\Pr(x \le k) = 0.5$. See Problem 13.14.

repaired original items can be repaired to good-as-new status forever. Repaired units can either return to their original location ("socket") or to a spares pool. If they are returned to a spares pool then this interpretation assumes that the repaired units have the same failure rate as the spares (there is no difference between the repaired units and the spares). These repairable items are referred to as "rotable." *Rotable* means that the component or inventory item can be repeatedly and economically restored to a fully serviceable condition. Rotable also refers to a servicing method in which an already repaired component is exchanged for a failed component, which in turn is repaired and kept for another exchange.

12.1.1 *Multi-Unit Spares for Repairable Items*

Equation (12.7) represents spares for a single fielded unit. If there are n identical units in service, the probability that k spares are sufficient to survive for repair times of t is given by [Ref. 12.3]

$$PL = \Pr(x \leq k) = \sum_{x=0}^{k} \frac{(n\lambda\, t)^x e^{-n\lambda\, t}}{x!} \tag{12.8}$$

where

k = the number of spares.

n = the number of unduplicated (in series, non-redundant) units in service.

λ = the constant failure rate (exponential distribution of time to failure assumed) of the unit <u>or</u> the average number of maintenance events expected to occur in time t.

t = the time interval.

PL, $\Pr(x \leq k)$ = the probability that k are enough spares <u>or</u> the probability that a spare will be available when needed ("protection level" or "probability of sufficiency").

$n\lambda t \cong$ Unavailability.

As an example, consider the following case. We need spare parts to keep a population of systems operational while failed original parts are repaired. The population consists of $n = 2000$ units; the spare part has $\lambda =$ 121.7 failures/million hours; it takes $t = 4$ hours to repair the failed parts; and we require a 90% confidence that there are a sufficient number of spares. How many spares (k) do we need? Substituting the numbers into Equation (12.8) we obtain

$$0.9 \le \sum_{x=0}^{k} \frac{\left(2000\dfrac{121.7}{1\times10^{6}}4\right)^{x} e^{-\left(2000\frac{121.7}{1\times10^{6}}4\right)}}{x!} \tag{12.9}$$

We need to solve Equation (12.9) for k. When $k = 1$, 0.9 is not less than or equal to the right-hand side of Equation (12.9), which is 0.7454, so the required confidence level is not satisfied. When $k = 2$, 0.9 is less than 0.9244, indicating that we need 2 or more spares to satisfy the required confidence level.

12.1.2 *Sparing for a Kit of Repairable Items*

A *kit* is a conglomeration of different items required to create a system of separate serviceable units. The protection level for a kit consisting of m rotable items is given by

$$PL_{kit} = \prod_{i=1}^{m} PL_i \tag{12.10}$$

where PL_i is the protection level for item i and Equation (12.10) assumes the independence of the failures of the m rotable items. If PL_{kit} is evenly apportioned to each of the m items in the kit,

$$PL_{kit} = \prod_{i=1}^{m} PL_i = PL_{item}^{m} \tag{12.11}$$

which gives,

$$\left(PL_{kit}\right)^{1/m} = PL_{item} = \sum_{x=0}^{k} \frac{(n\lambda t)^{x} e^{-n\lambda t}}{x!} = \sum_{x=0}^{k} PL_x \tag{12.12}$$

As a simple kit example, consider the following case. Assume that the required $PL_{kit} = 0.96$, and there are $m = 300$ items in the kit; that there are 4 units/system, 35 systems/fleet, 8 operational hours/day, a 12-day turnaround time to repair the original part (for every part in the kit); and that the *MTBUR* (mean time between unit removals) = 13,000 operational hours.[5]

$$n = (4)(35) = 140 \text{ (number of units in service).}$$
$$\lambda = 1/13{,}000 = 7.69\text{x}10^{-5} \text{ per operational hour (removal rate).}$$
$$t = (8)(12) = 96 \text{ operational hours.}$$
$$n\lambda t = 1.034 \text{ (expected number of unit removals in } t\text{).}$$

From Equation (12.11), the protection level for each item in the kit is

$$PL_{item} = (0.96)^{1/300} = 0.999864 \tag{12.13}$$

Solving Equation (12.12) for different values of x we obtain the results shown in Table 12.1. Searching the table for the smallest number of spares (k) that results in a PL_{item} that is greater than or equal to the PL_{item} (computed in Equation (12.13)), gives $k = 6$ spares. So it takes 6 or more spares for each item in the kit.

Table 12.1. Calculated Protection Levels.

x	PL_x		k	PL_{item}
0	0.355636494		0	0.355636494
1	0.367673422		1	0.723309916
2	0.190058876		2	0.913368792
3	0.065497213		3	0.978866005
4	0.01692851		4	0.995794515
5	0.003500295		5	0.99929481
6	0.000603128		6	0.999897938
7	8.90773E-05		7	0.999987015
8	1.15115E-05		8	0.999998527
9	1.32235E-06		9	0.999999849
10	1.36711E-07		10	0.999999986

[5] We will use *MTBUR* instead of *MTBF* because *MTBUR* includes all unit removals, not just the failures. For example, it includes misdiagnosis.

12.1.3 *Sparing for Large k*

When k is large, the Poisson distribution can be approximated by the normal distribution with a mean of $n\lambda t$ and a standard deviation of $\sqrt{n\lambda t}$ [Ref. 12.4],

$$k \cong \left\lceil n\lambda t + z\sqrt{n\lambda t} \right\rceil \tag{12.14}$$

where z is the number of standard deviations from the mean of a standard normal distribution (the standard normal deviate from 1-α, where α is 1 minus the desired confidence level).[6] The approximation in Equation (12.14) is independent of the underlying time-to-failure distribution and is valid when t and k are large.

For the kitting example in the previous section, using the *PL* given in Equation (12.13) we get,

$$z = 3.6405$$

the right-hand side of Equation (12.14) omitting the ceiling function = 4.74

$$k = \lceil 4.74 \rceil = 5$$

In this example, Equation (12.14) underestimates the number of spares because k is relatively small. Figure 12.1 shows a comparison of Equations (12.7) and (12.14).

---

[6] This is a single-sided z score. Note, the z that appears in Equation (9.12) is a two-sided z-score. z = NORMINV(*PL*,0,1) in Excel, where *PL* is the required protection level.

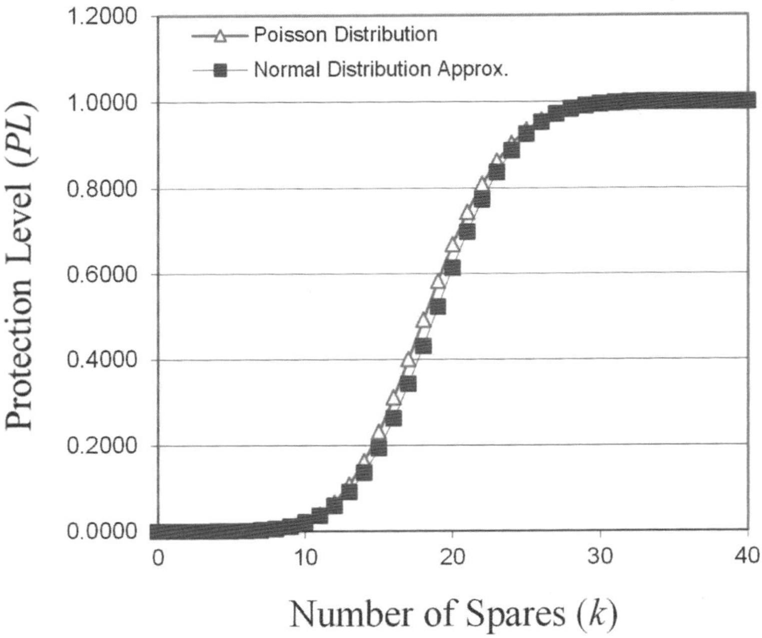

Fig. 12.1. Comparison of Poisson model (Equation (12.7)) and normal distribution approximation (Equation (12.14)), where $n = 25{,}000$, $t = 1500$ hours, $\lambda = 5\text{x}10^{-7}$ failures per hour.

12.2 The Cost of Spares

The protection level computed in Section 12.1 is the probability of having a spare available when required. The protection level is a hedge against the risk of a stock-out situation. While maximizing the spares will minimize this risk, the risk has to be traded off against cost — the more spares you have and the longer you hold them, the more it costs.

The costs associated with spares come from several sources. The total cost of spares in the jth period of time for one spared item is given by

$$C_{Totalj} = PD_j + \frac{C_p D_j}{Q} + \frac{C_h Q}{2} \tag{12.15}$$

where

P = the purchase price of the spare.

D_j = the number of spares needed in period j for one spared item.

C_p = the cost per order (setup, processing, delivery, receiving, etc.).
Q = the quantity per order.
C_h = the holding (or carrying) cost per period per spare (cost of storage, insurance, taxes, etc.).

The first term in Equation (12.15) is the purchase cost (the cost of purchasing D_j spares); the second term is the ordering cost (the cost of making D_j/Q orders in the time period); and the third term is the holding cost (the cost of holding the spares in the time period). In the third term, $Q/2$ is the average quantity in stock — this term does not use $D_j/2$ because the maximum number of spares that are held at any time is Q (not D_j).

Equation (12.15) can be used to solve for the economic order quantity (EOQ), which is the quantity per order (Q) that minimizes the total cost of spares in a period of time. To solve for the optimal order quantity, minimize the total cost:

$$\frac{dC_{Totalj}}{dQ} = -\frac{C_p D_j}{Q^2} + \frac{C_h}{2} = 0 \tag{12.16}$$

Solving for Q we obtain

$$Q = \sqrt{\frac{2C_p D_j}{C_h}} \tag{12.17}$$

Equation (12.17) is known as the Wilson EOQ Model or Wilson Formula.[7]

The basic EOQ model in Equation (12.17) only applies under the following conditions: (a) when the demand for spares is constant over the time period, (b) when each order is delivered in full when the inventory reaches zero, (c) when the cost per order is a constant that does not depend on the number of units ordered, and (d) when the time period (often referred to as the "review time" or "review period") is short.

One variation on the EOQ model is called the economic production quantity (EPQ) [Ref. 12.6]. The EOQ assumes that 100% of the order arrives instantaneously upon ordering when the inventory reaches zero. This assumption in the EOQ model is reflected in the third term in

---

[7] The model was developed by F. W. Harris in 1913 [Ref. 12.5]; however, R. H. Wilson, a consultant who applied it extensively, is given credit for it.

Equation (12.15). If instead, each order is delivered incrementally when the inventory reaches zero, Equation (12.15) becomes,

$$C_{Totalj} = PD_j + \frac{C_p D_j}{Q} + \frac{C_h Q}{2}\left(1 - \frac{u_r}{d_r}\right) \tag{12.18}$$

where

u_r = usage rate.
d_r = demand (production or delivery) rate.

Similar to Equation (12.16), we minimize the total cost of spares with respect to Q and then solve for Q to obtain

$$Q = \sqrt{\frac{2C_p D_j}{C_h}}\sqrt{\frac{d_r}{d_r - u_r}} \tag{12.19}$$

There are many other variations on the basic EOQ model. Some of these include volume discounts, loss of items in inventory (physical loss or shelf life issues), accounting for the ratio of production to consumption to more accurately represent the average inventory level, and accounting for the order cycle time.

12.2.1 *Spares Cost Example*

Consider the support of a system that contains a critical non-repairable item that has an $MTBUR$ = 13,000 operational hours. There are n = 300 systems to support (each has one instance of the item in it). A protection level of PL = 0.99 is desired. The purchase price of the item is P = \$5000, C_p = \$1000 per order, and C_h = \$150 per year per part. We wish to determine the optimum quantity per order (Q) and the total cost of spares (C_{Total}) for a one year period.

Using Equation (12.14), the number of spares necessary in a t = 8760 hour (one calendar year) period is, k = 236. The optimum order quantity from Equation (12.17) is given by

$$Q = \sqrt{\frac{2(1000)(236)}{(150)}} = 56.1 \tag{12.20}$$

Rounding Q up to 57 (since we cannot buy fractional parts) and using Equation (12.15),

$$C_{Total} = (5000)(236) + \frac{(1000)(236)}{57} + \frac{(150)(57)}{2} = \$1{,}188{,}415 \qquad (12.21)$$

Equation (12.21) is the cost of spares to support one year of the operation of the 300 systems.

12.2.2 *Extensions of the Cost Model*

We did not include the cost of money in Equation (12.15) because we have assumed that the time period of interest is relatively short. However, the total cost of spares over the entire support life of a system should include the cost of money. The total cost of spares (for a single spared item) over the entire life of a system is given by

$$C_{Total} = \sum_{j=0}^{n_t-1} \frac{C_{Totalj}}{(1+r)^j} \qquad (12.22)$$

where r is the discount rate per time period (assumed to be constant over time) and the support life of the system is n_t time periods.

If the 300 systems considered in Section 12.2.1 have to be supported for $n_t = 15$ years and the discount rate is $r = 6.5\%$/year (constant for all the years), the total cost (in year 0 dollars) is given by Equation (12.22) as

$$C_{Total} = \sum_{j=0}^{14} \frac{1{,}188{,}415}{(1+0.065)^j} = \$11{,}900{,}604 \qquad (12.23)$$

Several other effects can impact the cost of the spares. Two different types of obsolescence impact inventories. First, inventory or sudden obsolescence refers to the situation when the system that the spare parts were purchased for is changed (or retired) before the end of the projected support period, making the spares inventory obsolete [Ref. 12.7]. This represents a cost because the investment in the spare parts may not be recoverable. The opposite problem, which is common to sustainment-dominated systems, is DMSMS (diminishing manufacturing sources and material shortages) obsolescence, which represents the inability to continue to purchase spares over the life of the system--that is, the needed part is discontinued by its manufacturer and may become unprocurable at some point prior to the end of the need to support the system. DMSMS obsolescence is the topic of Chapter 16. The result in Equation (12.23) assumes that the needed spares can be procured as needed for the entire support time (i.e., for 15 years).

Other issues that are common to the management of inventories for sustainment-dominated systems include the inventory lead times (the time between spare replenishment orders and when the spares are delivered). Also, repair times for original units that have failed can be lengthy and are usually modeled using lognormal distributions (see Section 15.2). In fact, as repairable systems age, the electronic parts become obsolete and there may be delays in obtaining the parts necessary to repair repairable systems.

12.3 Summary and Comments

It should be stressed that much of the development in this chapter is based on the time-to-failure distribution given in Equation (12.2), which is an exponential distribution that assumes a constant failure rate, λ. Equations (12.3) through (12.8) and Equation (12.12) are specific to the constant failure rate assumption. Determining the number of spares for other time-to-failure distributions requires the calculation of renewal functions, which will be addressed in Chapter 13.

The cost of spares is a very important contributor to the life-cycle costs of many systems. In addition to the direct costs discussed in Section 12.2, many additional logistics costs must be considered, including costs to transport spares to the locations where they are needed, holding costs (which may vary by location), and the costs to transport failed systems to places where they can be repaired. See [Ref. 12.8] for a discussion of holding costs.

As mentioned in the introduction, spares exist because availability is important to many systems. Besides assessing the number of spares needed, sparing analysis also focuses on how to distribute the spares among multiple locations in order to have them available when needed (it does no good to have the correct number of spares to support a system stored in Oklahoma City if the system that needs the spares is in Germany).

Distribution of spares directly impacts system availability. Geographic distribution of spares may also influence spare quantity if spares cannot be easily or quickly transported between locations.

The development in this chapter implicitly assumes that spares can be replenished (that more can be purchased) whenever needed. This may not be the case. Original manufacturers often discontinue making parts at some point (this is especially problematic for electronic parts, some of whose procurement lifetimes are measured in months). See Chapter 16 for the cost ramifications of obsolescence.

Sparing is potentially about more than just hardware. Although the context of the spares calculations presented in this chapter has focused on hardware components, products or units, the spared item could also be trained personnel or a maintenance team.

References

12.1 Louit, D., Pascual, R., Banjevic, D. and Jardine, A. K. S. (2011). Optimization models for critical spare parts inventories – A reliability approach, *Journal of the Operational Research Society*, 62, pp. 994-1004.

12.2 Cox, R. (1962). *Renewal Theory* (Methuen, London).

12.3 Myrick, A. (1989). Sparing analysis – A multi-use planning tool, *Proceedings of the Reliability and Maintainability Symposium,* pp. 296-300.

12.4 Coughlin, R. J. (1984). Optimization of spares in a maintenance scenario, *Proceedings of the Reliability and Maintainability Symposium*, pp. 371-376.

12.5 Harris, F. W. (1913). How many parts to make at once, *Factory, The Magazine of Management*, 10(2), pp. 135-136, 152.

12.6 Taft, E. W. (1918). The most economical production lot, *The Iron Age*, 101, pp. 1410-1412.

12.7 Brown G., Lu J. and Wolfson, R. (1964). Dynamic modeling of inventories subject to obsolescence, *Management Science*, 11(1), pp. 51-63.

12.8 Lambert, D. M. and La Londe, B. J. (1976). Inventory carrying costs, *Management Accounting*, 58(2), pp. 31-35.

Bibliography

Sparing is also treated in many engineering reliability texts and engineering logistics texts, including the following:

Elsayed, E. A. (1996). *Reliability Engineering* (AddisonWesley Longman, Inc., Reading MA).

Blanchard, B. S. (1992). *Logistics Engineering and Management*, 4th Edition (Prentice Hall, Englewood Cliffs, NJ).

Gopalakrishnan, P. and Banerji, A. K. (1991). *Maintenance and Spare Parts Management* (PHI Learning Private Limited, New Delhi).

Problems

12.1 For a single non-repairable system defined by *MTBUR* = 8,000 operational hours, what is the probability that the system will survive 9,500 operational hours with 6 spares?

12.2 A customer requires a protection level of 0.96 and owns 8 spares for a single repairable system that has an *MTBUR* of 1 calendar month. What is the maximum amount of time that the repair of failed units can take?

12.3 Rework Problem 12.2 if the customer owns 4 identical systems.

12.4 If the system in Problem 12.2 actually consists of a kit consisting of 134 items (with evenly apportioned protection level), what is the protection level required for each item in the kit?

12.5 An organization has been supporting a product for several years. The product is repairable and spares are only used to maintain the product while repairs are made. The repair time is 1.2 months and 512 identical systems are supported. Experience has shown that 9 spares results in a protection level of 0.9015. What is the failure rate?

12.6 Assume you are supporting a product. You are going to order 450 spares and the $n\lambda t = 420.2983$. Assume the time to failure is exponentially distributed and that the large k assumption is valid. NOTE: to make life easier you may ignore all "ceiling functions" in the solution of this problem. Hint: you need the table at the end of this exam for this problem.

a) What confidence do I have that I have that 450 spares will be sufficient to support the product?

b) An engineer proposes some process improvements that will decrease the failure rate (λ) of this product by 7.5%. If spares cost \$1300 each, how much money can be saved by this improvement? Hint: you do not need to know n or t to solve this problem. Hint: the improved $\lambda_{improved} = (1 - 0.075)\ \lambda_{original}$.

c) If the process improvements cost a total of \$50,000 and all the return on the investment is in the reduction of the number of spares, what is the return on investment (ROI) of the process change? See Chapter 17 for a treatment of ROI.

12.7 A system supporter expects to need 200 parts per year to support a system. The storage space taken up by one part is costed at £20 per year. If the cost associated with ordering is £35 per order, what is the economic order quantity, given that the interest rate you have to pay on the money used to buy the spare parts is 10% per year and the cost of one part is £100? What is the total cost? Hint: Treat the 10% interest as a holding cost.

12.8 Suppose in Problem 12.7 a budget was only available to order 15 spare parts per order. What is the cost penalty associated with this budget limitation?

12.9 If the purchase price of the spares is a function of the quantity per order, such that $P = P_1(1\text{-}q(Q\text{-}1))$, what is the optimum order quantity? P_1 and q are constants.

12.10 For a particular part, the order cost is represented by a triangular distribution with a mode of \$595 per order (low = \$500, high = \$633). The holding cost is represented by a triangular distribution with a mode of \$13.54 per year (low = \$9, high = \$22). If 25 spares are needed per year and the purchase price is \$91 per spare, what is your confidence that the total cost of spares per year (if the optimum order quantity is used) will be less than \$3850?

12.11 Your company supports an electronic product. Demand for a particular integrated circuit (IC) to repair the product is 10,000 units per year (constant throughout the year). You have two choices for your repair operation: (1) You can provide resources that are capable of repairing at a rate of 15,000 units per year, at a cost of \$10.00 per repair; or (2) you can provide resources that are capable of repairing at a rate of 11,000 units per year, at a cost of \$10.10 per repair. You figure your holding cost per IC per year to be C_h = \$2 + (5%)(unit repair cost) and the repair operation set-up cost (C_p) is \$500 in both cases. Which choice should you use for your repair operation? Hint: this is an economic production quantity (EPQ) problem.

Chapter 13

Warranty Cost Analysis

The total cost of warranties for computer and related high-technology US companies is now about $8B per year [Ref. 13.1]. For many companies, warranty costs approach what they spend on new product development and often rival their net profit margins; this is particularly true for commodity-type businesses making products like PCs or personal printers.

Fundamentally, a warranty is a manufacturer's assurance to a buyer that a product or service is or shall be as it is represented. Warranties are considered to be a contractual agreement between the buyer and the manufacturer entered into upon sale of the product or service. In broad terms, the purpose of a warranty is to establish liability among two parties (manufacturer and buyer) in the event that an item fails. This contract specifies both the performance that is expected and the redress available to the buyer if a failure occurs.[1]

From a buyer's perspective, warranties are protectional — the warranty provides a form of compensation if the item, when properly used, fails to perform as intended or as specified by the manufacturer. From the manufacturer's perspective, warranties are both protectional and promotional. They are protectional in the sense that the warranty terms specify the conditions of use for which the product is intended and provide for limited or no coverage in the event of product misuse. They are promotional in the sense that buyers often infer that they are purchasing a more reliable product if it has a longer warranty than its competition, and the warranty can be used to differentiate the product from competing items in the marketplace.

---

[1] These definitions were adapted from [Ref. 13.2].

The exact historical origin of warranties[2] is difficult to pinpoint; however, concepts of product liability appeared in the Hammurabi code of laws as early as 1800 B.C., when penalties were imposed on craftsmen for making defective products. Notions of compensating the customer for the failure of products also appear in the Hammurabi code in the form of money-back guarantees — if a defect was discovered in a slave, the seller would return the money paid. Warranties evolved through Roman, middle European Jewish, and old English law over the next four thousand years, and approached the form we are familiar with today at the end of the nineteenth century, when the courts began to make exceptions to the concept of *caveat emptor* ("let the buyer beware") for common products. Modern U.S. laws governing warranties and guarantees are contained in the Uniform Commercial Code (UCC) of 1952 and the Magnuson-Moss Warranty Act of 1975.[3] An excellent summary of the history of warranties is provided in [Ref. 13.3].

How Warranties Impact Cost

Warranties are one mechanism by which companies that manufacture and support products are effectively charged (or penalized) for the lack of initial quality and, later, the reliability of their products.[4] Servicing warranty claims is not free; costs can include providing telephone or web-based support to customers, repairing products, or replacing defective products. It is important to be able to estimate the future costs of servicing warranty claims when setting the sales price of a product. For example, if a product costs $10 to manufacture, and an additional $2 to market and sell, selling the product for $15 results in a profit of $3 per product sold only if there are no warranty returns to address. If 25% of these products

---

[2] The word "warranty" comes from the French words "warrant" and "warrantie," and the German word "werēnto," which mean "protector" [Ref. 13.3].

[3] Note that there were no warranties on weapons systems in the United States until the Defense Procurement Reform Act of 1985 required the prime contractor for the production of weapons systems to provide a written guarantee.

[4] Other mechanisms by which companies are penalized include liability (lawsuits) and reductions in customer satisfaction that lead to the loss of future sales. These additional mechanisms are not addressed in this book.

are returned by the customers during the warranty period and need to be replaced with new products, then the effective cost per product to the manufacturer is approximately

$$\$10 + \$2 + 0.25(\$10) = \$14.50$$

This effectively cuts the $3 profit per product to $0.50, and this simple calculation does not account for the costs of shipping the replacement product to the customer or the possibility that some fraction of the replacement products could themselves also fail prior to the end of the warranty period.

This very simple example points out that the cost of servicing the warranty needs to be figured into the cost of the product when the selling price is established. Companies often establish warranty reserve funds for their products to cover the expected costs of warranty claims — this is usually implemented by adding a fraction of each product sale to the reserve fund for covering warranty costs.

The cost of servicing the warranty on a product is considered a liability in accounting. Generally, revenue recognition policies do not include the warranty reserve fund as revenue — that is, a company can't report as revenue the money paid to them by customers to support a warranty until the money goes unused (when the warranty period expires). For example, it would be misleading for a public company to report on their earnings statement a $3 profit for the product described above. In this case, the company should contribute $2.50 per product sold to a warranty reserve fund to cover future warranty claims, and only report a profit of $0.50 per product sold to its shareholders. Underestimation of warranty costs results in companies having to restate profits (causing stock value drops and potential shareholder lawsuits); overestimating warranty costs potentially results in overpricing a product, with an associated loss in sales. Therefore, accurate estimation of warranty costs is very important.

Consider the following warranty cost example. After the initial release of the Microsoft Xbox 360 video game console in May 2005, Microsoft claimed that the failure rate matched a consumer electronics industry average of 3 to 5%; however, representatives of the three largest Xbox 360 resellers in the world at the time (EB Games, GameStop and Best Buy) claimed that the failure rate of the Xbox 360 was between 30% and 33%

[Ref. 13.4].[5] According to the German computer magazine *c't*, in an article titled "Jede dritte stirbt den Hitzetod" ("Every third one dies of heat"), the main reason for the problems was that "the wrong type of lead free solder was used, a type that when exposed to elevated temperatures for a long time becomes brittle and can develop cracks" [Ref. 13.4]. Because of inadequate thermal management, the ball grid array solder joints of the CPU and GPU can break. On July 9, 2007, CRN Australia published an article claiming that Microsoft admitted there was a design flaw in Xbox 360 that could cause a failure of all Xbox 360 consoles produced to date [Ref. 13.6]. A few days before, the vice president of Microsoft's Interactive Entertainment Business division had published an open letter recognizing the problem and announcing a three-year warranty extension for every Xbox 360 console that experienced a general hardware failure [Ref. 13.7]. According to Bloomberg [Ref. 13.8], Microsoft created an internal account of more than one billion dollars dedicated to addressing this problem. A simple warranty reserve fund calculation, assuming that the replacement cost of an Xbox 360 was $300, suggests that the fund was sufficient to replace $1 billion/$300 = 3.3 million units. Microsoft had sold 11.6 million units as of June 30, 2007, meaning that the expected replacement rate was 3.3/11.6 = 28%.

The warranty servicing costs were only a portion of the effective long-term cost of the Xbox 360's reliability problems. What about the damage to the brand name? "It's a pretty big black eye," said Matt Rosoff, an analyst at the research firm Directions on Microsoft. "It's certainly not going to help the Xbox compete against Nintendo, and it may be the stumble" that PlayStation 3 maker Sony Corp. needs to win sales [Ref. 13.8]. On the day that Microsoft announced that it would be incurring over $1 billion in pre-tax costs to cover the Xbox warranty problems, its stock dropped 8 cents per share, or 0.25%.

---

[5] More recently, some have claimed that the failure rate may have been as high as 54.2% [Ref. 13.5].

13.1 Types of Warranties

Warranties are usually divided into two broad groups. Implicit warranties are assumed, not explicitly stated. Implicit warranties are inferred by customers from industry standards, advertising and sales implications. The second type of warranty is the explicit or express warranty. Explicit warranties contain a contractual description of the warranty in the "small print" in a user's manual or on the back of the product packaging. The remainder of this chapter addresses particular types of explicit warranties and their cost ramifications.

Based on the definition of a warranty given, a warranty agreement should contain three fundamental characteristics [Ref. 13.9]: a coverage period (usually called the warranty period), a method of compensation, and the conditions under which that compensation can be obtained. The various explicit warranty types differ in respect to one or more of these characteristics.

Generally, three types of warranties are common for consumer goods: ordinary free replacement warranties, unlimited free replacement warranties, and pro-rata warranties. In the first two types, the seller provides a free replacement or good-as-new repair.[6] In the case of an ordinary free replacement warranty (also called a non-renewing free replacement warranty), the warranty on the replacement is for the remaining duration of the original warranty, while for the unlimited free replacement warranty (also called renewing free replacement warranties) the warranty on the replacement is for the same duration as the original warranty. Unlimited free replacement warranties may be offered on inexpensive items with lifetime warranties, such as a surge protector. Ordinary free replacement warranties are offered for items that have warranties that last for a limited period, such as a laptop computer. In the case of a pro-rata warranty, the customer receives a rebate that depends on the age of the item at the time of failure. Examples of pro-rata warranty items include batteries, lighting systems, and tires.

---

[6] Many references do not draw a distinction between ordinary and unlimited free replacement warranties. In this case, they are usually just discussing ordinary free replacement warranties and referring to them as free replacement warranties, or FRWs.

Free replacement warranties favor the customer and pro-rata warranties favor the seller; therefore, mixed (or "combined") warranty policies that are a compromise between the two are common. In this type of warranty, there might be an initial period of free replacement, followed by a period of pro-rata coverage.

There are many variations on the basic warranties described above for repairable and non-repairable products; however, all of these warranties are "one-dimensional," meaning that the warranty period depends only on a single variable. Warranties can also be two-dimensional where the warranty is characterized by two variables — for example, time and/or mileage (say, 3 years or 36,000 miles, whichever comes first). Two-dimensional warranties will be discussed in Section 13.4.

13.2 Renewal Functions

Evaluating the cost of providing a product warranty requires predicting the number of failures the product will have during the warranty period. Renewals are defined as replacement of equipment or components.

Consider a product that is placed in operation at time 0. When the product fails at some later time it is immediately replaced with a new version of the product (a spare) that has a reliability identical to the original unit at time 0. The replaced product fails after a time and is similarly replaced by a good-as-new version of the product. The expected number of failures and associated renewals per product instance within a population of the product in the interval $(0,t]$ is denoted by a renewal function, $M(t)$:

$$M(t) = \mathrm{E}[N(t)] \tag{13.1}$$

where $N(t)$ is the total number of failures in the time interval $(0,t]$. If we account for only the first failure, $M(t) = F(t) = 1 - R(t)$, where $F(t)$ is the unreliability and $R(t)$ is the reliability. This estimation of $M(t)$ assumes that repaired or replaced products never fail. The difference between $M(t)$ and $F(t)$ is that $M(t)$ accounts for more than the first failure, including the possibility that the repaired or replaced product may fail again during the warranty period.

To determine $M(t)$, let T_1, T_2, ... be the sequence of failure times associated with a system and $t_i = T_i - T_{i-1}$ be the times between failures, as shown in Figure 13.1.[7] From the figure, the total time to the nth renewal is

$$S_n = \sum_{i=1}^{n} t_i \tag{13.2}$$

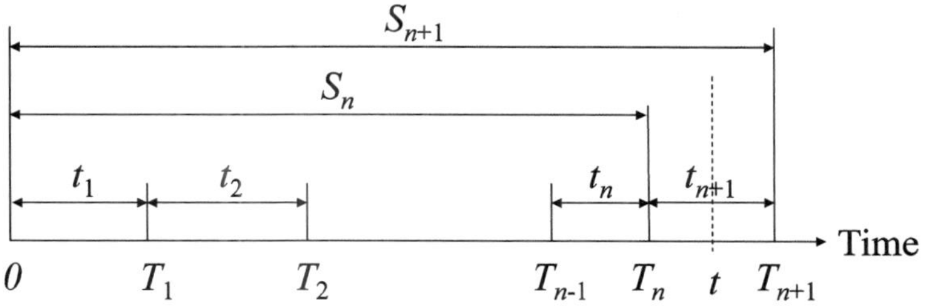

Fig. 13.1. Renewal counting process.

If $N(t)$ is the total number of failures in the interval $(0,t]$, then the probability that $N(t) = n$ is the same as the probability that t lies between the n and n+1 failures in Figure 13.1 which is

$$\begin{aligned}\Pr(N(t) = n) &= \Pr(N(t) \geq n) - \Pr(N(t) \geq n+1) \\ &= \Pr(S_n \leq t) - \Pr(S_{n+1} \leq t)\end{aligned} \tag{13.3}$$

If $F_n(t)$ represents the cumulative distribution function of S_n, then $F_n(t) = \Pr(S_n \leq t)$ and Equation (13.3) becomes

$$\Pr(N(t) = n) = F_n(t) - F_{n+1}(t) \tag{13.4}$$

The expected value of $N(t)$, which is called the renewal function is given by

$$M(t) = \mathrm{E}[N(t)] = \sum_{n=0}^{\infty} n \Pr(N(t) = n) \tag{13.5}$$

---

[7] If the inter-occurrence times t_1, t_2, ... are independent and identically distributed, then the counting process is called an ordinary renewal process. If t_1 is distributed differently than the other inter-occurrence times, the counting process is called a delayed renewal process. In this case the first event is different from the subsequent events.

Substituting Equation (13.4) into Equation (13.5) we get

$$M(t)=\sum_{n=0}^{\infty} n\left[F_n(t)-F_{n+1}(t)\right]=\sum_{n=1}^{\infty} F_n(t) \tag{13.6}$$

Equation (13.6) can be rewritten as,

$$M(t)=F_1(t)+\sum_{n=1}^{\infty} F_{n+1}(t) \tag{13.7}$$

$F_{n+1}(t)$ in Equation (13.7)[8] can be obtained from $F_n(t)$ and $f(t)$ (the PDF of $F(t)$) using

$$F_{n+1}(t)=\int_0^t F_n(t-x)f(x)dx \tag{13.8}$$

Substituting Equation (13.8) into Equation (13.7) and switching the order of the integral and the sum we get,

$$M(t)=F_1(t)+\int_0^t\left[\sum_{n=1}^{\infty} F_n(t-x)\right]f(x)dx \tag{13.9}$$

The term in the brackets in Equation (13.9) is $M(t\text{-}x)$, giving

$$M(t)=F_1(t)+\int_0^t M(t-x)f(x)dx \tag{13.10}$$

The integral equation in Equation (13.10) is commonly known as the fundamental renewal equation.

Taking the Laplace transform of both sides of Equation (13.10), assuming that all the $F(t)$ are the same and using the convolution theorem,[9] we get

$$\hat{M}(s)=\hat{F}(s)+\hat{M}(s)\hat{f}(s) \tag{13.11}$$

---

[8] $F_{n+1}(t)$ is the convolution of $F_n(t)$ and $f(t)$.

[9] The convolution theorem is, $\mathcal{L}\left[\int_0^t X(t-\tau)Y(\tau)d\tau\right]=\hat{X}(s)\hat{Y}(s)$.

Since $F_n(t)=\int_0^t f_n(\tau)d\tau$ from Equation (11.5) and $\mathcal{L}[F_n(t)]=\hat{f}_n(s)/s$ solving for $\hat{M}(s)$ gives

$$\hat{M}(s)=\frac{1}{s}\left[\frac{\hat{f}(s)}{1-\hat{f}(s)}\right] \tag{13.12}$$

the renewal density function is given by

$$m(t)=\frac{dM(t)}{dt} \tag{13.13}$$

The renewal density function is the mean number of renewals expected in a narrow interval of time near *t*. The Laplace transform of the renewal density function follows from Equations (13.12) and (13.13),

$$\hat{m}(s)=\frac{\hat{f}(s)}{1-\hat{f}(s)} \tag{13.14}$$

13.2.1 *The Renewal Function for Constant Failure Rate*

For a constant failure rate of λ, the $f(t)$ is given by Equation (11.14):

$$f(t)=\lambda e^{-\lambda t} \tag{13.15}$$

The Laplace transform of $f(t)$ is

$$\hat{f}(s)=\frac{\lambda}{s+\lambda} \tag{13.16}$$

Substituting Equation (13.16) into Equation (13.12) gives

$$\hat{M}(s)=\frac{\lambda}{(s+\lambda)s\left(1-\dfrac{\lambda}{s+\lambda}\right)}=\frac{\lambda}{s^2} \tag{13.17}$$

and taking the inverse Laplace transform,

$$M(t)=\lambda t \tag{13.18}$$

The renewal density function from Equation (13.14) is $m(t)=\lambda$.

If, for example, a system with a constant failure rate of $1\text{x}10^{-5}$ failures per hour of continuous operation has a one-year warranty, and if 10,000 of these systems are fielded, what is the expected number of legitimate warranty claims during the warranty period? From Equation (13.18), $M(t) = (1\text{x}10^{-5})(24)(365) = 0.0876$ expected failures per unit. So the expected number of claims is (0.0876)(10,000) = 876 claims.

13.2.2 *Asymptotic Approximation of M(t)*

Often it is difficult or impossible to determine the Laplace transform of the PDF, $f(t)$. This may be due to the distribution chosen or simply to a lack of knowledge of what the failure distribution is. There are several approximations for renewal functions. The following non-parametric renewal function estimation for large t is commonly used [Ref. 13.10]:

$$M(t) \cong \frac{t}{\mu} + \frac{\sigma^2}{2\mu^2} - \frac{1}{2} \tag{13.19}$$

where μ and σ^2 are the mean and variance of the failure distribution given by,

$$\mu = -\frac{d\hat{f}(s)}{ds} \text{ and } \sigma^2 = \frac{d^2\hat{f}(s)}{ds^2} - \mu^2 \tag{13.20}$$

both evaluated at $s = 0$.

Equations (13.19) and (13.20) are valid for any distribution. For example, for exponentially distributed failures, $\mu = 1/\lambda$ (the MTBF) and $\sigma^2 = 1/\lambda^2$, which from Equation (13.19) gives, $M(t) = \lambda t$, which is the same result derived from Equation (13.18).

A commonly used time-to-failure distribution for electronic systems is the 2-parameter Weibull distribution:

$$f(t) = \frac{\beta}{\eta}\left(\frac{t}{\eta}\right)^{\beta-1} e^{-\left(\frac{t}{\eta}\right)^{\beta}} \tag{13.21}$$

where β is the shape parameter and η is the scale parameter. The mean and variance are given by

$$\mu = \eta\Gamma\left(1+\frac{1}{\beta}\right) \text{ and } \sigma^2 = \eta^2\left[\Gamma\left(1+\frac{2}{\beta}\right) - \Gamma^2\left(1+\frac{1}{\beta}\right)\right] \tag{13.22}$$

where Γ( ) denotes a gamma function. Using Equations (13.22) and (13.19), an approximation to the renewal function for a Weibull distribution can be found.

13.3 Simple Warranty Cost Models

In this section we will construct cost models for simple (one-dimensional) warranty reserve funds. The models in this section are idealized in the sense that they assume that the time that the unit is out of service undergoing warranty repair or replacement is effectively zero (or at least much smaller than the warranty period). The models in this section do not necessarily assume good-as-new replacement or repair; however, if the form of the renewal functions derived in Section 13.2 is used, then good-as-new replacement or repair is implicitly assumed.

It is not uncommon for warranty cost models to replace $M(t)$ with $F(t)$, the unreliability. This is an approximation that is valid only if the warranty period is short relative to the mean of the time-to-failure distribution — that is, if units rarely fail more than once during the warranty period. In the following we will define warranty reserve fund costs in terms of the renewal function, which is more accurate.

This section focuses on "non-renewing" warranties. A non-renewing warranty means that the warranty period starts on the product sale date and ends after the specified warranty period is reached regardless of how many renewals are performed on the product. Alternatively, a renewing warranty (not treated in this section) means that each renewal gets a new warranty period equal to the original warranty period.

13.3.1 *Ordinary (Non-Renewing) Free-Replacement Warranty Cost Model*

The basic model for an ordinary free replacement warranty's cost (total warranty cost for the product — i.e., the warranty reserve fund) is given by

$$C_{rw} = C_{fw} + \alpha M\left(T_W\right)C_{cw} \tag{13.23}$$

where

C_{fw} = the fixed cost of providing warranty coverage.
α = the quantity of products sold.
$M(T_W)$ = the renewal function — the expected number of renewal events per product during the interval $(0,T_W]$.
T_W = the warranty period.
C_{cw} = the average cost of servicing one warranty claim (manufacturer's cost).

Note, this model could be cast in terms of something other than time, e.g., miles. C_{fw} represents the cost of creating a warranty system for the product (toll-free telephone number, web site, training people, and so on) and C_{cw} is the recurring cost of each individual warranty claim (replacement, repair or a combination of replacement and repair as well as administrative costs).

As a simple example of the application of Equation (13.23), consider the manufacturer of a new television who is planning to provide a 12-month ordinary free replacement warranty. The lifetimes of the televisions are independent and exponentially distributed with $\lambda = 0.004$ failures per month. Assume that all failures result in replacements (no repairs and no denied claims). The manufacturer's recurring cost per television plus additional warranty claim resolution costs is \$112. Assume that C_{fw} = \$10,000 and that 500,000 televisions are sold. What warranty reserve should be put in place — that is, how much money should the manufacturer of the television budget to satisfy the promised warranty? In this case,

$$M(T_W) = \lambda T_W = (0.004)(12) = 0.048$$

$$C_{rw} = 10{,}000 + (500{,}000)(0.048)(112) = \$2{,}698{,}000$$

Since 500,000 televisions are sold, the customers should pay \$2,698,000/500,000 = \$5.40 per television for the warranty. Note, if we had used the unreliability instead of the renewal function,

$$F(T_W) = 1 - e^{-\lambda T_W} = 1 - \mathrm{e}^{-(0.004)(12)} = 0.04687$$

$$C_{rw} = 10{,}000 + (500{,}000)(0.04687)(112) = \$2{,}634{,}720$$

$M(T_w) > F(T_w)$ because a small number of televisions fail more than once during the warranty period, which results in a warranty reserve fund that is \$63,280 larger (\$0.13 more per television).

Not all warranty returns result in a repair or replacement. Failed products also include items damaged through use not covered by the warranty, items that are beyond their warranty period, and fraudulent claims. However, all the warranty claims, whether legitimate or not, cost money to resolve. A more complete model for the total warranty cost is given by

$$C_{rw} = C_{fw} + \alpha\left[M(T_W)C_{cw} + D(T_W)C_{dw}\right] \tag{13.24}$$

where

C_{dw} = the cost of resolving a denied warranty claim.
$D(T_W)$ = the expected number of denied warranty claims per product.

13.3.2 *Pro-Rata (Non-Renewing) Warranty Cost Model*

In the case of a pro-rata warranty, the customer receives a rebate that depends on the age of the item at the time of replacement (the warranty terminates when the rebate occurs). The pro-rated customer rebate at time t is given by

$$Rb(t) = \theta\left(1 - \frac{t}{T_W}\right) \tag{13.25}$$

where

θ = the product price (including warranty).
T_W = the warranty period duration.

Since the cost of servicing a warranty claim in this case is a function of t, we can't just substitute Rb for C_{cw} in Equation (13.23). The expected number of first-time warranty claims in the interval $(0,t]$ is $\alpha F(t)$;[10] if we assume a constant failure rate then this becomes $\alpha(1-e^{-\lambda t})$. Therefore, the expected number of warranty claims in an incremental time, dt, is $\alpha\lambda e^{-\lambda t}dt$.

[10] $\alpha F(t)$ is used instead of $\alpha M(t)$ because only the first-time warranty claims count in this case. There are no subsequent claims because the warranty makes a pro-rata payment at the first failure at which point the warranty ends.

Combining this result with Equation (13.23) and substituting Equation (13.25) for C_{cw}, we get

$$d(C_{rw} - C_{fw}) = Rb(t)\alpha\lambda e^{-\lambda t}\,dt = \theta\left(1 - \frac{t}{T_W}\right)\alpha\lambda e^{-\lambda t}\,dt \quad (13.26)$$

Integrating both sides of Equation (13.26) gives us the total warranty reserve cost during the warranty period T_w:

$$C_{rw} = C_{fw} + \int_0^{T_w}\theta\left(1 - \frac{t}{T_W}\right)\alpha\lambda e^{-\lambda t}\,dt = C_{fw} + \theta\alpha\left[1 - \frac{1 - e^{-\lambda T_w}}{\lambda T_w}\right] \quad (13.27)$$

Therefore, the effective warranty cost per product instance is

$$C_{pw} = \frac{C_{rw}}{\alpha} = \frac{C_{fw}}{\alpha} + \theta\left[1 - \frac{1 - e^{-\lambda T_w}}{\lambda T_w}\right] \quad (13.28)$$

Assuming that $\theta = \theta' + C_{pw}$, where θ' is the unit price without the warranty, then

$$\theta' = \theta\left(1 - \frac{C_{pw}}{\theta}\right) \quad (13.29)$$

Consider again the example at the end of Section 13.3.1, but assume that the manufacturer is going to provide a pro-rata warranty instead of an ordinary free replacement warranty. In this case what size warranty reserve fund should be put in place? Using Equation (13.28),

$$C_{pw} = \frac{\$10{,}000}{500{,}000} + \theta\left[1 - \frac{1 - e^{-0.004(12)}}{0.004(12)}\right]$$

In this case, $\theta' = \$200 = \theta - C_{pw}$, so $C_{pw} = \theta - \$200$.[11] Substituting for C_{pw} above we get

$$\theta = \frac{0.004(12)}{1-e^{-0.004(12)}}\left[\frac{\$10{,}000}{500{,}000} + \$200\right] = \$204.86$$

and therefore C_{pw} = \$4.86. The total warranty reserve fund in this case is C_{rw} = (500,000)(\$4.86) = \$2,430,000. Note the warranty cost per television when an ordinary free replacement warranty is used is 10% higher at \$5.40/unit (because it has to continue to provide a warranty to the end of the warranty period on the replaced televisions, whereas the pro-rata warranty pays off one time (on the first failure).

13.3.3 *Investment of the Warranty Reserve Fund*

The warranty reserve fund is usually collected when a product is sold and held until needed to fund warranty actions. During this holding period the warranty reserve fund can be invested to generate a return for the manufacturer. The investment return effectively reduces the amount of money that needs to be collected per product.

If the warranty reserve fund is invested, the average cost of servicing one warranty claim for an ordinary free replacement warranty (C_{cw}) is time-dependent. From Equation (13.23), the total recurring cost of warranty claims at time t is given by

$$X(t) = \alpha C_{cw}(t)M(t) \tag{13.30}$$

---

[11] Why isn't θ'=\$112? This is because \$112 is the cost to the manufacturer to replace a television; it is not the price of the television. The pro-rated payment to the customer is based on the price the customer paid, not on the cost to the manufacturer to make the television. The \$112 includes the manufacturing cost and other recurring costs associated with servicing the warranty claim (packing and shipping of the television to and from the manufacturer, administrative paper work, claim verification, etc.). The price of the television will likely be significantly larger than the cost of the television to the manufacturer due to marketing and sales costs, profit, and other factors.

The expectation value of the total recurring cost of warranty claims through the entire warranty period is

$$\mathrm{E}[X(t)] = \int_0^{T_w} \alpha C_{cw}(t) m(t) dt \tag{13.31}$$

If we assume that failures are exponentially distributed, $m(t) = \lambda$, then Equation (13.31) becomes

$$\mathrm{E}[X(t)] = \int_0^{T_w} \alpha C_{cw}(t) \lambda dt \tag{13.32}$$

Using the present value of $C_{cw}(t)$ from Equation (II.1), we obtain

$$\mathrm{E}[X(t)] = \int_0^{T_w} \alpha \frac{C_{cw}(0)}{(1+r)^t} \lambda dt \tag{13.33}$$

where r is the discount rate. Equation (13.33) implicitly assumes that all of the α products are sold (and their subsequent warranty periods start) at the same time. When $(1+r)^t \cong e^{rt}$ Equation (13.33) becomes[12]

$$\mathrm{E}[X(t)] = \int_0^{T_w} \alpha C_{cw}(0) e^{-rt} \lambda dt = \frac{\alpha C_{cw}(0)\lambda}{r}\left(1 - e^{-rT_w}\right) \tag{13.34}$$

For the example in Section 13.3.1, the total warranty cost if there is a 5% per year discount rate becomes

$$C_{rw} = 10{,}000 + \frac{(500{,}000)(112)(0.004)}{0.05}\left(1 - e^{-(0.05)(12)}\right) = \$2{,}031{,}323$$

This result is 25% less than the warranty reserve fund when there is no investment of the warranty reserve fund.

Similarly, for the pro-rata warranty, Equation (13.34) becomes

$$\mathrm{E}[X(t)] = \int_0^{T_W} \alpha\theta\left(1 - \frac{t}{T_W}\right) e^{-rt} \lambda e^{-\lambda t} dt = \frac{\alpha\theta\lambda}{\lambda + r}\left[1 - \frac{\left(1 - e^{-(\lambda+r)T_w}\right)}{(\lambda + r)T_w}\right] \tag{13.35}$$

---

12 Equation (II.1) assumes discrete compounding; alternately if continuous compounding is assumed (i.e., k compoundings per year in the limit as $k \to \infty$) then the *Present value* $= V_n e^{-rn_t}$.

which reduces to the second term in Equation (13.28) when $r = 0$ (and $\alpha = 1$). Investment of the warranty reserve fund can make a significant difference when either T_w is long and/or the discount rate, r, is high.

13.3.4 *Other Warranty Reserve Fund Estimation Models*

There are many warranty models based on various different assumptions about how a product is replaced or repaired to satisfy a warranty claim. For example, there are models for minimally repaired failed products, where minimal repair means that the unit is repaired to a state that is as good as other units fielded at the same time as the original unit. Lump-sum rebate models pay a fixed or lumped-sum rebate to customers for any failure occurring in the warranty period. Mixed warranty policies provide 100% of the purchase price as compensation upon failure during a specified period of time, followed by a pro-rata compensation to the end of the warranty period. These and other variations in warranty models are discussed in [Refs. 13.11, 13.12 and 13.13].

13.4 Two-Dimensional Warranties

The models discussed so far are one-dimensional warranties that are characterized by an interval called the warranty period, which is defined in terms of a single variable that defines the warranty's limits — for example, time, age, mileage, or some other usage measure. Two-dimensional warranties are characterized by a region in a two-dimensional plane with one axis representing time or age and the other representing usage. The shape of the resulting warranty coverage region defines the two-dimensional warrant policy.

Fundamentally, two-dimensional warranties differ from one-dimensional warranties in two ways [Ref. 13.12]. First, the warranty is defined by a two-dimensional region instead of a one-dimensional interval; and second, the failures are events that occur randomly in the two-dimensional region.

The left side of Figure 13.2 defines the warranty coverage region for a two-dimensional warranty in which the manufacturer agrees to repair or replace failed units up to a time or age, *W*, or up to a usage, *U*, whichever comes first. W is the warranty period and U is the usage limit in this case. Any failure that falls inside the region on the left side of Figure 13.2 is covered by this warranty. An example of this type of warranty is the warranty on a new car: "3 years or 36,000 miles, whichever comes first."

An alternative warranty policy is shown on the right side of Figure 13.2. In this policy the manufacturer agrees to repair or replace failed units up to a minimum time or age, W, and up to a minimum usage, U. Other two-dimensional warranty models have been proposed [Ref. 13.12].

To estimate the cost of supporting a two-dimensional warranty, we have to determine the expected number of warranty claims, E[$N(W,U)$], where $N(W,U)$ is the number of failures under the warranty defined by W and U.

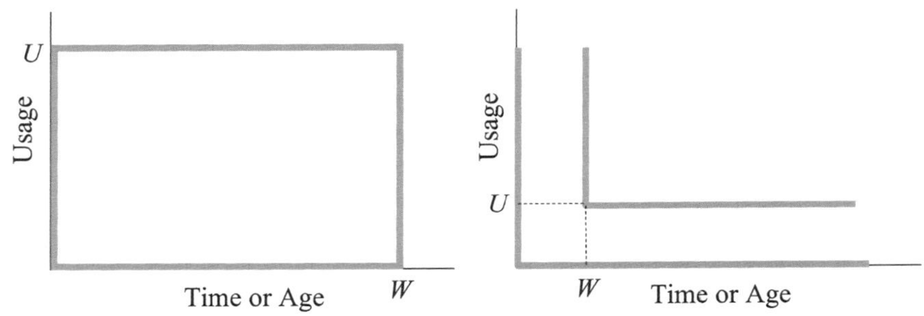

Fig. 13.2. Warranty regions defined for two different two-dimensional warranty policies.

Consider the construction shown in Figure 13.3. In Figure 13.3, u is the usage rate (usage per unit time) and $\gamma_1 = U/W$. When $u < \gamma_1$ the warranty ends at time W; when $u \geq \gamma_1$ the warranty ends at usage U, which corresponds to time U/u. The number of failures under the warranty defined by W and U conditioned on the usage rate u is given by

$$N(W,U|u) = \begin{cases} N(W|u), & \text{if } u < \gamma_1 \\ N\left(\dfrac{U}{u}|u\right), & \text{if } u \geq \gamma_1 \end{cases} \tag{13.36}$$

where $N(t)$ is the number of failures in the interval $(0,t]$ and $N(t|u)$ is the number of failures in the interval $(0,t]$ conditioned on u.

As in Equation (13.4),

$$\Pr(N(t\,|\,u) = n) = F_n(t\,|\,u) - F_{n+1}(t\,|\,u) \tag{13.37}$$

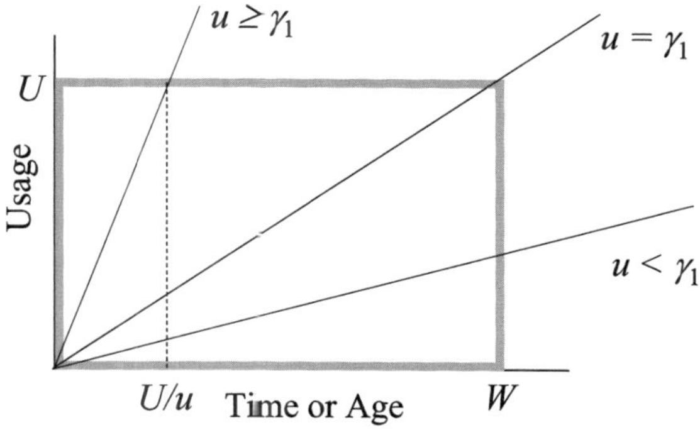

Fig. 13.3. Definition of usage rate (u).

Therefore,

$$\mathrm{E}\left[N(W,U|u)\right] = \begin{cases} M(W \mid u) & \text{if } u < \gamma_1 \\ M\left(\dfrac{U}{u} \mid u\right) & \text{if } u \geq \gamma_1 \end{cases} \tag{13.38}$$

where $M(t|u)$ is the conditioned renewal function associated with $F(t|u)$.

From Equation (13.38),

$$\mathrm{E}[N(W,U)] = \int_0^{\gamma_1} M(W \mid u)dG(u) + \int_{\gamma_1}^{\infty} M\left(\frac{U}{u} \mid u\right)dG(u) \tag{13.39}$$

where $G(u)$ is the cumulative distribution of the usage rates, u — that is, $G(u) = \Pr(\text{usage rate} \leq u)$.

The renewal functions in Equation (13.39) can be defined as

$$M(t \mid u) = \int_0^t \lambda(\tau \mid u)d\tau \tag{13.40}$$

The λ that appears in the Poisson Equation (Equation (12.5)) is called the intensity function of the process. In a "stationary" process, λ is a constant — for example, a constant failure rate. In a nonstationary process, λ varies with time. When failures are rectified via replacement (non-repairable), the intensity function has the general form [Ref. 13.12]

$$\lambda(t|u) = \theta_0 + \theta_1 u \tag{13.41}$$

Using Equation (13.41), Equation (13.39) becomes

$$\mathrm{E}[N(W,U)] = \int_0^{\gamma_1} (\theta_0 + \theta_1 u) W dG(u) + \int_{\gamma_1}^{\infty} (\theta_0 + \theta_1 u) \frac{U}{u} dG(u) \qquad (13.42)$$

$G(u)$ can take many different forms. One common form is a gamma function:

$$G(x,p) = \int_0^x \frac{y^{p-1} e^{-y}}{\Gamma(p)} dy \qquad (13.43)$$

Using Equation (13.43) in Equation (13.42) we get

$$\begin{aligned} \mathrm{E}[N(W,U)] &= \theta_0 W G(\gamma_1, p) + \theta_1 W p G(\gamma_1, p+1) \\ &\quad + \theta_1 U[1 - G(\gamma_1, p)] + \frac{\theta_0 U}{p-1}[1 - G(\gamma_1, p-1)] \end{aligned} \qquad (13.44)$$

As an example of the use of Equation (13.44), consider a non-repairable system for which the usage rate follows a gamma distribution with a mean of 3 (similar examples are presented in [Ref. 13.12]). In this case, $\theta_0 = 0.004$, $\theta_1 = 0.0006$, and several different values of W and U are assessed in Table 13.1.

Table 13.1. Expected Number of Failures in the Warranty Period.

		W (years)			
		0.5	1.0	2.0	3.0
U (10^4 miles)	0.9	0.001983	0.002490	0.002754	0.002833
	1.8	0.002570	0.003965	0.004979	0.005337
	2.7	0.002711	0.004747	0.006676	0.007469
	3.6	0.002742	0.005140	0.007931	0.009246

In Table 13.1 the units on W are years and on U are 10^4 miles; therefore the units on u are 10^4 miles/year. In Table 13.1, $W = 3.0$ and $U = 3.6$ corresponds to 3 years or 36,000 miles, whichever comes first. For this case, the expected number of failures is $(0.009246)(10^4) = 92.46$ warranty claims per 10,000 units. Moving from left to right and top to bottom in Table 13.1, the number of warranty claims increases because the region shown in Figure 13.3 increases.

The cost of the warranty claims in this example can be calculated as described in Section 13.3.1 using $E[N(W|U)]$ as the renewal function.

13.5 Warranty Service Costs — Real Systems

Analysis of real warranty problems usually reveals a range of warranty claims containing a mixture of different types of problems. Real warranty claims contain various types of failures, which are qualitatively presented in Figure 13.4. The failure rate curves shown in Figure 13.4 reflect the general trends in automotive electronics warranty observed at Delphi Electronics & Safety [Ref. 13.14], but do not represent any particular set of data. The typical automotive warranty mix includes:

A: initial performance or quality.
B: manufacturing or assembly-related failure.
C: design-related failure or unacceptable performance degradation due to applied stresses (environment, usage, shipping, etc.).
D: service damage, misdiagnosis, etc.
E: software-related problems.

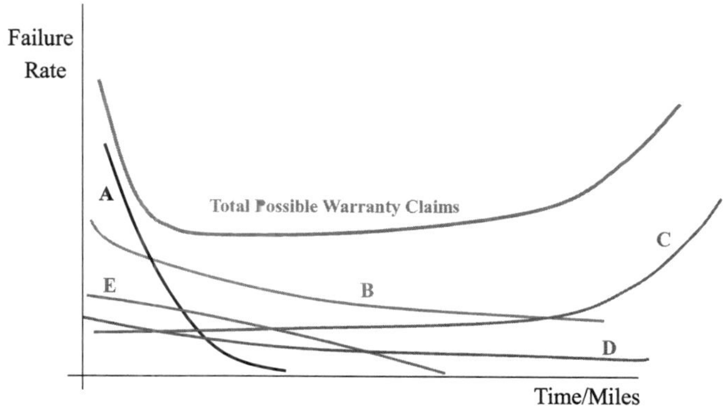

Fig. 13.4. Warranty claim content from Delphi [Ref. 13.14].

The sum of these failures makes up the total warranty claims (top curve in Figure 13.4). Based on the collected data for automotive electronics presented in Figure 13.5, the total warranty curve approximately follows the first two sections of the bathtub curve (Figure 11.2).

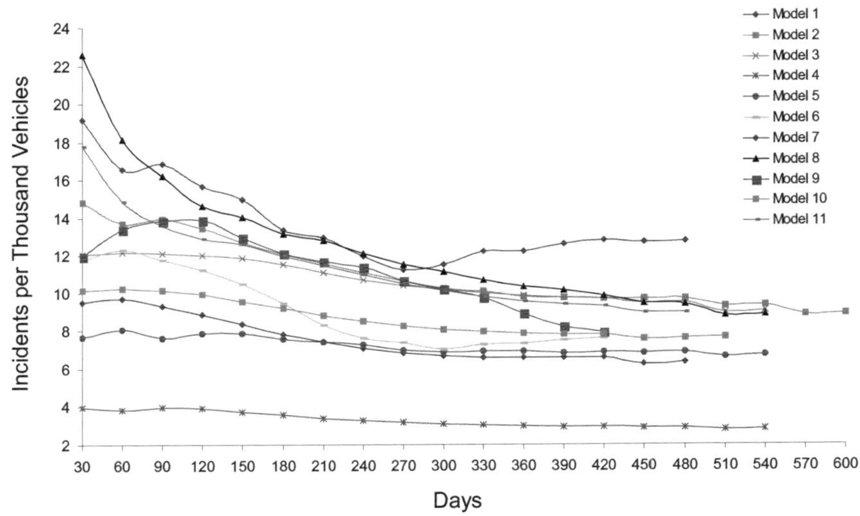

Fig. 13.5. Failure rates for selected passenger compartment mounted electronic products (models) from Delphi [Ref. 13.14].

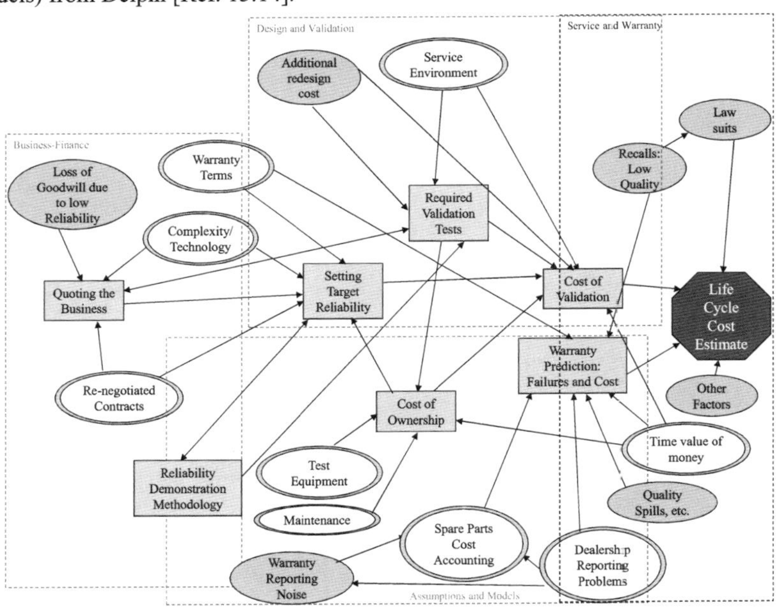

Fig. 13.6. Complete life-cycle cost influence diagram [Ref. 13.14]. Rectangles are decision nodes where decisions must be made. Filled ovals are chance nodes that represent a probabilistic variable. Unfilled ovals are deterministic nodes that are determined from other nodes or non-deterministic variables. Arrows denote the influence among modes and the direction of the decision process flow.

The influence diagram in Figure 13.6 shows all the factors affecting this life-cycle cost decision-making process. Those factors include the variety of inputs affecting the process from the new business quoting event through design, validation, and warranty. All the influence factors fall under the following major categories: (1) business-finance, (2) design and validation, (3) service and warranty, and (4) assumptions and models. The first three represent the flow of product development from business contract to design, validation, and consequent repair/service. The fourth group (assumptions and models) influences categories (1) through (3), since the modeling process incorporates a number of engineering assumptions, utilized models, and equations. Each of the four categories has at least one major decision-making block and a variety of probabilistic and deterministic node inputs. All of these inputs will directly and indirectly affect the outcome value node, where the final dependability-related portion of the life-cycle cost is calculated and minimized.

References

13.1 Arnum, E. (2007). *Warranty Week*, May.

13.2 Murthy, D. N. P. and Djamaludin, I. (2002). New product warranty: A literature review, *International Journal of Production Economics*, 79(3), pp. 231-260.

13.3 Loomba, A. P. S. (1995). Chapter 2: Historical perspective on Warranty, *Product Warranty Handbook*, W. R. Blischke and D. N. P. Murthy, Editors, (Marcel Dekker, New York).

13.4 c't (2007). Xbox 360: Jede dritte stirbt den Hitzetod, *c't*, 16, p. 20.

13.5 Thorsen, T. (2009). Xbox 360 failure rate = 54.2%?, *GameSpot*, August 18. http://www.gamespot.com/articles/xbox-360-failure-rate-542/1100-6215590/. Accessed April 25, 2016.

13.6 Sanders, T. (2007). Microsoft facing US$1.15bn Xbox 360 repair bill, *CRN*, July 9. http://www.crn.com.au/News/85600,microsoft-facing-us115bn-xbox-360-repair-bill.aspx. Accessed April 25, 2016.

13.7 Open Letter from Peter Moore, https://xbl10kclubnews.wordpress.com/2007/07/07/open-letter-from-peter-moore/. Accessed April 25, 2016.

13.8 Bass, D. (2007). Microsoft to incur Xbox cost of up to $1.15 billion, Bloomberg.com, July 5. http://www.bloomberg.com/apps/news?pid=20601087&sid=aOrvYZ2gPwZk&refer=home. Accessed June 2013.

13.9 Pham, H. (2006). Chapter 7 Promotional warranty policies: Analysis and perspectives, *Springer Handbook of Engineering Statistics* (Springer Verlag, London).

13.10 Smith, W. L. (1954). Asymptotic renewal theorems, *Proceedings of the Royal Society*, 64, pp. 9-48.
13.11 Elsayed, E. A. (1996). *Reliability Engineering* (AddisonWesley Longman, Inc., Reading, MA).
13.12 Blischke, W. R. and Murthy, D. N. P. (1994). *Warranty Cost Analysis* (Marcel Dekker, New York).
13.13 Thomas, M. U. (2006). Reliability and Warranties, Methods for Product Development and Quality Improvement (CRC Press, Boca Raton, FL).
13.14 Kleyner, A. V. (2005). Determining Optimal Reliability Targets Through Analysis of Product Validation Cost and Field Warranty Data, Ph.D. Dissertation, University of Maryland.

Problems

13.1 If 20 legitimate warranty claims are made in a 12-month period, there are 5000 fielded units, and the product is believed to have a constant failure rate, what is the failure rate? Express your answer to 6 significant figures.

13.2 In Problem 13.1, if a Weibull distribution is believed to represent the reliability, what are the values of β and η? Hint: make a graph of valid β versus η values.

13.3 The company in Problem 11.8 created a $2 million warranty reserve fund for the GPS chip. Assuming an ordinary free replacement warranty, if 1 million GPS chips are sold, the fixed cost of warranty is $100,000, and the average cost per warranty claim is $13, what should the warranty period be?

13.4 For a product with a failure time probability density given by $f(t) = a\eta e^{-at} + b(1-\eta)e^{-bt}$ for $t \geq 0$ find $M(t)$. Assume that $a = 4$ failures/year, $b = 3$ failures per year, $C_{cw} = \$80$, $C_{fw} = 0$, and $\eta = 0.3$. If the warranty period is 3 years, how much money should be set aside for each product instance? Assume an ordinary free replacement warranty.

13.5 Derive Equation (13.19).

13.6 The manufacturer of a part quotes an *MTBF* of 32 months. The cost of repairing the part is estimated to be $22.50/repair. Assuming a constant failure rate and an ordinary free replacement warranty, what is the length of the warranty period and average warranty cost per part that will ensure that the reliability during the warranty period is at least 0.96? Assume that the fixed cost of providing the warranty is negligible.

13.7 An electronic instrument is sold for $2500 with a 1-year ordinary free replacement warranty (however, the instruments are never replaced; they are always repaired). The *MTBF* is 2.5 years; the average cost of a warranty claim is $40. Customers are given the option of extending the warranty an additional year for $20. Assuming that the failures are exponentially distributed, if it costs $50/repair out of warranty does it make sense for the customer to spend $20 for the extended warranty? Assume that the fixed cost of providing the warranty is negligible.

13.8 A manufacturer currently produces a product that has a *MTBF* of 2 years. The product has an 18-month ordinary free replacement warranty. The warranty claims cost an average of $45 per claim to resolve. Assuming the failure rate is constant, if the manufacturer wishes to reduce its warranty costs by 25%, how much does the reliability of the product have to improve? Assume that the fixed cost of providing the warranty is negligible.

13.9 The manufacturer of an electronic instrument offers a pro-rata warranty that gives customers the option of obtaining a new instrument at a discounted price if their original instrument fails. The period of the pro-rata warranty is 20 years. The purchase price of the instrument has changed over the last 20 years according the schedule below (due to inflation). The price of a new instrument today is $2500. What would be a fair (linear) discount for each of the following instruments?

Age (years)	Original Retail Price	Discount Off New Instrument
0	$2500	$2500
5	$2375	?
10	$2250	?
15	$2125	?
19	$2025	?
20	$2010	$0

13.10 In the limit at r approaches zero, show that Equation (13.34) approaches the form used in Section 13.3.1.

13.11 Rework the example in Section 13.3.2 with a 5% discount rate.

13.12 Derive Equation (13.44) using Equations (13.42) and (13.43).

13.13 Customers value a product's warranty relative to the perceived quality of the product, e.g., if the customer thinks that the quality of an item is high; they will not require as much warranty. Alternatively, for products of lesser or unknown quality, the customer will require more warranty coverage (e.g., a longer warranty period). Your company makes a non-repairable product that costs you $1000 to replace if it fails during the warranty period. The product fails at a rate of 0.5/year (assume this is a constant failure rate). The cost of marketing the product varies depending on the length of the warranty offered according to the following relation:

$$B(w) = (b_0 - b_1 w)^2$$

where w is the warranty length in years. Assume that $b_0 = 50$, $b_1 = 10$, the fixed cost of providing the warranty (per product) = $3, and an unlimited free replacement warranty is offered. What is the optimum warranty period (w) from the manufacturer's perspective? Optimum means minimum total cost.

13.14 Prove or demonstrate that $\Pr(x \leq k) = 0.5$ in Equation (12.7) predicts the same number of spares as a renewal function for the constant failure rate assumption.

Chapter 14

Burn-In Cost Modeling

Burn-in is the process by which units are stressed prior to being placed in service (and often, prior to being completely assembled). The goal of burn-in is to identify particular units that would fail during the initial, high-failure rate infant mortality phase of the bathtub curve shown in Figure 11.2. The goal is to make the burn-in period sufficiently long (or stressful) that the unit can be assumed to be mostly free of further early failure risks after the burn-in.

A precondition for a successful burn-in is a bathtub-curve failure rate, meaning that there is a non-negligible number of early failures (infant mortality), after which failure rate decreases. Stressing all units for a specified burn-in time causes the units with the highest failure rate to fail first so they can be taken out of the population. The units that survive the burn-in will have a lower failure rate thereafter.

The strategy behind burn-in (see Figure 14.1) is that early in-use system failures can be avoided at the expense of performing the burn-in and a reduction in the number of units shipped to customers.[1]

---

[1] The view of burn-in has changed significantly in the past twenty years. Twenty years ago, burn-in was an important process in the electronics industry due to high infant mortality rates. Back then, you had to make a case NOT to include a burn-in in your process. These days the opposite is true — in many industries the case must be made for burn-in due to the cost implications and reasonably low infant mortality rates.

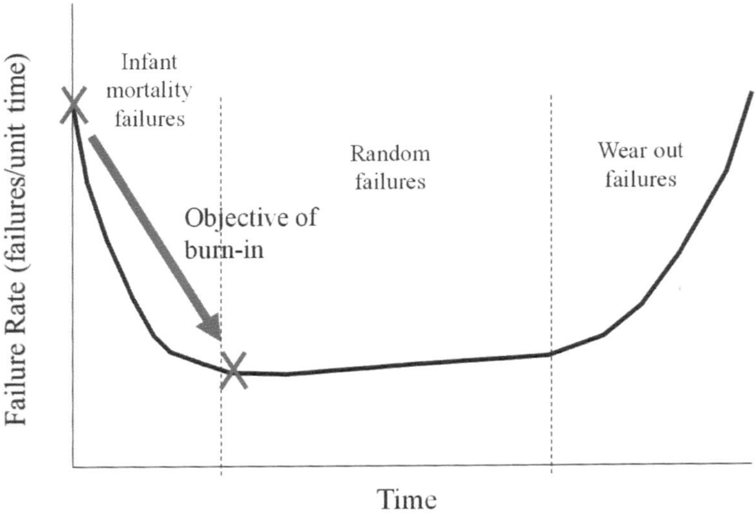

Fig. 14.1. The goal of burn-in is to reach the random failures portion of the bathtub curve before sending the product to the customers.

The Cost Tradeoffs Associated with Burn-In

Burn-in is not free and neither are its benefits clear. Evaluating whether burn-in makes sense requires an application-specific cost analysis (discussed in the next section). The cost of performing burn-in is a combination of the following factors:

- the cost of the development of the burn-in tests.
- the cost of performing the burn-in (fixed and variable).
- the cost of units that are failed in burn-in.
- the opportunity cost associated with units failed in burn-in.
- the value of the life removed from units that pass burn-in testing.

The potential value of burn-in is a combination of a

- reduction in warranty claims (or field repairs) during field use.
- improved availability of the product.
- customer satisfaction improvement (market share retention or growth).

The next section constructs a model that incorporates many of the factors listed above.

14.1 Burn-In Cost Model

For burn-in modeling, we will assume all units are non-repairable (see Section 14.4 for a discussion of repairable units). Even if the units are technically repairable, in this section we are assuming that if they fail during burn-in, the units will not be repaired or replaced; they are discarded. The assumption is that every manufactured unit is burned-in (burn-in is not a test performed on a "sample" from the manufactured units — it is part of the manufacturing process for all units). Everything in this chapter is presented in terms of time; however, an alternative unit of environmental stress could be used, e.g., thermal cycles.

14.1.1 *Cost of Performing the Burn-In*

Equivalent burn-in time (t_{bd}), sometimes called time under operating conditions, can be measured in calendar time or operational time and is given by

$$t_{bd} = A_F t_s \tag{14.1}$$

where

A_F = the acceleration factor associated with the burn-in test.

t_s = the actual time under stress (burn-in test time).

The cost of performing burn-in (C_{BI}) on all units can be expressed as

$$C_{BI} = C_{BD} + C_{BNR} + n_u C_B + C_{LR} \tag{14.2}$$

where

C_{BD} = the fixed cost of burn-in development.

C_{BNR} = the non-recurring burn-in cost (includes the cost of qualifying, calibrating and maintaining the burn-in equipment and facilities, and training people).

n_u = the number of units being burned-in.

C_B = the recurring burn-in cost per unit (energy costs, etc.).

C_{LR} = the cost associated with life removed by the burn-in from non-failed units.

The recurring burn-in cost per unit (C_B) is given by

$$C_B = C_{TB}(t_{bd}) + F(t_{bd})[C_P + C_O] \tag{14.3}$$

where

$C_{TB}(t_{bd})$ = the cost of burning-in one unit for the equivalent of t_{bd}.
$F(t_{bd})$ = the unreliability in the interval $(0, t_{bd}]$.
C_P = the unit cost.
C_O = the opportunity cost associated with the unit (profit that could have been made by selling the unit that failed at burn-in) assuming all manufactured units could be sold.

The second term on the right side of Equation (14.3) is the cost (per unit) of units that fail the burn-in. Note that the unreliability is used instead of a renewal function because units that fail burn-in are not repaired and not replaced, so there is no replaced or repaired version of the unit to fail at a later time.

The cost associated with the life removed by the burn-in from non-failed units, C_{LR}, is 0 if the warranty period, $t_{bd}+T_W$, does not reach wear-out for the units, where T_W is the warranty period as shown in Figure 14.2.

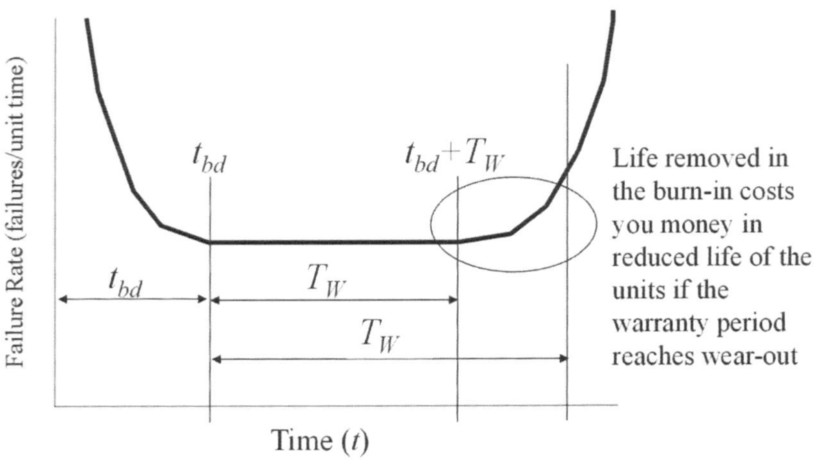

Fig. 14.2. Life removed by burn-in.

The model may be equipment-capacity-limited — that is, the facilities and equipment (C_{BNR}) cannot support burning-in an infinite number of units concurrently and can probably only be expanded in discontinuous steps (i.e., the capacity of the equipment only increases in steps). The burn-in facility/equipment has both a depreciation life over which its investment cost can be spread, and a facility life after which it must be replaced.

There may be cost factors associated with the length (in elapsed time) of the burn-in. For example, burn-in could impact delivery/program schedules ("schedule slip" cost) that have not been accounted for in this model. There will also be escapes from the burn-in that are not accounted for here, i.e., some fraction of infant mortality units are not detected.

14.1.2 *The Value of Burn-In*

The value (per unit that survives the burn in) of performing a burn-in is given by

$$V_B = [M(T_W) - (M(t_{bd} + T_W) - M(t_{bd}))]C_{cw} + C_{CS} \tag{14.4}$$

where

$M(t)$ = the renewal function, mean number of renewal events (warranty claims) that occur in the interval $(0,t]$ (see Section 13.2).

C_{cw} = the average cost of servicing one warranty claim on the unit.

C_{CS} = the customer satisfaction value (allocated per unit).

The term in brackets in Equation (14.4) is the decrease in the number of renewals (warranty claims) assuming an ordinary non-renewing free replacement warranty. A renewal function is used here (instead of the unreliability) because failed units are replaced and can fail again before the end of the warranty is reached.

Equation (14.4) represents the value of units that will be put into the field. If a unit is removed due to another defect that is not associated with burn-in, then the value in Equation (14.4) is not realized for that unit (this also impacts the number of units appearing in Equation (14.5)). For a constant failure rate in all periods of the product's life (including the infant mortality region), $M(t) = \lambda t$ and the term multiplying C_{cw} goes to zero —

that is, for a constant failure rate there are the same number of renewals in any interval of length T_W in the part's life.

The return on investment (see Chapter 17) associated with the burn-in is given by

$$ROI = \frac{Return - Investment}{Investment} = \frac{n_u\left[1\text{-}F\left(t_{bd}\right)\right]V_B - C_{BI}}{C_{BI}} \qquad (14.5)$$

Note that C_{BI} includes the cost of units that do not survive burn-in. The quantity multiplying V_B is the number of units surviving burn-in assuming that n_u units start burn-in. $ROI = 0$ is break-even ($ROI < 0$ means there is no economic return and $ROI > 0$ means that there is an economic return).

14.2 Example Burn-In Cost Analysis

As an example, consider a product characterized in Figure 14.3, with a Weibull failure distribution during the first 20 operational hours: $\beta = 0.95$, η = 3,200,000 operational hours, $\gamma = 0$; and a constant failure rate: λ = 0.000986 failures/operational year assumed after 20 operational hours. We are assuming for simplicity that there is only one failure mechanism, that our burn-in conditions accelerate that mechanism, and that the units are non-repairable (units that fail during burn-in are discarded and have no salvage value). The remaining inputs are given in Table 14.1.

Using the values in Table 14.1 and Figure 14.3,

$$C_O = (0.25)C_P = \$75.$$
$$A_F = t_{bd} / t_s = 20/1 = 20.$$
$$t_{bd} = 20/365/5 = 0.010959 \text{ operational years.}$$
$$C_{TB} = (C_{OBF})(t_s)/(\text{burn-in facility capacity}).$$
C_{OBF} = the operational cost of the burn-in facility per hour (varied in the results that follow).

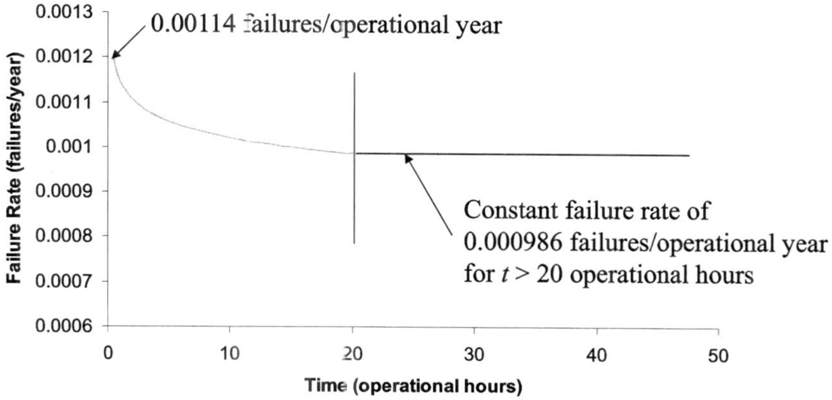

Fig. 14.3. Failure rate example.

Table 14.1. Example Input Data.

Quantity	Symbol	Value
Burn-in development cost	C_{BD}	$100,000
Non-recurring equipment and facilities cost	C_{BNR}	$250,000
Number of units that start the burn-in process	n_u	1,700,000
Cost per unit	C_P	$300
Profit per unit (fraction of C_P)		0.25
Time under stress	t_s	1 hour
Warranty period	T_W	2 operational years
Burn-in facility capacity		300 units
Life removed cost	C_{LR}	$0
Customer satisfaction cost	C_{CS}	$0 per unit
Warranty fixed cost	C_{fw}	$100,000
Average replacement/repair cost per warranty claim	C_{cw}	$400
Operational hours per day		5

In this example, different portions of the product's life are characterized by different renewal functions. In order to determine the value using Equation (14.4), we need to determine $M(t_{bd} + T_W)$. Using the diagram in Figure 14.4, we get

$$M(t_{bd} + T_W) = M_1(t_{bd}) + [M_2(t_{bd} + T_W) - M_2(t_{bd})] \qquad (14.6)$$

For this example, $M_1(t)$ is given by Equations (13.19) and (13.22), and $M_2(t) = \lambda t$.

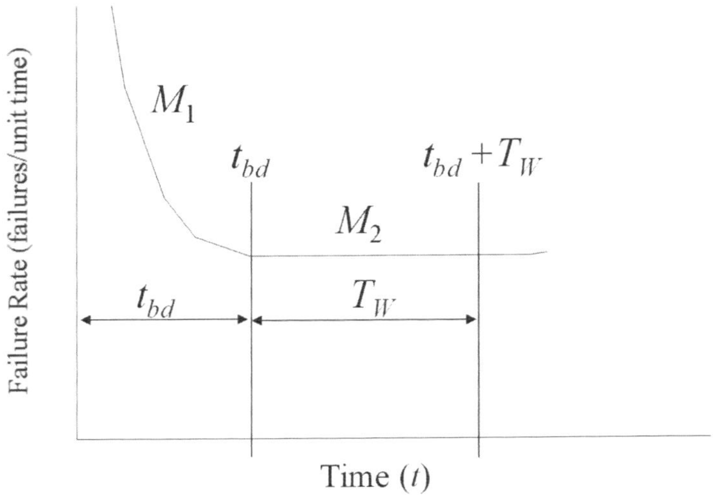

Fig. 14.4. Renewal functions for different periods of time.

The ROI computed using Equation (14.5) is shown in Figure 14.5 as a function of the operational cost of the burn-in facility. Obviously, as the cost of operating the facility goes down, the ROI associated with the burn-in process increases.

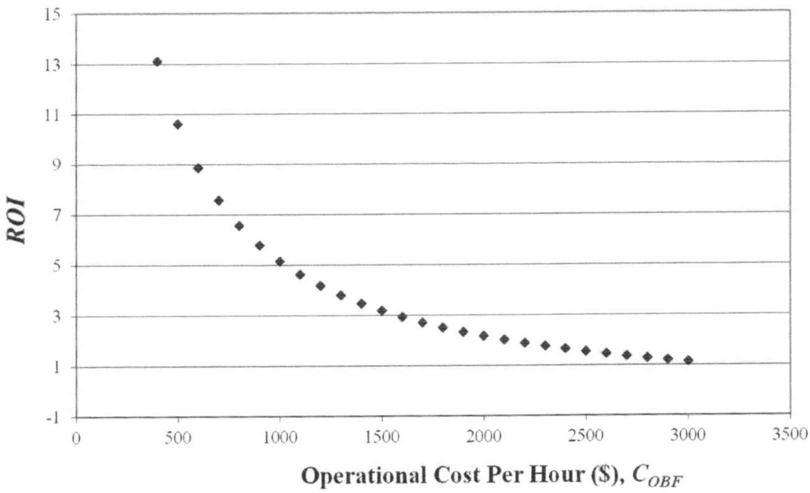

Fig. 14.5. Return on investment (ROI) as a function of operational cost of the burn-in facility.

14.3 Effective Manufacturing Cost of Units That Survive Burn-In

In this section we present an alternative model for the manufacturing cost of units that survive burn-in. This model was developed by Nguyen and Murthy [Ref. 14.1]. The model makes one key simplifying assumption: $t_{bd} = t_s$ (i.e., $A_F = 1$, there is no acceleration of the stress conditions in the burn-in). Under this assumption the burn-in cost per unit is given by

$$C_{BI/unit}(t) = \begin{cases} C_1 + C_{Bt}t & \text{for } t < t_{bd} \\ C_1 + C_{Bt}t_{bd} & \text{for } t \geq t_{bd} \end{cases} \tag{14.7}$$

where C_1 is a combination of the fixed and non-recurring costs per unit and C_{Bt} is the recurring burn-in cost per unit per time. The first item in Equation (14.7) is for units that fail during burn-in and the second is for units that survive burn-in. From Equation (14.7), the expected burn-in cost per unit is given by

$$\mathrm{E}\left[C_{BI/unit}(t)\right] = \int_0^{t_{bd}} \left(C_1 + C_{Bt}t\right) f(t)dt + \int_{t_{bd}}^{\infty} \left(C_1 + C_{Bt}t_{bd}\right) f(t)dt \tag{14.8}$$

where $f(t)$ is the failure time distribution (PDF). Equation (14.8) reduces to

$$\mathrm{E}\left[C_{BI/unit}(t)\right] = C_1 + C_{Bt}\int_0^{t_{bd}} \left(1 - F(t)\right)dt \tag{14.9}$$

where $F(t)$ is given by Equation (11.5).

The burn-in process is part of the manufacturing process, so the final effective manufacturing cost of units that survive the burn-in is given by

$$C_{manuf+burn-in} = \frac{C_{manuf} + C_1 + C_{Bt}\int_0^{t_{bd}} \left(1 - F(t)\right)dt}{1 - F(t_{bd})} \tag{14.10}$$

where C_{manuf} is given in Equation (2.5). In Equation (14.10), $1\text{-}F(t_{bd})$ is the probability of survival through the burn-in process (to $t = t_{bd}$), which means that Equation (14.10) assumes that units that do not survive the burn-in process are discarded and have no salvage value.

14.4 Burn-In for Repairable Units

All the previous formulations in this chapter assume that we are burning-in non-repairable units. If we are burning-in repairable units, then the following modifications must be made:

(1) Replace *F*( ) with *M*( ), the renewal function, in the calculation of the burn-in costs (this assumes that parts that fail are replaced and the burn-in continues).
(2) Diagnosis costs must be included — when a repairable unit fails during burn-in or in the field, you must determine what portion of the unit failed (see Section 8.1).
(3) Some failures result in a replacement of the unit (the unit is scrapped) and some result in a repair of the unit.
(4) Part-level burn-in (stress screening) may be used in addition to unit-level burn-in.

14.5 Discussion

Different failure mechanisms have different reliability distributions, failure rates and renewal functions. Burn-in may accelerate more than one mechanism and not others. It does little good to apply a burn-in that accelerates a non-relevant failure mechanism.

Investment costs in developing a burn-in process or in burn-in equipment may be made today, but the value in the form of reduced warranty costs happens in the future. Depending on the size of the effective discount rate and the length of the warranty period, it may be necessary to include cost of money in the calculations.

There may be a disconnect between what the customer perceives as defects and what the manufacturer thinks is a defect; not all the defects that the burn-in removes will necessarily result in warranty claims.

References

14.1 Nguyen, D. G. and Murthy, D. N. P. (1982). Optimal burn-in time to minimize cost for products sold under warranty, *IIE Transactions*, 14(3), pp. 167-174.

Bibliography

The following references include cost models for burn-in of electronic equipment:

Yan, L. and English, J. R. (1997). Economic cost modeling of environmental stress-screening and burn-in, *IEEE Transactions on Reliability*, 46(2), pp. 275-282.

Chan, H. A. (1994). A formulation to optimize stress testing, *Proceedings of the Electronic Components and Technology Conference*, pp. 1020-1027.

Alani, A., Dislis, C. and Jalowiecki, I. (1996). Burn-in economics model for multi-chip modules, *Electronics Letters*, 32(25), pp. 2349-2351.

Mok, Y. L. and Xie, M. (1996). Planning and optimizing environmental stress screening, *Proceedings of the Reliability and Maintainability Symposium (RAMS)*, pp. 191-198.

Sheu, S-H. and Chien, Y-H. (2004). Minimizing cost-functions related to both burn-in and field-operation under a generalized model, *IEEE Transactions on Reliability*, 53(3), pp. 435-439.

Problems

14.1 Why is $F(\,)$ used in Equation (14.5) instead of $M(\,)$?

14.2 In the example provided in Section 14.2, if C_{OBF} = \$2500/hour, what value of burn-in facility capacity causes the ROI to be 0?

14.3 Derive Equation (14.9).

14.4 Explain why Equations (14.7) through (14.10) assume that $A_F = 1$.

Chapter 15

Availability

Availability is the ability of a service or a system to be functional when it is requested for use or operation. The concept of availability accounts for both the frequency of failure (reliability) and the ability to restore the service or system to operation after a failure (maintainability). The maintenance ramifications generally translate into how quickly the system can be repaired upon failure and are usually driven by logistics management. Availability only applies to systems that are either externally maintained or self-maintained.

Availability has been a critical design parameter for the aerospace and defense communities for many years, but more recently it is beginning to be recognized, quantified, and studied for other types of systems. Many real world systems are significantly impacted by availability. A failure — the decrease of availability — of an ATM machine causes inconvenience to customers; poor availability of wind farms can make them non-viable; the unavailability of a point-of-sale system to retail outlets can generate a huge financial loss; the failure of a medical device or of hospital equipment can result in loss of life. For web-based business services, the availability of a web site and the data to support it may depend on the reliability and maintainability of servers. In these example systems, insuring the availability of the system becomes the primary interest and the owners of the systems are often willing to pay a premium (purchase price and/or support) for higher availability.

15.1 Time-Based Availability Measures

Reliability is the probability that an item will not fail; maintainability is the probability that a failed item can be successfully restored to operation.

Availability is the probability that an item will be able to function (i.e., not be failed or undergoing repair) when called upon to do so over a specific period of time under stated conditions. Measuring availability provides information about how efficiently a system is supported.

In general, availability is computed as the ratio of the accumulated uptime and the sum of the accumulated uptime and downtime:

$$A = \frac{uptime}{uptime + downtime} \tag{15.1}$$

where *uptime* is the total accumulated operational time during which the system is up and running and able to perform the tasks that are expected from it; *downtime* is the period for which the system is down and not operating when requested due to repair, replacement, waiting for spares, or any other logistics or administrative delays. The sum of the accumulated uptimes and downtimes represents the total operation time for the system. Equation (15.1) implicitly assumes that uptime is equal to operational time, whereas in reality, not all of the uptime is actually operational time; some of it corresponds to time the system spends in standby mode waiting to operate.

Many different types of availability can be measured. Availability measures are generally classified by either the time interval of interest or the collection of events that cause the downtime [Ref. 15.1].

15.1.1 *Time-Interval-Based Availability Measures*

If the primary concern is the time interval of interest, then we consider instantaneous, average, and steady-state availability.

Instantaneous (also called point or pointwise) availability is the probability that an item will be able to perform its required function at the instant it is required. Instantaneous availability is given by:

$$A(t) = R(t) + \int_0^t R(t-\tau)m(\tau)d\tau \tag{15.2}$$

where

$R(t)$ = the reliability at time t, (the probability that the item functioned without failure from time 0 to t).

$R(t\text{-}\tau)$ = the probability that the item functioned without failure since the last repair time τ.

$m(\tau)$ = the renewal density function.

Equation (15.2) represents a sum of probabilities. The first term is the probability of no failure occurring from time 0 to t, the second term is the probability of no failure since the last repair time (τ).

A renewal function, $M(t)$, (see Chapter 13) is the expected number of failures in a population. The renewal density function is the mean number of renewals expected in a narrow interval of time near t: $m(t) = dM(t)/dt$. In general, the renewal density function in Equation (13.14) can be written as

$$\hat{m}(s) = \frac{\hat{w}(s)\hat{g}(s)}{1 - \hat{w}(s)\hat{g}(s)} \tag{15.3}$$

where $\hat{m}(s)$ is the Laplace transform of $m(t)$, and $\hat{w}(s)$ and $\hat{g}(s)$ are the Laplace transforms of the time-to-failure distribution and time-to-repair distributions, respectively.[1] Using Equation (15.3) in Equation (15.2), the Laplace transform of the availability becomes

$$\hat{A}(s) = \frac{1 - \hat{w}(s)}{s\left(1 - \hat{w}(s)\hat{g}(s)\right)} \tag{15.4}$$

Instantaneous availability is a useful measure for systems that are idle for periods of time and then are required to perform at a random time, such as a defibrillation unit in a hospital or a torpedo in a submarine.

---

[1] $f(t)$ is the convolution of $w(t)$ and $g(t)$, $f(t) = \int_0^t w(t-\tau)g(\tau)d\tau$, and therefore $\hat{f}(s) = \hat{w}(s)\hat{g}(s)$. $f(t)$ is the time derivative of the probability of failure <u>or</u> repair: $f(t) = w(t)$ only if the time to repair is zero.

The average (also called mean, average uptime, or interval) availability is given by

$$\overline{A(t)} = \frac{1}{t}\int_{0}^{t} A(\tau)d\tau \tag{15.5}$$

The average availability in Equation (15.5) is the proportion of time in the interval (0,*t*] that the system is available. Average availability is used for systems whose usage is defined by a duty cycle, like a commercial airliner or construction equipment at a job site.

The steady-state (or limiting) availability is given by

$$A(\infty) = \lim_{t\to\infty} A(t) \tag{15.6}$$

where $A(t)$ is the instantaneous availability. Equation (15.6) is only valid if the limit exists. Steady-state availability is often applied to systems that operate continuously — for example, an air traffic control radar system or a computer server.

15.1.2 *Downtime-Based Availability Measures*

Availability measures that focus on the various mechanisms that result in downtime include inherent availability, achieved availability, and operational availability. The relevant time measures are summarized in Table 15.1. Availability measures in this category are differentiated based on what activities are included in the downtime and have the general form shown in Equation (15.1). All of these availability measures assume a steady-state condition.

Inherent availability is defined as

$$A_i = \frac{MTBF}{MTBF + MTTR} \tag{15.7}$$

where *MTBF* is the mean time between failures and *MTTR* is the mean time to repair (or mean corrective maintenance time). Inherent availability only includes downtime due to corrective maintenance actions (excluding preventative maintenance, logistics, and administrative downtimes). Inherent availability is used to model an ideal support environment.

Table 15.1. Summary of Relevant Maintenance Time Measures.

Symbol	Name	Content
MTBF	Mean time between failures	Mean time between corrective maintenance activities.
MTTR ($\overline{M}ct$)	Mean time to repair (Mean corrective maintenance time)	Corrective maintenance (as a result of failure): failure detection, diagnosis (fault isolation), disassembly, repair, reassembly, verification, etc.
MTBM	Mean time between maintenance	Mean time between all (corrective and preventative) maintenance activities.
MTPM	Mean time to perform preventative maintenance	
$\overline{M}$	Mean active maintenance time	Corrective and preventative maintenance (weighted sum of $\overline{M}ct$ and $\overline{M}pt$).
MDT	Mean maintenance downtime	$\overline{M}$ with *LDT* and *ADT* included
$\overline{M}pt$	Mean preventative maintenance time	Preventative maintenance: scheduled maintenance, periodic inspection, servicing, calibration, overhaul, etc. Can overlap with $\overline{M}ct$ and operational time.
LDT	Logistics delay time	Time spent waiting for spares, test equipment, and/or facilities; transportation time.
ADT	Administrative delay time	Time spent waiting for personnel assignments, prioritization, organizational delays, etc.
MSD	Mean supply delay	*LDT* + *ADT*

Achieved availability is given by

$$A_a = \frac{MTBM}{MTBM + \overline{M}} \tag{15.8}$$

where *MTBM* is the mean time between maintenance activities and $\overline{M}$ is the mean active maintenance time. Sometimes inherent and achieved availability are referred to as intrinsic availability. Achieved availability is also used to model an ideal support environment.

Operational availability is the availability that the customer actually experiences in a real operational environment:

$$A_o = \frac{MTBM}{MTBM + MDT} \tag{15.9}$$

The denominator of Equation (15.9) is the overall operational time period. Operational availability is used to model an actual (non-ideal) support environment.

A common availability metric used in inventory analysis is supply availability, which is defined as

$$A_s = \frac{MTBM}{MTBM + MSD} \tag{15.10}$$

The denominator of Equation (15.10) specifically excludes the time associated with diagnosing or making a repair — that is, it is independent of the maintenance policy and only depends on the sparing policy for stocking spares [Ref. 15.2].

As an example of availability estimation using downtime-based availability measures, consider an electronic system with the following characteristics ("op hours" = operational hours):

- Operational cycle = 2000 op hours/year
- Support life = 5 years
- Failures that require corrective maintenance = 2/year
- Repair time per failure = 40 op hours
- Preventative maintenance activities = 1/year
- Preventative maintenance time per preventative maintenance action = 8 op hours
- Average wait time for repair materials for corrective maintenance = 10 op hours

From the given information, $MTTR = 40$ op hours, $MTPM = 8$ op hours, $LDT = 10$ op hours, and the following quantities can be calculated:

$$\text{Total number of maintenance actions} = (2)(5)+(1)(5) = 15 \tag{15.11a}$$

$$\overline{M} = \frac{(40)(2)(5) + (8)(1)(5)}{15} = 29.333 \text{ op hours} \tag{15.11b}$$

$$MDT = \frac{(40+10)(2)(5) + (8)(1)(5)}{15} = 36 \text{ op hours} \tag{15.11c}$$

$$MTBF = \frac{(5)(2000)}{(2)(5)} = 1000 \text{ op hours} \tag{15.11d}$$

$$\text{Total operational cycle} = (5)(2000) = 10{,}000 \text{ op hours} \tag{15.11e}$$

$$\text{Total downtime} = (15)(36) = 540 \text{ op hours} \tag{15.11f}$$

$$\text{Total uptime} = 10{,}000 - 540 = 9460 \text{ op hours} \tag{15.11g}$$

$$MTBM = \frac{9460}{15} = 630.667 \text{ op hours} \tag{15.11h}$$

Using the quantities in Equation (15.11), we can calculate the availabilities as:

$$A_i = \frac{1000}{1000 + 40} = 0.9615 \tag{15.12a}$$

$$A_a = \frac{630.667}{630.667 + 29.333} = 0.9556 \tag{15.12b}$$

$$A_o = \frac{630.667}{630.667 + 36} = 0.9460 \text{ or } A_o = \frac{9460}{10{,}000} = 0.9460 \tag{15.12c}$$

Notice that the same operational availability is computed different ways in Equation (15.12c).

15.1.3 *Application-Specific Availability Measures*

Several additional specialized types of time-based availability also exist. These availability measures represent the availability for specific applications.

Mission availability — the probability that each individual failure occurring in a mission of a specific total operating time can be repaired in a time that is less than or equal to some specified time length. Mission availability is applicable to situations when only a finite amount of repair time is acceptable.

Work-mission availability — the probability that the sum of all the repair times for all the failures occurring in a mission of a specified total operating time is less than or equal to some specified time length.

Joint availability — the probability of finding the system operating at two distinct times during a mission.

Random-request availability — incorporates the performance of several tasks arriving randomly during the fixed mission period. Random-request availability includes both the system state and random task arrival rates.

Computation availability — the mean performance level at a given time, which is the weighted sum of state probabilities.

15.2 Maintainability and Maintenance Time

Maintenance refers to the measures taken to keep a product in operable condition or to repair it to an operable condition [Ref. 15.3]. The term maintainability is used to denote the study and improvement of the ability to maintain products, primarily focused on reducing the amount of time required to diagnose and repair failures. Quantitatively, maintainability is the probability that a failed unit will be repaired (restored to an operable state) within a given amount of time. The time associated with this definition is the downtime in Equation (15.1). For example, a system with a maintainability of 95% in one day has a 95% probability of being restored to operability within one day of its failure. The maintainability, $M_a(t)$, is the probability of completing maintenance in a time T, which is less than t and is given by

$$M_a(t) = \Pr(T < t) = \int_0^t f(\tau)d\tau \tag{15.13}$$

where $f(\tau)$ is the repair time probability density function. If $f(t)$ is given by

$$f(t) = \mu e^{-\mu t} \tag{15.14}$$

where μ is the constant repair rate and t is the time to repair (downtime), then the maintainability becomes

$$M_a(t) = 1 - e^{-\mu t} \tag{15.15}$$

Under the assumption of a constant repair rate, which is assumed in Equation (15.14), the mean time to repair is given by

$$MTTR = \frac{1}{\mu} \tag{15.16}$$

A more common distribution for repair times for electronics is the lognormal distribution:

$$f(t) = \frac{1}{t\sigma\sqrt{2\pi}} e^{-\frac{1}{2}\left(\frac{\ln(t)-\mu}{\sigma}\right)^2} \tag{15.17}$$

where

μ = the mean of ln(t), location parameter.
σ = the standard deviation of ln(t), scale parameter.

Substituting Equation (15.17) into Equation (15.13), the maintainability corresponding to lognormally distributed repair times becomes

$$M_a(t) = \int_0^t \frac{1}{\tau\sigma\sqrt{2\pi}} e^{-\frac{1}{2}\left(\frac{\ln(\tau)-\mu}{\sigma}\right)^2} d\tau = \Phi\left(\frac{\ln(t)-\mu}{\sigma}\right) \tag{15.18}$$

where Φ is the standard normal CDF.[2] In this case the *MTTR* is given by[3]

$$MTTR = e^{\left(\mu+\frac{\sigma^2}{2}\right)} \tag{15.19}$$

In general, the time to repair should include the time to diagnose, disassemble, and transport the failed unit to a place it can be repaired; obtain replacement parts and other necessary materials; make the repair; perform functional testing; reassemble the unit; and verify and test the unit in the field.

There are many other maintenance metrics that can be computed; see [Refs. 15.3 and 15.4].

---

[2] The standard normal CDF is given by

$$\Phi(x) = \frac{1}{\sqrt{2\pi}} \int_{-\infty}^{x} e^{-t^2/2} dt = \frac{1}{2}\left[1 + \operatorname{erf}\left(\frac{x}{\sqrt{2}}\right)\right]$$

[3] Note, the units on *MTTR* will be the same as the units on t since μ is the ln(t).

15.3 Monte Carlo Time-Based Availability Calculation Example

Given constant failure rates and constant repair rates, it is simple to apply the relations in Section 15.2 to compute time-based availabilities. However, when general distributions of failures and repair times are used, how can we solve for the availability? If the distributions are defined by known probability distribution forms, closed-form solutions may be obtainable. However, this may not always be the case, and we need to be able to also numerically solve for the availability. This can be accomplished, in general, by using the Monte Carlo method described in Chapter 9.

Consider the following simple inherent availability example. Assume that both the time to failure and time to repair are exponentially distributed with $MTBF = 1$ and $MTTR = 1$. Using Equation (15.7), $A_i = 0.5$, which is exactly correct. If we numerically determine the availability using the actual distributions for time to failure and time to repair in Equation (15.7), we should get the same answer. Figure 15.1 shows the input exponential distributions and the output inherent availability distribution that results from a Monte Carlo analysis applied to Equation (15.7).

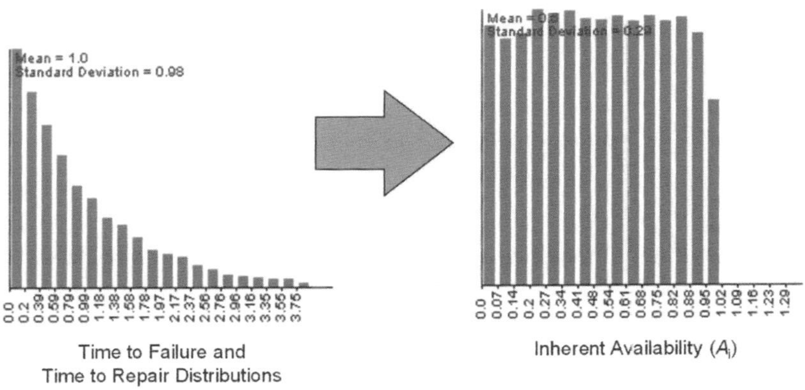

Fig. 15.1. Monte Carlo analysis to determine inherent availability, 10,000 samples used.

The mean of the resulting distribution of inherent availability is 0.5. In general, the distribution of availability when failure and repair times are exponentially distributed is a Beta distribution; the uniform distribution in Figure 15.1 is a special case of the Beta distribution.

Figure 15.1 demonstrates a very important point. Just because *MTBF* = 1 and *MTTR* = 1 and the mean $A_i = 0.5$, this does not imply that every instance of the system has $A_i = 0.5$. The right side of Figure 15.1 is a histogram of the inherent availabilities of the population of systems. Some individuals in this population have availabilities far less than 0.5 and some have availabilities far greater than 0.5. The average availability of the systems in the population is 0.5.

Consider a case where *MTBF* = 600 and *MTTR* = 34 (exponential distributions assumed). Running 10,000 samples in our Monte Carlo analysis of Equation (15.7) results in the histogram of inherent availabilities shown in Figure 15.2. In this case, the mean is 0.8786.

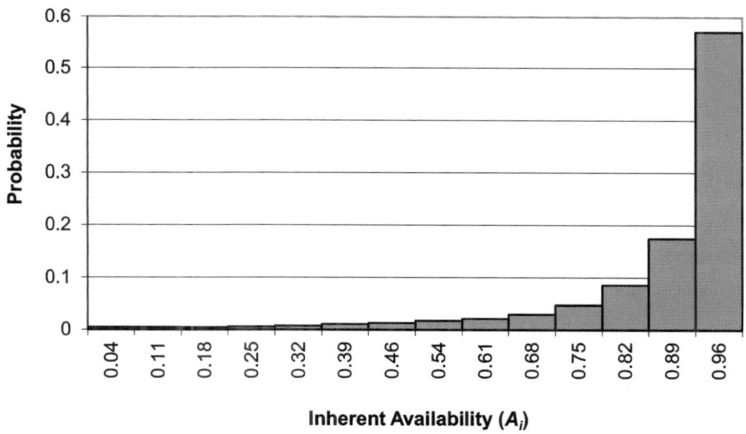

Fig. 15.2. Monte Carlo analysis to determine inherent availability, 10,000 samples used.

Simply plugging the mean values of the failure rate and the repair time into Equation (15.7) only provides an approximation to the correct value of A_i, because in general,

$$\overline{\left[\frac{X_i}{X_i + Y_i}\right]} \neq \frac{\overline{X}_i}{\overline{X}_i + \overline{Y}_i} \qquad (15.20)$$

The left side of Equation (15.20) represents the correct way to assess the mean value of the availability.

15.4 Markov Availability Models

Markovian approaches to the formulation of availability models have also been widely used. The simplest Markov model is the Markov chain, which models the state of a system with a random variable that changes over time. In this context, the Markov property suggests that the distribution for this variable depends only on the distribution of the previous state.[4]

Let $X(T)$ represent the status of the system (S) at time T. $X(T) = 0$ means the system is down (not available) at time T, and $X(T) = 1$ means the system is up (available) at time T. The state transition diagram for our system S is shown in Figure 15.3.

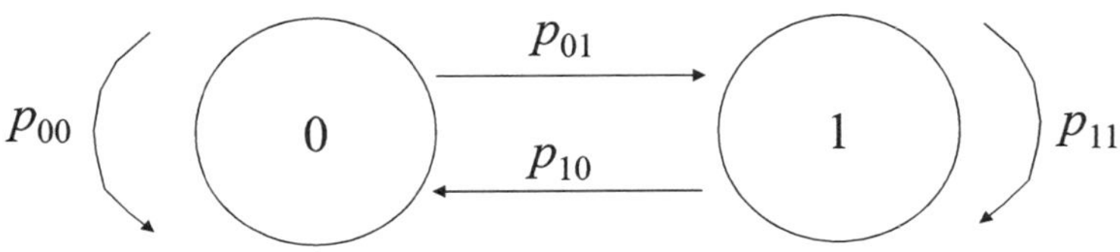

Fig. 15.3. State transition diagram for system S.

The state transition probabilities in Figure 15.3 are given by p_{ij}, which is the probability that the state is j at T, given that it was i at time T-1. The state transition probabilities in Figure 15.3 are given by

$$p_{01} = \mathrm{P}[X(T) = 1|X(T\text{-}1) = 0] = q$$

$$p_{10} = \mathrm{Pr}[X(T) = 0|X(T\text{-}1) = 1] = p$$

$$p_{00} = \mathrm{Pr}[X(T) = 0|X(T\text{-}1) = 0] = 1\text{-}q$$

$$p_{11} = \mathrm{Pr}[X(T) = 1|X(T\text{-}1) = 1] = 1\text{-}p$$

where $p_{00} + p_{01} = 1$ and $p_{10} + p_{11} = 1$, since there are only two states the system can be in.

Markov chains can be represented using a state transition probability matrix like the one constructed in Figure 15.4.

---

[4] Markov processes are "memoryless", i.e., the probability distribution of the next state depends only on the current state and not on the sequence of events that preceded it.

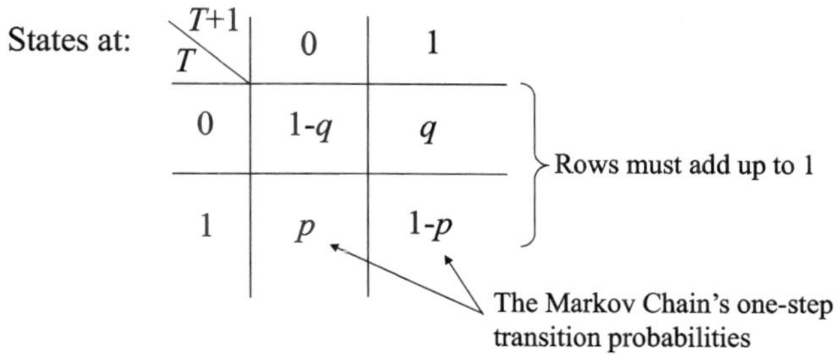

Fig. 15.4. State transition matrix construction.

The state transition probability matrix for our simple system represents the probabilities of moving from one state to any other state, and is given by

$$\begin{bmatrix} 1-q & q \\ p & 1-p \end{bmatrix} \tag{15.21}$$

If we need to determine the probabilities of moving from one state to another state in two steps, all we have to do is raise Equation (15.21) to the second power:

$$\begin{bmatrix} 1-q & q \\ p & 1-p \end{bmatrix}^2 = \begin{bmatrix} 1-q & q \\ p & 1-p \end{bmatrix}\begin{bmatrix} 1-q & q \\ p & 1-p \end{bmatrix}$$
$$= \begin{bmatrix} (1-q)^2 + qp & (1-q)q + q(1-p) \\ p(1-q) + (1-p)p & pq + (1-p)^2 \end{bmatrix} = \begin{bmatrix} p_{00}^2 & p_{01}^2 \\ p_{10}^2 & p_{11}^2 \end{bmatrix} \tag{15.22}$$

Note that a matrix multiplication is used in Equation (15.22). For example, the probability p_{10}^2 in Equation (15.22) represents the probability that system S is down after operating for $T = 2$ time steps if it was initially up (in state 1). Note that the rows of the state transition probability matrix in Equation (15.22) still add up to one.

For large n, the state transition matrix has quasi-identical rows and the results are interpreted as "long run averages" or "limiting probabilities" of S being in the state corresponding to column i:

$$\begin{bmatrix} 1-q & q \\ p & 1-p \end{bmatrix}^n = \frac{1}{p+q}\begin{bmatrix} p & q \\ p & q \end{bmatrix} + \frac{(1-p-q)^n}{p+q}\begin{bmatrix} q & -q \\ -p & p \end{bmatrix} \tag{15.23}$$

In the limit as n approaches infinity,

$$\lim_{n\to\infty}\begin{bmatrix}1-q & q\\ p & 1-p\end{bmatrix}^n = \frac{1}{p+q}\begin{bmatrix}p & q\\ p & q\end{bmatrix} \tag{15.24}$$

For the example considered in Section 15.3 with an *MTBF* = 600 and an *MTTR* = 34,

$p = p_{10} = 1/600 = 0.00167$ (probability of failing is $1/MTBF$)
$q = p_{01} = 1/34 = 0.0294$ (probability of being repaired is $1/MTTR$)

The transition probabilities are given by

$$p_{11}^n = p_{01}^n = \frac{q}{p+q} = 0.9464$$

$$p_{00}^n = p_{10}^n = \frac{p}{p+q} = 0.0536$$

Thus p_{11}^n and p_{00}^n are state occupancy rates, which can also be interpreted as the fraction of time that the system will spend in the "up" and "down" states respectively — that is, the expected availability and unavailability of the system. In this case the inherent availability is p_{11}^n, note, 600/(600+34) = 0.9464.

15.5 Spares Demand-Driven Availability

Not all availability measures are directly based on time.[5] One way to view availability is operational (time based), while an alternative view is through the lens of demand. Viewing availability as the ability to support a system when the demand for the system arrives, leads us to the consideration of availability as an inventory problem. *MDT* discussed in Section 15.1.2 depends on both the time to perform a repair and the availability of spare parts (the spare part stocking or inventory level).

[5] However, to the extent that demand is a function of time, the availability measures discussed in this section are also obviously dependent on time. In fact, supply availability appeared in Section 15.1.2 and appears again in this section.

Sections 15.5.1 and 15.5.2 address the challenge of determining the minimum number of spares (and in the real world, their physical distribution) necessary to meet an availability requirement. Section 15.5.3 is also an inventory view of availability, but one in which the inventory is the fielded systems (not spare parts); and Section 15.5.4 is a discussion of energy availability used for energy generation sources.

15.5.1 *Backorders and Supply Availability*

A backorder is an unfulfilled demand due to lack of spares. Equation (12.5) is the probability of an item system having exactly x failures in time t. If k spares exist for a population of n items, then the probability of needing $k+ m_b$ spares resulting in a backorder of m_b is given by Equation (12.8):

$$\Pr(k+m_b)=\frac{(n\lambda\, t)^{k+m_b}\, e^{-n\lambda\, t}}{(k+m_b)!} \tag{15.25}$$

The expected number of backorders for the population of items with k available spares is

$$EBO(k)=\sum_{x=k+1}^{\infty}(x-k)\Pr(x) \tag{15.26}$$

where Pr(x) is given by Equation (15.25). Each of the terms in the sum in Equation (15.26) is the probability of needing 1, 2, 3, … , ∞ more spares than you have multiplied by that number of spares.

As an example, if there are $n\lambda t = 20$ demands for spares and you have $k = 10$ spares, then the expected number of backorders from Equation (15.26) is $EBO(10) = 10.01$.

Now we can relate the expected number of backorders to the supply availability (A_s) using [Ref. 15.2]:

$$A_s=\prod_{i=1}^{l}\left[1-\frac{EBO_i(k_i)}{NZ_i}\right]^{Z_i} \tag{15.27}$$

where

l = the number of unique repairable items in the system.
N = the number of instances of the system.
Z_i = the number of instances of item i in each system.

$EBO_i(k_i)$ = the expected number of backorders for the ith item if k_i spares exist (this is the total expected backorders for all instances of the ith item in N systems).

In Equation (15.27), the product NZ_i is n, which is the number of sockets for the ith item in the N systems (number of places that the ith repairable item occupies). Sockets are the places in a system where the items go. The ratio $EBO_i(k_i)/NZ_i$ is the probability of an unfulfilled spare demand for the entire population of the ith item. Then, $1\text{-}EBO_i(k_i)/NZ_i$ is the probability that there are no unfulfilled spare demands in the entire population of the ith item. Raising this quantity to the power Z_i gives the probability of no unfulfilled spare demands for the ith item in one instance of the system. That is, the system is assumed to be available only if there are no unfulfilled spares in the Z_i items of the ith type in the system. The product in Equation (15.27) assumes that all l unique repairable items that make up one instance of the system have to function for the system to be available, so A_s represents the supply available for the system.

Equation (15.27) assumes that all the i items have independent failures and that the N systems are independent as well. Also, there is no cannibalization (i.e., no failed systems are robbed for parts to fix other systems). Equation (15.27) only applies if $EBO_i(k_i) \leq NZ_i$ for all i.

Consider an example, if there are 1000 systems, each containing 2 unique repairable items (one instance of item 1 and three instances of item 2), that must be spared for 60 days, and item 1 experiences twenty demands during the time period and has ten spares, while item 2 experiences seventeen demands during the time period and has twelve spares, what is the supply availability for each system in the fleet? In this case,

$$N = 1000 \qquad Z_1 = 1 \qquad Z_2 = 3$$
$$l = 2 \qquad n\lambda_1 t = 20 \qquad n\lambda_2 t = 17$$
$$k_1 = 10 \qquad k_2 = 12$$

From Equation (15.26) $EBO_1(10) = 10.1$ and $EBO_2(12) = 5.18$. Using Equation (15.27), the supply availability is given by

$$A_s = \left[1 - \frac{10.1}{(1000)(1)}\right]^1 \left[1 - \frac{5.18}{(1000)(3)}\right]^3 = 0.9848$$

15.5.2 *Erlang-B*

One way to relate availability to spares is to use the Erlang-B (also known as the Erlang loss formula), [Ref. 15.5]. This formula was originally developed for planning telephone networks, and it is used to estimate the stock-out probability for a single-echelon repairable inventory:[6]

$$1 - \overline{A} = \frac{a^k / k!}{\sum_{x=0}^{k} \left(a^x / x!\right)} \tag{15.28}$$

where

$\overline{A}$ = the steady-state availability (1-$\overline{A}$ is the unavailability).
a = the number of units under repair.
k = the number of spares.

In Equation (15.28) 1-$\overline{A}$ is the stock-out probability.[7] The number of units under repair can be computed from

$$a = NF_t \mu_r \tag{15.29}$$

where

N = the number of fielded units.

---

[6] Single-echelon repairable inventory means that the members of the lowest echelon are responsible for their own stocking policies, independent of each other and independent of a centralized depot. Single-echelon means we are basically dealing with a single inventory (or stocking point) of spares. Multi-echelon inventory considers multiple stocking points coupled together (multiple distribution centers and layers) — e.g., a centralized depot that provides common stock to multiple lower stocking points.

[7] For telephone networks, 1-$\overline{A}$ is called the blocking probability, the probability of all k servers being busy and a call being blocked (lost). a is the traffic offered to the group measured in Erlangs, and k is the number of trunks in the full availability group. Equation (15.28) is used to determine the number of trunks (k) needed to deliver a specified service level (1-$\overline{A}$), given the traffic intensity (a). In general, this formula describes a probability in a queuing system.

F_t = the failures that need to be repaired per unit per unit time.
μ_r = the mean repair time (mean time to repair one unit).

The product NF_t is the arrival rate, or the number of repair requests per unit time. Equation (15.28) assumes that a follows a Poisson process and is derived assuming that the number of spares (k) is equal to the number of fielded systems requesting a spare (see [Ref. 15.6]).

As an example of the usage of Equation (15.28), consider a population of 3000 systems where each system has a failure rate of λ = 7x10^{-6} failures/hour; 50% of the failures require repair (the other 50% are assumed to either result in system retirement or are resolved with permanent spares taken from another source outside the scope of this problem); the mean repair time is 72 hours. We want a 99.9% availability. How many spares are needed?

F_t = 0.5λ=3.5x10^{-6} failures per unit per hour.
a = (3000)(3.5x10^{-6})(72) = 0.756 the number of units under repair at any one time (this unit of measure is referred to as an Erlang).
$1-\overline{A}$ = 0.001.

Applying Equation (15.28), we find that when k = 5, $1-\overline{A}$ = 0.00097 (which is less than 0.001), 5 or more spares are needed.

15.5.3 *Materiel Availability*

Materiel or matériel is equipment, apparatus, and supplies used by an organization or institution, often specifically associated with a military application. Materiel availability is the fraction of the total inventory of a system that is operationally capable (ready for tasking) for performing a required mission at a specific point in time governed by the condition of the materiel. The key word in this definition is "inventory". If I have an inventory of 10 helicopters and 8 are currently operational and ready for use, then my materiel availability is 0.8 or 80%.

The point or instantaneous materiel availability is expressed as the fraction of end items that are operational, which can be calculated using either of the following relations,

$$A_m = \frac{\text{Number of Operational End Items}}{\text{Total Population of End Items Fielded (in Inventory)}} \tag{15.30a}$$

$$A_m = \frac{\text{Active Inventory}}{\text{Active Inventory + Inactive Inventory}} \tag{15.30b}$$

Materiel availability is distinguished from time-based availability measures by the fact that it depends on the total population of systems (end items) fielded (in inventory) and it considers the total life cycle of the system (end item).[8]

The materiel availability can be calculated using Equation (15.1), however, the uptime and downtime have different definitions and the materiel availability is <u>not</u> interchangeable with the operational availability. The materiel availability must apply to the entire fielded inventory of systems, apply to the entire life cycle of the system, and incorporate all categories of downtime. Operational availability always applies to a limited number of systems and frequently incorporates only unscheduled maintenance downtimes. A_m is a function A_o and other factors that do not impact A_o, including technology insertion. While A_o is an operational measure, A_m is a programmatic measure that spans a larger timeframe, additional sources of downtime, and additional sources of unscheduled maintenance.

15.5.4 *Energy-Based Availability*

Specific applications have discovered that time-based availability measures do not always adequately represent their needs. For example in the renewable energy generation domain, time-based availability does not account for the fact that the system is not producing efficiently all the time, i.e., just because the system is operating does not mean it is operating efficiently. Conversely, just because the system is not operating does not mean that energy could be produced if it was operational. For example, for a wind farm, 3% unavailability when there isn't much wind could

---

8 Since the definition of materiel availability mandates that it consider the entire fielded population of systems and the entire system life cycle, technically it is impossible to measure until after a system has completed its entire field life.

represent very little energy loss. While the same unavailability could represent a loss of up to 10% during high-wind periods [Ref. 15.7].

While time-based availability[9] is used for renewable energy applications, energy-based availability measures like the following are also widely used,

$$A_E = \frac{\text{Available Energy}}{\text{Available Energy} + \text{Energy Lost}} \tag{15.31a}$$

$$A_E = \frac{E_{real}}{E_{theoretical}} \tag{15.31b}$$

15.6 Availability Contracting

Customers of avionics, large scale production lines, servers, and infrastructure services with high availability requirements are increasingly interested in buying the availability of a system, instead of actually buying the system itself, resulting in the introduction of "availability-based contracting." Availability-based contracts are a subset of outcome-based contracts [Ref. 15.8], through which the customer pays for the delivered outcome, instead of paying for specific logistics activities, system reliability management, or other tasks. Basically, in this type of contract, the customer pays the service or system provider to ensure that their specific availability requirement is met. For example, the Availability Transformation: Tornado Aircraft Contract (ATTAC) [Ref. 15.9] is an availability contract; BAE Systems has agreed to support the Tornado GR4 aircraft fleet at a specified availability level throughout the fleet service life for the UK Ministry of Defence. The agreement implements a new cost-effective approach to improving the availability of the fleet while minimizing the life-cycle cost [Ref. 15.9].

Before providing background on relevant outcome-based contracts, it is useful to clearly distinguish availability-based contracts from other common contract mechanisms that are applied to the support of products and systems (Table 15.2). Availability-contracts are not warranties, lease

---

[9] The term "availability factor" is often used to mean operational availability in power plants.

agreements or maintenance contracts, which are all break-fix guarantees. Rather these contracts are quantified "satisfaction guaranteed" contracts where "satisfaction" is a combination of outcomes received from the product, usually articulated as a time (e.g., operational availability), usage measure (e.g., miles), or energy-based availability.

Table 15.2. Common mechanisms that are applied to the support of products and systems.

Type of Contract Mechanism	Examples	Key Characteristics	Support Provider Commitment
Break-fix guarantee	Common warranties, leases and maintenance contracts	Definition of, or threshold for, failure	Replace or repair on failure
Satisfaction guarantee	Warranties and leases	Satisfaction is not quantified	Replace or repair if not satisfied
Outcome guarantee	Performance-based contracts (PBL, PBH, PPP, and PPA)	Carefully quantified "satisfaction"	Provider has autonomy to meet required outcomes any way they like

The evaluation of an availability requirement is a challenging task for both suppliers and customers. From a suppliers' perspective, it is not trivial to estimate the cost of delivering a specific availability. Entering into an availability contract is a non-traditional way of doing business for the suppliers of many types of safety- and mission-critical systems. For example, the traditional avionics supply chain business model is to sell the system, and then separately to provide the sustainment of the system. As a result, avionics suppliers may sell the system for whatever they have to in order to obtain the business, knowing that they will make their money on its long-term sustainment. From a customer's perspective, the amount of money that should be spent on a specific availability contract is also a mysterious quantity; if a choice has to be made between two offers of availability contracts for which the values of the promised availabilities are close (e.g., one contract offers an availability of 95%, and the other one offers 97%), then how much money should the customer be willing to spend for a specific availability improvement?

15.6.1 *Product Service Systems (PSS)*

Two common mechanisms that may include elements of availability contracting are product service systems (PSS) and leasing models. Figure 15.5 shows an example PSS spectrum that indicates the concept of outcome-based contracting models (of which availability contracting is an example). PSS provide both the product and its service/support based on the customer's requirements [Ref. 15.10], which could include an availability requirement. Lease contracts [Ref. 15.11] are use-oriented PSS, where the ownership of the product is usually retained by the service provider. A lease contract may indicate not only the basic product and service provided but also other use and operation constraints, such as the failure rate threshold. In leasing agreements the customer has an implicit expectation of a minimum availability, but the availability is generally not quantified contractually.[10]

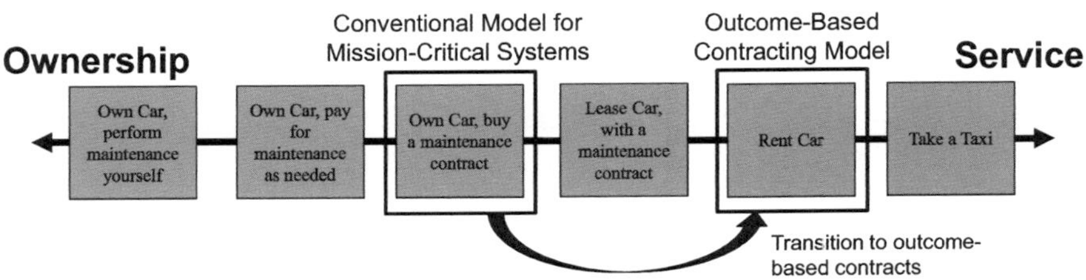

Fig. 15.5. Example PSS spectrum for a car.

15.6.2 *Power Purchase Agreements (PPAs)*

A PPA (also called Energy Performance Contracting (EPC)) is defined as a long-term contract to buy electricity from a power plant. PPAs secure

[10] Leases often have availability-like requirements; however, the primary difference is that the requirement is usually imposed by the owner of the system upon the customer, rather than the other way around. For example, a copy machine lease may require the customer to make 1000 copies per month or fewer; if they make more they pay a penalty; or there may be a maximum amount of data you can use per month on your mobile phone plan. Alternatively, if this was an availability contract, the copy machine user would tell the owner of the machine that it must be able to successfully make at least 1000 copies per month or they will pay the owner of the machine less for the lease.

the payment stream for a power producer and satisfy the purchaser's (often federal and state) regulations/requirements for long-term electricity generation. A PPA defines a price schedule for the electricity that is generated with optional annual escalation and a variety of time-of-delivery factors. The price schedule is based on several parameters that include: the levelized cost of energy (with/without state and federal incentives) — see Section 20.3, the length of the agreement, the internal rate of return, and various milestones.

As far as availability contracts are concerned, the salient attribute of PPAs is that the power purchaser does not own or operate the power producer's generation, and the power purchaser only cares about being delivered the promised power. It is up to the power producer to decide how to operate and manage the production. PPAs exist for all types of power generation, but are particularly useful for renewable power generation (i.e., solar and wind). In these cases, the PPA insulates the power purchaser from uncontrollable risks (e.g., too many cloudy days and if the wind doesn't blow) as well as the risks associated with maintaining the generation (e.g., weather problems for offshore wind farms).

15.6.3 *Performance-Based Logistics (PBLs)*

The form of outcome-based contracting that is used by the U.S. Department of Defense is called performance-based contracting (or Performance Based Logistics (PBL)). In PBL contracts the contractor is paid based on the results achieved, not on the methods used to achieve them [Refs. 15.12 and 15.13]. Availability contracts, and most outcome-based contracting, include cost penalties that may be assessed for failing to fulfill a specified availability requirement within a defined time frame (or a contract payment schedule that is based on the achieved availability).

15.6.4 *Public-Private Partnerships (PPPs)*

Public-private partnerships (PPPs) have been used to fund and support civil infrastructure projects, most commonly highways. Availability payment models for civil infrastructure PPPs require the private sector to take responsibility for designing, building, financing, operating, and maintaining an asset. Under the "availability payment" concept, once the asset is available for use, the private sector begins receiving an annual payment for a contracted number of years based on meeting performance requirements. The challenge in PPPs is to determine a payment plan (cost and timeline) that protects the public interest, i.e., does not overpay the private sector; but also, minimizes the risk that the asset will become unsupported.

15.7 Readiness

Readiness is the state of having been made ready or prepared for use or action. Quantitatively, readiness is determined using the same relationship as availability in Equation (15.1). In some definitions, readiness is distinguished from availability solely based on what is included in the downtime. For example, in [Ref. 15.14], downtime for readiness calculations includes free time and storage time in addition to operational downtimes.

However, readiness often has a broader scope than availability. Qualitatively, readiness includes

- the operational availability of the system
- the availability of the people who are needed to operate the system, and
- the availability of the infrastructure and other resources needed to support the operation of the system.

Consider the example of an aircraft that could have 100% operational availability, but less than 100% readiness because of lack of fuel or crew, damage to the runway it requires, or the unavailability of the items it has to transport. Therefore, readiness is really the cumulative (series) availability of a collection of individual system availabilities. The availability of a set of n systems in series is the product of their individual availabilities (if they are independent):

$$A = \prod_{i=1}^{n} A_i \tag{15.32}$$

Equation (15.32) is valid if the unavailability of any of the n systems causes the system to be inoperable.[11] Equation (15.32) assumes that all the non-failed systems continue to function during the time when a failed system is repaired.

Another approach to readiness is called "fleet readiness" [Ref. 15.15]. For a fleet, readiness is defined as the probability that there are at least k systems available at any random point in time:

$$\Pr(N \geq k) = \sum_{i=k}^{n} \binom{n}{i} A^i (1-A)^{n-i} \tag{15.33}$$

where

N = the number of available systems.

n = the number of identical systems in the fleet.

A = the availability of a single system.

$\binom{n}{i} = \dfrac{n!}{i!(n-i)!}$, the binomial coefficient.

If $A = 0.95$, $n = 20$, $k = 18$, then $\Pr(N \geq k) = 0.925$, or there is a 92.5% probability that at least 18 systems in the fleet are ready for operation at any time.

15.8 Discussion

What quality is to manufacturing costs, availability is to life-cycle costs. It makes little sense for many systems to evaluate or minimize life-cycle costs without a corresponding assessment of availability. Availability and

---

[11] Alternatively, if the unavailability of one of the n systems leads to one of the other systems taking over the operation of the unavailable system, the two systems are considered to be operating in parallel. The availability of a set of n systems in parallel is given by

$$A = 1 - \prod_{i=1}^{n} (1 - A_i)$$

See [Ref. 15.1] for a summary of methods for determining the availability of other system configurations.

readiness are both part of a broader concept called effectiveness, which is the extent to which an activity fulfils its intended purpose or function.

Availability can be evaluated at different levels. For example, it can be evaluated at the system level, as for an airplane, or at the subsystem level, the engine on the airplane. The availability described in this chapter really targets "sockets." Sockets are the places in a system where the objects (often called line replaceable units) are located. For example, when we talk about spares having an impact on availability (Section 15.5), we are really considering the availability of the socket for the object. ften the socket's availability is more important than the availability of a particular instance of an item that goes into the socket. The instance may occupy several different sockets as it fails, is repaired and goes back into a spares pool rather than just the original socket it came from.

Section 15.1 provides definitions for numerous different availability measures; however, exactly what is included in the uptimes and downtimes that define availability depends on what the user wants or is contractually required to measure.

Most availability predictions used during the design and support of real systems are performed either using Markov models or some form of a discrete-event simulator (see Appendix C). Discrete-event simulators track the current state of the system, and based on the present events, predict the occurrence of future events [Ref. 15.16]. Markov models do not explicitly embrace the concept of future events; rather, they track the model state at each time step and sample how long the model will be in the current state before it switches to the next state. Each event in a discrete-event simulator depends on the time spent in that event and the path that led to it; on the other hand, Markov models depend only on the current state of the model, regardless of the duration spent in the current state and the path that led to it. Discrete-event simulators accumulate the outcomes resulting from the type and duration of previous events, then use only the set of data inputs that are necessary at a specific point on the timeline to predict future events. Markov models incorporate all provided data to generate an analytical solution and use it to determine the current model state and to move to the next state.

Discrete-event simulators are generally more efficient than Markov models for modeling complex systems with large numbers of variables, specifically in data capturing without aggregation. In general, discrete-event simulators order the failure and maintenance events for a system temporally, and the durations associated with the failure and maintenance events can be readily accumulated to estimate availability. Thus, it is straightforward for a discrete-event simulation to compute the availability based on a particular sequence of failures, logistics and maintenance events.

While there is a significant body of literature that addresses availability optimization (maximizing availability), little work has been done on designing to meet a specific availability requirement, as would be done for an availability contract. Unlike availability optimization, in availability contracts there may be no financial advantage to exceeding the required availability. Recent interest in availability contracts that specify a required availability has created an interest in deriving system design and support parameters directly from an availability requirement. In general, determining design parameters from an availability requirement is a stochastic reverse simulation problem. While determining the availability that results from a sequence of events is straightforward, determining the events that result in a desired availability is not, and has not in general been done. See [Ref. 15.17] for a discussion of design for availability modeling.

References

15.1 Lie, C. H., Hwang, C. L. and Tillman, F. A. (1977). Availability of maintained systems: A state-of-the-art survey, *AIIE Transactions*, 9(3), pp. 247-259.

15.2 Sherbrooke, C. C. (2004). *Optimal Inventory Modeling of Systems*, 2nd Edition (Kluwer Academic Publishers, New York, NY).

15.3 Dhillon, B. S. (1999). *Engineering Maintainability* (Gulf Publishing Company, Houston TX).

15.4 Blanchard, B. (1992) *Logistics Engineering and Management*, 4th Edition (Prentice Hall, Englewood Cliffs, NJ).

15.5 Erlang, A. (1948). Solution of some problems in the theory of probabilities of significance in automatic telephone exchanges in *The Life and Works of A.K. Erlang*, E. Brockmeyer, H. Halstrom, and A. Jensen, eds., *Transactions of the Danish Academy of Technical Sciences*, No. 2.

15.6 Cooper, R. B. (1972). Introduction to Queuing Theory (MacMillan, New York).

15.7 Conroy, N., Deane, J. P. and Ó Gallachóir, B. P. (2011). Wind turbine availability: Should it be time or energy based? – A case study in Ireland. Renewable Energy, 36(11), pp. 2967–2971.

15.8 Ng, I. C. L., Maull, R. and Yip, N. (2009). Outcome-based contracts as a driver for systems thinking and service-dominant logic in service science: Evidence from the defence industry, *European Management Journal*, 27(6), pp. 377-387.

15.9 BAE (2008). *BAE 61972 BAE Annual Report.*

15.10 Bankole, O. O., Roy, R., Shehab, E. and Wardle, P. (2009). Affordability assessment of industrial product-service system in the aerospace defense industry, *Proceedings of the CIRP Industrial Product-Service Systems (IPS2) Conference*, p. 230.

15.11 Yeh, R. H. and Chang, W. L. (2007). Optimal threshold value of failure-rate for leased products with preventive maintenance actions, *Mathematical and Computer Modeling*, 46, pp. 730-737.

15.12 Beanum, R. L. (2006). Performance-based logistics and contractor support methods, *Proceedings of the IEEE Systems Readiness Technology Conference (AUTOTESTCON).*

15.13 Hyman, W. A. (2009). Performance-based contracting for maintenance, *NCHRP Synthesis 389*, Transportation Research Board of the National Academies.

15.14 Pecht, M. (2009). *Product Maintainability Supportability Handbook*, 2nd Edition (CRC Press, Boca Raton, FL).

15.15 Jin, T and Wang, P. (2011). Planning performance based logistics considering reliability and usage uncertainty, Working Paper from Ingram School of Engineering, Texas State University, San Marcos, TX.

15.16 Banks, J., Carson, J. S., Nelson, B. L. and Nicol, D. M. (2010). *Discrete-Event System Simulation*, 5th Edition (Prentice Hall, Upper Saddle River, NJ).

15.17 Jazouli, T. Sandborn, P. and Kashani-Pour, A. (2014). "A Direct Method for Determining Design and Support Parameters to Meet an Availability Requirement," *International Journal of Performability Engineering*, 10(2), pp. 211-225.

Problems

15.1 Derive Equation (15.3). Hint: See Section 13.2.

15.2 Derive Equation (15.4).

15.3 For the case of a constant failure rate (λ) and a constant repair rate (μ), what is the renewal density function?

15.4 For the conditions in Problem 15.3, what is the steady-state availability?

15.5 If the failure rate and the repair rate are exponentially distributed with $\lambda = 6\text{x}10^{-5}$ failures per hour and $\mu = 5\text{x}10^{-2}$ repairs per hour, what is the steady-state availability? Hint: You need to solve Problem 15.4 first.

15.6 What order (by magnitude) do the different availabilities described in Section 15.1.2 occur in?

15.7 If performing one more preventative maintenance activity per year in the example in Section 15.1.2 results in a reduction in the number of failures per year from 2 to 1.5 (i.e., 3 every two years), is there any improvement in the system's operational availability?

15.8 How do the availabilities in the example in Section 15.1.2 change if there is an additional administrative delay time (*ADT*) of 20 operational hours that has to be applied to only two of the preventative maintenance activities performed during the 5-year support life of the system?

15.9 For the example shown in Figure 15.2, what is the probability that inherent availability is greater than 90%? Hint: First write a Monte Carlo model to reproduce Figure 15.2.

15.10 Derive Equations (15.23) and (15.24).

15.11 Create the PSS spectrums (like Figure 15.5) for other types of systems.

15.12 Assuming that the times to failure and times to repair are exponentially distributed, what is the inherent availability of a system consisting of the following three components: Component 1: $\lambda = 0.05, \mu = 0.067$; Component 2: $\lambda = 0.033, \mu = 0.053$; and Component 3: $\lambda = 0.04, \mu = 0.045$. Assume that the components are connected in series and that all non-failed components continue to operate during the time when the failed component is repaired.

15.13 Why does Equation (15.32) assume that all non-failed systems continue to operate during the time when the failed system is repaired?

15.14 Rework Problem 15.12, assuming that all non-failed components are shut down (i.e., do not operate) during the time when the failed component is repaired.

Chapter 16

The Cost Ramifications of Obsolescence

Technology obsolescence is defined as the loss or impending loss of original manufacturers of items or suppliers of items or raw materials [Ref. 16.1]. The type of obsolescence addressed in this chapter is referred to as DMSMS (diminishing manufacturing sources and material shortages) , which is caused by the unavailability of technologies or parts[1] that are needed to manufacture or sustain a product. DMSMS means that due to the length of the system's manufacturing and support life and possible unforeseen life extensions to the support of the system, the necessary parts and other resources become unavailable (or at least unavailable from their original manufacturer) before the system's demand for them is exhausted. Part unavailability from the original manufacturer means an end of support for that particular part and an end of production of new instances of that part (i.e., the part is obsolete).[2]

The DMSMS-type obsolescence problem is especially prevalent in "sustainment-dominated" systems for which the cost of sustaining (maintaining) the system over its support life far exceeds the cost of manufacturing or procuring the system (see Section II.1). Sustainment-dominated systems have long enough design cycles that a significant portion of the electronics technology in them may be obsolete prior to the system being fielded for the first time, as shown in Figure 16.1. Once in

---

[1] In this chapter, "part" refers to the lowest management level possible for the system being analyzed. In some systems, the "parts" are laptop computers, operating systems, and cables, while in other systems the parts are integrated circuits (chips).

[2] Inventory or sudden obsolescence refers to the opposite problem from DMSMS obsolescence. Inventory obsolescence occurs when the product design or system part specifications changes such that existing inventories of components are no longer required [Ref. 16.2].

the field, the operational support for these systems can last for twenty, thirty or more additional years. A possibly more significant issue is that the end-of-support date for systems like the one shown in Figure 16.1 is not known and will likely be extended from the original plan one or more times before the system is retired.

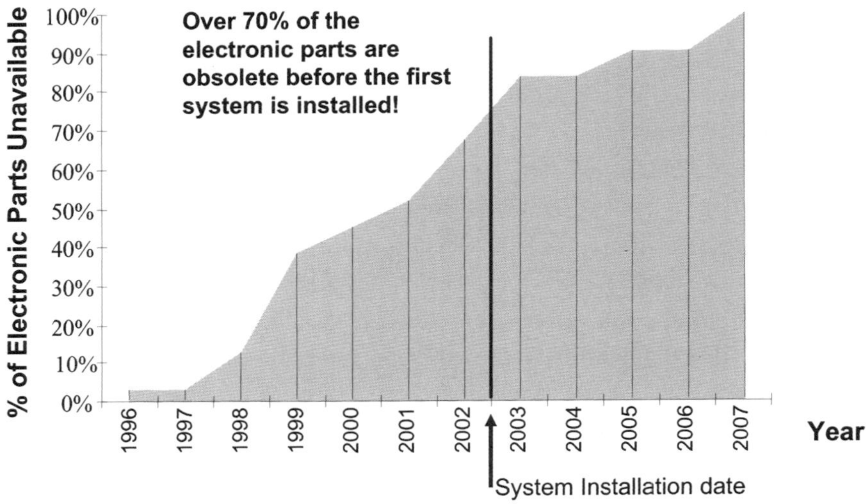

Fig. 16.1. Percent of commercial off-the-shelf (COTS) parts that are un-procurable versus the first 10 years of a surface ship sonar system's life cycle (Courtesy of NAVSURFWARCENDIV Crane).

For systems like the one shown in Figure 16.1, simply replacing obsolete parts with newer parts is often not a viable solution because of high re-engineering costs and the potentially prohibitive cost of system re-qualification and re-certification. For example, if an electronic part in the twenty-five-year old control system of a nuclear power plant fails, an instance of the original component may have to be used to replace it because replacement with a part with the same form, fit, function and interface that isn't an instance of the original part could jeopardize the "grandfathered" certification of the plant.

Sustainment-dominated products particularly suffer the consequences of electronic part obsolescence because they have no control over their electronic part supply chain due to their relatively low production volumes. DMSMS-type obsolescence occurs when long field life systems must depend on a supply chain that is organized to support high-volume products. Obsolescence becomes a problem when it is forced upon an organization; in response, that organization may have to involuntarily make a change to the product that it manufactures, supports or uses.[3]

Electronic Part Obsolescence

Electronic part obsolescence began to emerge as a problem in the 1980s when the end of the Cold War accelerated pressure to reduce military outlays and led to an effort in the United States military called *acquisition reform*. Acquisition reform included a reversal of the traditional reliance on military specifications ("Mil-Specs") in favor of commercial standards and performance specifications. One of the consequences of the shift away from Mil-Specs was that Mil-Spec parts that were qualified to more stringent environmental specifications than commercial parts and manufactured over longer periods of time were no longer available, creating the necessity to use commercial off-the-shelf (COTS) parts that are manufactured for non-military applications. Because their supply chains are driven by commercial and consumer products, the parts are usually procurable for much shorter periods of time. Although this history is associated with the military, the problem it has created reaches much further, since many non-military applications, such as commercial avionics, oil well drilling, power plant control systems, medical systems, and industrial equipment, depended on Mil-Spec parts.

---

[3] Researchers who study product development characterize different industries using the term "clockspeed," which is a measure of the dynamic nature of an industry [Ref. 16.3]. The type of industries that generally suffer from DMSMS problems would be characterized as slow clockspeed industries. In addition, because of the expensive nature of sustainment-dominated products (e.g., airplanes and ships) customers can't afford to replace these products with newer versions very often (slow clockspeed customers). DMSMS-type obsolescence occurs when slow clockspeed industries must depend on a supply chain that is organized to support fast clockspeed industries.

16.1 Managing Electronic Part Obsolescence

Effective long-term management of DMSMS in systems requires addressing the problem on three different management levels: reactive, pro-active and strategic. The reactive management level is concerned with determining an appropriate, immediate resolution to the problem of components becoming obsolete, executing the resolution process and documenting/tracking the actions taken.

Many mitigation strategies exist for reactively managing obsolescence once it occurs. Replacement of parts with non-obsolete substitute or alternative parts can be done as long as the burden of system re-qualification is not unreasonable. There are also aftermarket electronic part sources, ranging from original manufacturer-authorized aftermarket sources that fill part needs with a mixture of stored devices (manufactured by the original manufacturer) and new fabrication in original manufacturer-qualified facilities, to brokers and even eBay. However, buying obsolete parts on the secondary market from non-authorized sources carries its own set of risks — namely, the possibility of counterfeit parts [Ref. 16.4]. David Sarnoff Laboratories operates GEM and AME, which are electronic part emulation foundries that fabricate obsolete parts that meet original part qualification standards using newer technologies (BiCMOS gate arrays). Thermal uprating of commercial parts to meet the extended temperature range requirements of an obsolete Mil-Spec part is also a possible obsolescence mitigation approach [Ref. 16.5].

Most semiconductor manufacturers notify customers and distributors when a part is about to be discontinued, providing customers six to twelve months of warning and giving them the opportunity to place a final order for parts (a "lifetime buy"). Users of the part determine how many parts will be needed to satisfy manufacturing and sustainment of the system until the end of the system's life and place a last order for them.

Pro-active management of obsolescence means that critical components that (a) have a risk of going obsolete, (b) lack sufficient available quantity after obsolescence, and/or (c) will be problematic to manage if/when they become obsolete are identified and managed prior to their actual obsolescence. Pro-active management requires an ability to forecast the obsolescence risk for components. It also requires that there be a process for articulating, reviewing and updating the system-level DMSMS status.

Strategic management of DMSMS means using DMSMS data, logistics management inputs, technology forecasting, and business trending to enable strategic planning, life-cycle optimization, and long-term business case development for the support of systems. The most common approach to DMSMS strategic management is design refresh planning, which determines the set of refreshes that maximizes future cost avoidance.

All the obsolescence management approaches mentioned in this section cost money to perform. Being able to predict the life-cycle cost of managing obsolescence within a system is important for two reasons. First, it allows an estimation of the cost associated with managing a system in a specific way to be determined as part of the budgeting or bidding process for supporting the system. Secondly, it enables optimization of the management of a system by measuring and trading off the cost impact of multiple management approaches. The remainder of this chapter describes several cost modeling approaches that are applicable to managing obsolescence.

16.2 Lifetime Buy Costs

Lifetime buy is one of the most prevalent obsolescence mitigation approaches employed for DMSMS management. Purchasing sufficient parts to meet current and future demands is simpler in theory than in practice, due to many interacting influences and the complexity of multiple concurrent buys, as shown in Figure 16.2. The lifetime buy problem has two facets: demand forecasting, and optimizing the buy quantities based on the demands forecasted.

Forecasted demand depends on sales forecasts and sustainment expectations (spares) for fielded systems (we will not deal with this portion of the problem in this chapter, sparing is addressed in Chapter 12). The second aspect of the problem is determining how many parts should be purchased (lifetime buy quantity).

Given a demand forecast, the quantities of parts necessary to minimize life-cycle cost can be calculated (depending on the penalty for running short or running long on parts, these quantities could be different than what simple demand forecasting suggests). In general, this is an asymmetric problem, where the penalty for underbuying parts and overbuying parts are not the same; if they were the same, then the optimum quantity to purchase would be exactly the forecasted demand. For example, the penalty for underbuying parts is the cost to acquire additional parts long after they become obsolete; the penalty for overbuying parts is paying for extra parts and for the holding (inventory or storage) cost of those parts for a long period when you may lose all or some of that investment.[4] In general, for sustainment-dominated systems, the penalty for underbuying parts is significantly greater than the penalty for overbuying parts.

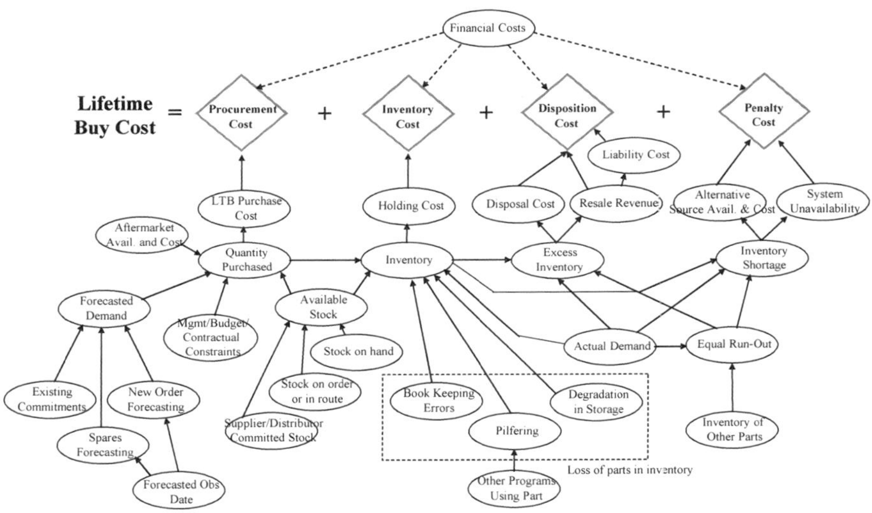

Fig. 16.2. Lifetime buy costs [Ref. 16.6].

[4] Additionally, you may need to pay to dispose of the extra parts. The cost of disposal could be negative (reselling the parts) or positive (ensuring that parts are destroyed so they can't enter the counterfeit parts supply stream is not free).

16.2.1 *The Newsvendor Problem*

Lifetime buy optimization is more generally referred to as the final-order problem, which is a special case of the newsvendor problem[5] from traditional operations management. Existing final-order models are intended for systems like complex manufacturing machinery that have long-term service contracts. To be able to provide long-term service, a manufacturer must be able to supply parts throughout the service period. However, the duration of the service period is typically much longer than the production period for the machine. The period after the machine has been taken out of production is called the end-of-life service period (EOL). To avoid out-of-stock situations during the EOL, an initial stock of spare parts is ordered at the beginning of the EOL. This initial stock is called the final order.

The factors relevant to solving this problem are:

C_O = the overstock cost – the effective cost of ordering one more unit than what you would have ordered if you knew the exact demand (i.e., the effective cost of one left-over unit that can't be used or sold).

C_U = the understock cost – the effective cost of ordering one fewer unit than what you would have ordered if you knew the exact demand (i.e., the penalty associated with having one less unit than you need or the loss of one sale you can't make).

Q = the quantity ordered.

D = Demand.

The newsvendor problem is a classic example of an optimal inventory problem. As an example, consider a newsvendor who purchases newspapers in advance for \$0.20/paper. The papers can be sold for \$1.00/paper. The demand was generated from a beta distribution with shape parameters: $\alpha = 2$ and $\beta = 5$ (lower bound 0, upper bound 40), which

[5] The newsvendor problem seeks to find the optimal inventory level for an asset, given an uncertain demand and unequal costs for overstock and understock. This problem dates back to an 1888 paper by Edgeworth [Ref. 16.7].

is shown in Figure 16.3.[6] How many papers should the newsvendor buy in order to maximize his profit?

In this case C_U = \$1.00-\$0.20 = \$0.80 (\$0.80 is lost for each sale that cannot be fulfilled) and C_O = \$0.20 (\$0.20 is lost for each paper purchased that cannot be sold).

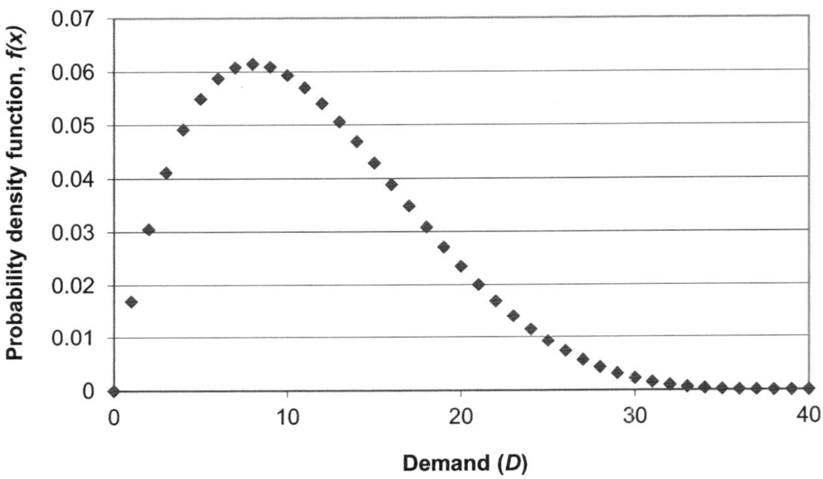

Fig. 16.3. Demand forecast.

Table 16.1 shows the calculations when Q = 10 (assuming discrete demand). The quantities in Table 16.1 are determined using:

$$\text{Overstock Cost} = (Q - D)C_O, \text{ when } D < Q \tag{16.1}$$

$$\text{E}[C_O] = f(x) \text{ (Overstock Cost)} \tag{16.2}$$

$$\text{Understock Cost} = (D - Q)C_U, \text{ when } D \geq Q \tag{16.3}$$

$$\text{E}[C_U] = f(x) \text{ (Understock Cost)} \tag{16.4}$$

[6] The analysis presented here can be done with any distribution. A Beta distribution was chosen because it has a defined lower bound (i.e., it does not go to $-\infty$).

Table 16.1. Newsvendor Problem Calculations for $Q = 10$.

Demand (D)	$f(x)$	Overstock Quantity (Q-D)	Overstock Cost	E[Co]	Understock Quantity (D-Q)	Understock Cost	E[Cu]
0	0	10	2	0			
1	0.0169441	9	1.8	0.030499			
2	0.030544	8	1.6	0.04887			
3	0.0411803	7	1.4	0.057652			
4	0.0492075	6	1.2	0.059049			
5	0.0549545	5	1	0.054955			
6	0.0587257	4	0.8	0.046981			
7	0.0608016	3	0.6	0.036481			
8	0.06144	2	0.4	0.024576			
9	0.0608766	1	0.2	0.012175			
10	0.0593262	0	0	0	0	0	0
11	0.0569831				1	0.8	0.045586
12	0.0540225				2	1.6	0.086436
13	0.0506011				3	2.4	0.121443
14	0.0468579				4	3.2	0.149945
15	0.0429153				5	4	0.171661
16	0.03888				6	4.8	0.186624
17	0.0348435				7	5.6	0.195124
18	0.0308834				8	6.4	0.197654
19	0.027064				9	7.2	0.194861
20	0.0234375				10	8	0.1875
21	0.0200445				11	8.8	0.176392
22	0.0169151				12	9.6	0.162385
23	0.0140697				13	10.4	0.146325
24	0.01152				14	11.2	0.129024
25	0.0092697				15	12	0.111237
26	0.0073155				16	12.8	0.093639
27	0.005648				17	13.6	0.076813
28	0.0042525				18	14.4	0.061236
29	0.0031098				19	15.2	0.047269
30	0.0021973				20	16	0.035156
31	0.0014897				21	16.8	0.025027
32	0.00096				22	17.6	0.016896
33	0.0005803				23	18.4	0.010678
34	0.0003227				24	19.2	0.006197
35	0.0001602				25	20	0.003204
36	0.0000675				26	20.8	0.001404

Table 16.1. (*Continued*)

Demand (*D*)	*f*(*x*)	Overstock Quantity (*Q-D*)	Overstock Cost	E[C_O]	Understock Quantity (*D-Q*)	Understock Cost	E[C_U]
37	2.195E-05				27	21.6	0.000474
38	4.453E-06				28	22.4	9.98E-05
39	2.856E-07				29	23.2	6.63E-06
40	0				30	24	0

The total expected loss in this case is given by

$$\text{Expected Total Loss} = \sum_{i=0}^{Q-1} \mathrm{E}\left[C_{O_i}\right] + \sum_{i=0}^{40} \mathrm{E}\left[C_{U_i}\right] = \$0.37 + \$2.64 = \$3.01 \tag{16.5}$$

The result in Equation (16.5) means that if the newsvendor purchases Q = 10 newspapers, he can expect to lose \$3.01.[7] If the analysis in Table 16.1 is repeated for Q = 16, the total loss = \$1.97, which indicates that buying 16 newspapers instead of 10 is better (a smaller loss). So, what is the value of Q that minimizes the expected total loss — that is, what is the optimum number of newspapers for the newsvendor to purchase?

If we let the expected total loss as a function of Q be denoted by $L(Q)$, and assume a continuous demand, then

$$L(Q) = C_O \int_0^Q (Q-x) f(x)dx + C_U \int_Q^\infty (x-Q) f(x)dx \tag{16.6}$$

Equation (16.6) expresses what was shown discretely in Table 16.1 and Equation (16.5), where $f(x)$ is the probability density function of the demand. The first term in Equations (16.5) and (16.6) is the expected cost of overstocking (having too many) and the second term is the expected

[7] Depending on the type of demand distribution used, the second sum in Equation (16.5) may go to ∞. In this example, a beta distribution with a fixed upper bound of 40 was used so the sums are complete (no terms are omitted).

cost of understocking (having too few). Taking the derivative of both sides of Equation (16.6) and setting it equal to zero to find a minimum gives

$$\frac{dL(Q)}{dQ} = C_O F(Q) - C_U\left[1 - F(Q)\right] = 0 \qquad (16.7)$$

where $F(Q)$ is the cumulative distribution function of the demand (the in-stock probability):

$$F(Q) = \int_0^Q f(x)dx \qquad (16.8)$$

The value of Q that satisfies Equation (16.7) is given by Q_{opt}, which is defined by

$$F(Q_{opt}) = \frac{C_U}{C_O + C_U} \qquad (16.9)$$

Equation (16.9) is called the critical ratio (or critical fractile) and is valid for any demand distribution (any $f(x)$). At Q_{opt}, the marginal cost of overstock is equal to the marginal cost of understock (marginal means just exactly break-even). At Q_{opt}, $F(Q_{opt}) = \Pr(D \le Q_{opt})$.

For the example given earlier in this section, $F(Q)$ is shown in Figure 16.4 and $F(Q_{opt}) = 0.8$ from Equation (16.9), which corresponds to $Q_{opt} =$ 16.9 from Figure 16.4.

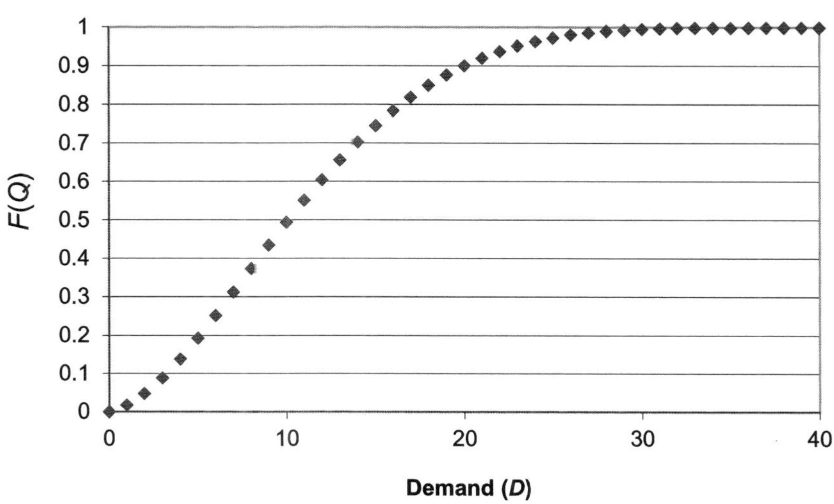

Fig. 16.4. $F(Q)$ vs. demand (D).

The solution discussed in this section assumes that backlogs are not allowed (i.e., unfulfilled demand is lost) and carryover is not allowed (i.e., leftover inventory has zero salvage value).

16.2.2 *Application of the Newsvendor Optimization Problem to Electronic Parts*

How can the newsvendor problem analysis in Section 16.2.1 be applied to lifetime-buying electronic parts? Assume you have to make a lifetime buy of an electronic part because it has become obsolete. Assume that the future demand for the part (to continue manufacturing and supporting the product) is given by a beta distribution with $\alpha = 2$ and $\beta = 5$ (lower bound 900, upper bound 1200); the parts can be purchased for \$2/part at the lifetime buy point. If the lifetime buy runs out the parts must be purchased from a broker for \$30/part. What is the optimum number of parts to buy?

For the example described above, the $C_O = \$2$ and $C_U = \$30\text{-}\$2 = \$28$. Satisfying Equation (16.9), the optimum quantity of parts to purchase is $Q = 1066$. However, in this simple treatment, there is an important implicit assumption and several key elements left out.

A "must support" assumption is implicit in lifetime buy problems, which can significantly increase the magnitude of the penalty associated with running out of parts. In the example above, you cannot choose not to support the product — that is, you are not allowed to fail to fulfill the demand and therefore you must pay the penalty to purchase extra parts from the broker if you run out.

Another significant assumption that is implicitly made in the classical newsvendor problem is that there is no time dependence. The examples given so far assume that time periods between purchasing newspapers (or parts) and selling, using, or running out of them are short. For the lifetime buy problem, this is not true. For lifetime buys of electronic parts to support sustainment-dominated systems, the parts are purchased, placed in inventory, and drawn from inventory over years, and if you run short of parts, the penalty is assessed at the end of the support period many years after the lifetime buy was made. In this case the cost of money (non-zero discount rate) and the cost of holding parts in inventory will play significant roles. The electronic part lifetime buy problem is analogous to the newspaper boy buying an inventory of papers in year 2010, paying to store the papers as he gradually sells them over a 10-year period, and then either having extra papers that can't be sold or customers that can't be satisfied in year 2021.

The inclusion of the cost of money in the discrete newsvendor problem solution does not affect the E[C_O] term in Equation (16.5) because the overbuy occurs at the beginning of the analysis (beginning of year 1) if money is in year 0 dollars. However, the E[C_U] term is impacted because the penalty for underbuying occurs after the order quantity, Q, runs out, which is at the end of the demand. The value of C_U depends on the year in which the understocking is rectified and the quantity that needs to be purchased in that year. In this case, Equation (16.3) for the ith demand in Table 16.1 becomes

$$\text{Understock Cost}_i = \sum_{j=1}^{y} q_{i,j} \frac{C_U}{(1+r)^{j-1}} \tag{16.10}$$

where y is the number of years the part needs to be supported for, and the quantity for the ith discrete demand in the jth year is given by

$$q_{i,j} = \begin{cases} \left\lceil \frac{D_i - Q}{y} \right\rceil & \text{if } (D_i - Q) \geq (y\text{-}j+1)\left\lceil \frac{D_i - Q}{y} \right\rceil \\ (D_i - Q) - (y\text{-}j)\left\lceil \frac{D_i - Q}{y} \right\rceil & \text{if } (y\text{-}j)\left\lceil \frac{D_i - Q}{y} \right\rceil < (D_i - Q) < (y\text{-}j+1)\left\lceil \frac{D_i - Q}{y} \right\rceil \\ 0 & \text{if } 0 < (D_i - Q) \leq (y\text{-}j)\left\lceil \frac{D_i - Q}{y} \right\rceil \text{ or } 0 > (D_i - Q) \end{cases} \tag{16.11}$$

Equation (16.11) simply says that the amount $\lceil (D_i - Q)/y \rceil$ is purchased in every year (starting with the last year and working backwards) until the entire understock has been purchased. Equation (16.11) assumes that parts are consumed uniformly over time and that the distribution of demand represents the total demand for the part over the whole life cycle of the system. The second term in Equation (16.5) is computed using

$$\text{E}[C_{U_i}] = f(x)_i\left(\text{Understock Cost}_i\right) \tag{16.12}$$

Now let's rework the discrete demand example presented at the beginning of this section. Assume a 5%/year discount rate and that the demand distribution represents the total demand over a 30-year period. In this case $r = 0.05$, $y = 30$, $C_O = \$2$ (year 0 dollars), and $C_U = \$28$ (year 0 dollars). The results from the analysis are shown in Figure 16.5. The minimum total loss with 0 discount rate is at Q = 1066 (as solved for previously). With the 5%/year discount rate, the minimum total loss corresponds to a significantly smaller buy of Q = 1022. The total loss is smaller because money is cheaper in the future, making the effective underbuy penalty smaller. The optimum buy size is less because the future underbuy penalty is less (i.e., the solution is to buy fewer because an underbuy isn't penalized as severely).

There are many extensions to the classical newsvendor formulation that accommodate a variety of different situations. Other, more detailed, discrete-event simulators have also been developed that include detailed penalty models and time-dependent effects, e.g., [Ref. 16.6].

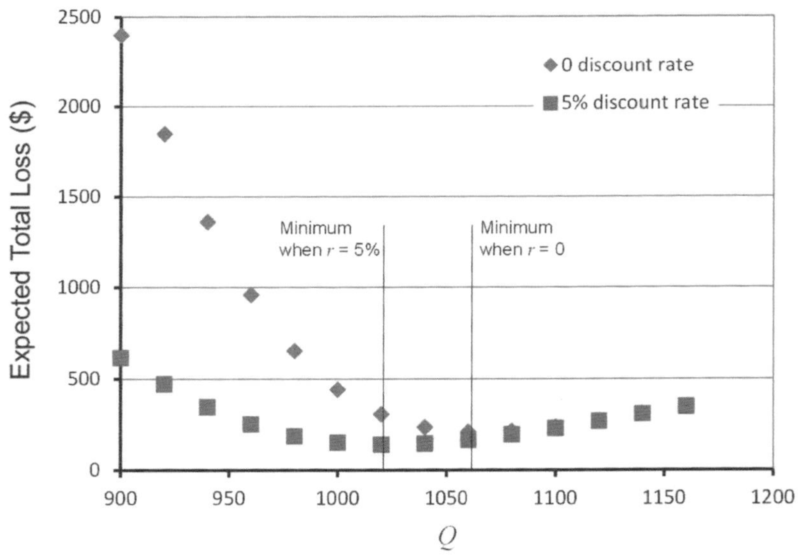

Fig. 16.5. Total loss as a function of Q.

16.3 Strategic Management of Obsolescence

All the obsolescence mitigation approaches discussed in Section 16.1 are *reactive* in nature, focused on minimizing the costs of obsolescence mitigation — that is, minimizing the cost of resolving the problem after it has occurred. While reactive solutions always play a major role in obsolescence management, ultimately, higher payoff is possible through *strategic* management approaches.

Planning strategic management activities requires life-cycle cost estimation in order to determine the magnitude of cost avoidance (see Section II.2). Because of the long manufacturing and field life associated with sustainment-dominated systems, they are usually refreshed or redesigned one or more times during their life to update functionality and manage obsolescence. Unlike high-volume commercial products for which redesign is driven by improvements in manufacturing, equipment or technology, for sustainment-dominated systems, design refresh[8] is often driven by obsolescence that would otherwise render the product un-producible and/or un-sustainable.

Ideally, a methodology that determines the best dates for design refreshes and the optimum reactive management approaches to use between the refreshes is needed. The next three subsections describe life-cycle cost modeling focused refresh planning solutions.

16.3.1 *Porter Design Refresh Model*

The simplest model for performing life-cycle planning associated with technology obsolescence (specifically, electronic part obsolescence) was developed by Porter [Ref. 16.9]. Porter's approach focuses on calculating the present value (PV) of last-time (bridge) buys[9] and design refreshes as a function of the design refresh date. As a design refresh is delayed, its PV decreases and the quantity (and thus, cost) of parts that must be purchased in the last-time buy required to sustain the system until the design refresh takes place increases. Alternatively, if design refresh is scheduled

---

[8] Refresh refers to changes that "have to be done" in order for the system functionality to remain usable. Redesign or technology insertion implies "want to be done" system changes, which include adopting new technologies to accommodate system functional growth and/or to replace and improve the existing functionality of the system [Ref. 16.8].

[9] A last-time or bridge buy means buying a sufficient number of parts to last until the part can be designed out of the system at a design refresh. Last-time buys become lifetime buys when there are no more planned refreshes of the system.

relatively early, then last-time buy cost is lower, but the PV of the design refresh is higher. In a Porter model, the cost of the last-time buy (C_{LTB}) is given by

$$C_{LTB} = \begin{cases} 0 & \text{when } i = 0 \text{ or if } Y_R = 0 \\ P_0 \sum_{i=1}^{Y_R} Q_i & \text{if } Y_R > 0 \end{cases} \quad (16.13)$$

where

i = the year.

P_0 = the price of the obsolete part in the year of the last-time buy (beginning of year 1 in this case).

Y_R = the year of the design refresh (0 = year of the last-time buy, 1 = one year after the last time buy, etc.).

Q_i = the number of parts needed in year i.

Equation (16.13) assumes that the part becomes obsolete at the beginning of year 1 and that the last-time buy is made at the beginning of year 1. Equation (16.13) also ignores holding costs — since the parts are purchased at the beginning of year 1, they must be held in inventory until they are needed. Holding costs for electronic parts (depending on the type of part) may not be negligible.

The design refresh cost for a refresh in year Y_R (in year 0 dollars), C_{DR}, is given by

$$C_{DR} = \frac{C_{DR_0}}{(1+r)^{Y_R}} \quad (16.14)$$

where

C_{DR_0} = the design refresh cost in year 0.

The total cost for managing the obsolescence with a year Y_R design refresh is given by

$$C_{Total} = C_{LTB} + C_{DR} \quad (16.15)$$

Figure 16.6 shows a simple example using the Porter model. In this case C_{DR_0} = \$100,000, r = 12%, Q_i = 500 (for all i from year 1 to 20, Q_i = 0 thereafter), and P_0 = \$10. In this simple example, the model predicts that the optimum design refresh point is in year 7.

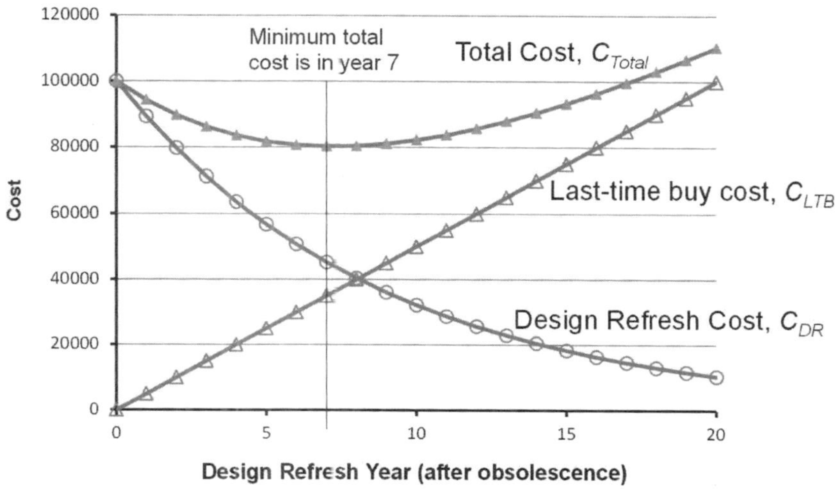

Fig. 16.6. Example application of Porter's design refresh costing model.

The optimum refresh year from the Porter model can be solved for directly for a simplified case. Substituting Equations (16.13) and (16.14) into Equation (16.15) and assuming that the demand quantity is the same in every year, we get

$$C_{Total} = P_0 \sum_{i=1}^{Y_R} Q_i + \frac{C_{DR_0}}{(1+r)^{Y_R}} = P_0 Q Y_R + C_{DR_0} e^{-rY_R} \tag{16.16}$$

Equation (16.16) assumes that $Q = Q_i$ for all $i = 1$ to Y_R and that $1/(1+r)^{Y_R} \cong e^{-rY_R}$ (see footnote 12 in Chapter 13). The minimum value of C_{Total} can be found by setting the derivative of Equation (16.16) with respect to Y_R equal to zero:

$$\frac{dC_{Total}}{dY_R} = P_0 Q - rC_{DR_0} e^{-rY_R} = 0 \tag{16.17}$$

Solving Equation (16.17) for Y_R we get[10]

$$Y_R = \frac{1}{-r} \ln\left(\frac{P_0 Q}{rC_{DR_0}}\right) \tag{16.18}$$

Equations (16.17) and (16.18) are only applicable when $r > 0$ (non-zero discount rate) and $rC_{CRo} \geq P_0Q$. For cases where $r = 0$ or $rC_{CRo} < P_0Q$ the optimum design refresh date is at $Y_R = 0$. It should be pointed out that the Y_R appearing in Equations (16.16) - (16.18) is the Y_R that minimizes life-cycle cost, whereas the Y_R appearing in Equations (16.13) and (16.14) is a selected refresh year. For the example given earlier, Equation (16.18) gives $Y_R = 7.3$ years.

The Porter model only treats the cost of supporting the system up to the design refresh, i.e., there is no accommodation for costs incurred after the design refresh. In the Porter model, the analysis terminates at Y_R. This means that the time span between the refresh (Y_R) and the end of support of the system is not modeled, i.e., the costs associated with buying parts after the design refresh to support the system to some future end-of-support date are not included and are not relevant for determining the optimum design refresh date. In order to treat multiple design refreshes in a product's lifetime, Porter's analysis can be reapplied after a design refresh to predict the next design refresh. Thus effectively optimizing each individual refresh, but the coupled effects of multiple design refreshes (coupling of decisions about multiple parts and coupling of multiple refreshes) in the lifetime of a product are not accounted for, which is a significant limitation for the application of the Porter approach to real systems.

---

[10] At its simplest level, the conceptual basis for the construction of the basic Porter model is similar to the construction of EOQ (Economic Order Quantity) models, (see Section 12.2). In the case of EOQ models, the sum of the part cost (purchase price and holding/carrying cost) and the order cost is minimized to determine the optimum quantity per order. The Porter model has a similar construction where the part cost is the same as in the EOQ model (with the addition of the cost of money) and the order cost is replaced by the cost of design refreshing the system to remove the obsolete part.

The Porter model performs its tradeoff of last-time buy costs and design refresh costs on a part-by-part basis. While the simple Porter approach can be extended to treat multiple parts, and a version of Porter's model has been used to plan design refreshes in conjunction with lifetime buy quantity optimization in [Ref. 16.10], it only considers a single design refresh at a time.

16.3.2 *MOCA Design Refresh Model*

A more complete optimization approach to design refresh planning, the mitigation of obsolescence cost approach (MOCA), has been developed that optimizes over multiple design refreshes (removes the single design refresh constraint in the Porter model), accommodates multiple obsolescence mitigation approaches (the Porter model only considers last-time buys), and includes appropriate holding costs for last-time buys (the Porter model assumes these are zero) [Ref. 16.11]. Using a detailed cost analysis model, the MOCA methodology determines the optimum design refresh plan during the field-support life of the product. The design refresh plan considers the number of design refresh activities, their content, and their respective calendar dates that minimize the life-cycle sustainment cost of the product.

MOCA is a discrete-event simulator that stochastically models a timeline (Figure 16.7). Fundamentally, the model supports a design through periods of time when no parts are obsolete, followed by multiple part-specific obsolescence events. When a part becomes obsolete, some type of mitigation approach must take effect immediately: either sufficient inventory exists, a lifetime buy of the part is made, or some other short-term mitigation strategy is used that only applies until the next design refresh. Next, there are periods of time when one or more parts are obsolete, and short-term mitigation approaches are in place on a part-specific basis. When design refreshes are encountered, the change in the design at the refresh is determined and the costs associated with performing the design refresh are computed. At a design refresh, a long-term obsolescence mitigation solution is applied (until the end of the product life or possibly until some future design refresh), and non-recurring, recurring, and re-qualification costs are computed. Re-qualification may be required depending on the impact of the design change on the application. The necessity for re-qualification depends on the role that the particular part(s) play and/or the quantity of non-critical changes made. The last activity appearing on the timeline is production. Systems often have to be produced after parts begin to become obsolete due to the length of the initial design/manufacturing process, additional orders for the system, and replenishment of spares.

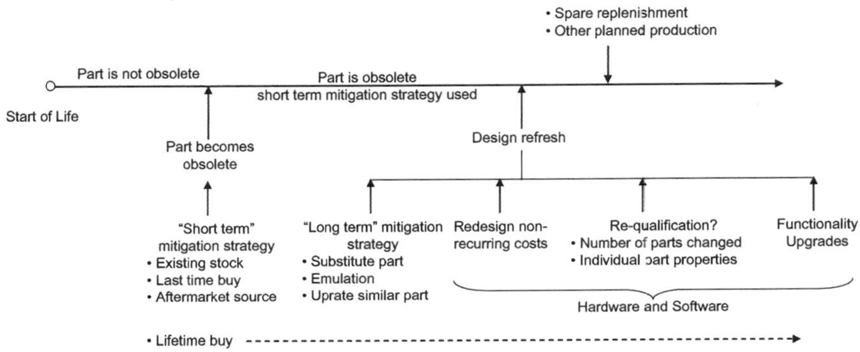

Fig. 16.7. Design refresh planning analysis timeline (presented for one part only, for simplicity; however, in reality, there are coupled parallel timelines for many parts, and design refreshes and production events can occur multiple times and in any order).

The MOCA methodology can be used during either the original product design process, or to make decisions during system sustainment, (to determine the best set of changes to make given an existing history of the product and forecasted future obsolescence and future design refreshes). See [Ref. 16.11] for design refresh planning analyses using MOCA.

16.3.3 *Material Risk Index (MRI)*

The idea of an MRI is to evaluate the time-dependent risk of a particular function or subsystem within a system being impacted by obsolescence to specific degrees that require specific actions. This evaluated risk can then be mapped to life-cycle cost or sustainment dollars at risk.

To perform an MRI on a system, first, a catalog of functions, subsystems, or specific part profiles is created. For example, the catalog could contain memory modules, processor boards, and so on. Each profile is characterized by a set of time-dependent obsolescence risk impacts. The periods can represent whatever timeframe is relevant to the function or subsystem (usually 3 or 5 years). The obsolescence risk (*OR*) can be interpreted using any one of the following:

- Period-independent model = The *OR* is used to determine the average number of items of a particular profile that are impacted by obsolescence to the extent that some action is required during a period.
- Fractional sum model = *OR* is the % of "up-to-date" items that experience obsolescence problems severe enough in the present period to require some action in the present period.
- Probabilistic model = *OR* is the probability of an item encountering obsolescence problems in the period.

Each of the risk models above represents a different interpretation (and thereby a different accumulation) of the obsolescence risk values.

Once a catalog has been created, cost models for each type of action that appears in the catalog are developed. Activity-based cost (ABC) models for organizations are an appropriate source of data to characterize the costs of activities. Application-specific results are obtained from the MRI model in the ith time period using

$$C_i = \sum_{p=1}^{n} (N_p)(OR_{p_i})(C_{p_i}) \tag{16.19}$$

where a profile can represent a function, subsystem, or part type, and,

n = the total number of profiles in the application.
N_p = the number of instances of the profile p in the application.
OR_{p_i} = the *OR* for profile p in period i.
C_{p_i} = the cost of the action defined in profile p in period i.

MRI models require significant resources to create and calibrate, but once created they are very easy and quick to use.

16.4 Discussion

Electronic part obsolescence is a growing (and expensive issue) for sustainment-dominated products. There are several other topics associated with obsolescence that impact system cost.

16.4.1 *Budgeting/Bidding Support*

Methods have been developed in [Ref. 16.12] to facilitate accurate budgeting or bidding. These methods perform two actions: first, they determine the probabilities of using specific resolution activities, and then they predict an application-specific cost of performing the predicted group of resolution activities. Both actions are performed based on practitioner surveys, expert opinions, and historical information. The result is an estimation of the obsolescence management costs for a defined contract period using commonly defined resolution approaches. For organizations that wish to estimate management costs for systems based on their own or the industry's prior system management history, this approach is valuable. It may also be possible to use this approach to perform tradeoffs associated with shifting the resolution approach focus within organizations.

16.4.2 *Value of DMSMS Management*

Determining the value of DMSMS management activities is an important metric for establishing the value of DMSMS management organizations. The most common cost avoidance approach used by DMSMS management organizations is based on a bookkeeping approach first articulated in a DMEA report written by ARINC from 1999 [Ref. 16.13]. In this approach, the cost avoidance associated with the chosen mitigation solution is equal to the difference between the cost of your solution and the next most expensive mitigation option.

Requesting resources to create cost avoidance is not as persuasive as making a return on investment (ROI) argument. Because of the problems with the conventional cost avoidance calculation and the need for more persuasive arguments to offer management, ROI-based evaluation methods have been developed [Ref. 16.14].

16.4.3 *Software Obsolescence*

Obsolescence also impacts system software. The applicable definition of software obsolescence varies depending on the system that uses the software, and where and how that system is being used. Commercial software has both end-of-sale dates and end-of-support dates that can be separated by long periods of time. For many mainstream commercial software applications (e.g., PC operating systems), both the end-of-sale and end-of-support dates may be published by the software vendors. For applications that have a connection to the public web (e.g., servers and communications systems), the relevant software obsolescence date for both the deployment of new systems and the continued use of fielded systems is often the end-of-support date, because that is the date on which security patches for the software terminate, making continued use of the software a security risk. For other embedded or isolated applications, the relevant software obsolescence date is governed by either an inability to obtain the necessary licenses to continue using it or changes to the system that embeds it (functional obsolescence issues). See [Ref. 16.15] for more discussion of software obsolescence.

16.4.4 *Human Skills Obsolescence*

Obsolescence isn't confined to just hardware and software. Many types of systems that have to be supported for long periods of time lose critical portions of their workforce before the support for the system ends. The loss of critical workforce does not refer to the normal turnover of unskilled labor, but rather, the loss of highly-skilled engineers that have unique experience and are either non-replenishable or would take very long periods of time to reconstitute.

While there is lots of existing research on "skills obsolescence," — people who have obsolete skills and need to be retrained in order to be employable [Ref. 16.16]. The type of obsolescence referred to here is the opposite of skill obsolescence, it is "critical skills loss," which is a special case of "organizational forgetting" , it is the loss of knowledge gained through learning-by-doing [Ref. 16.17].

In [Ref. 16.18] a model is constructed that uses historical workforce data to forecast the size and experience of the workforce pool as a function of time. The workforce experience pool is then used to determine the cost of supporting a system as a function of time. The model is used to determine what today's skills pool look like in the future, and what impact the future skills pool will have on the organization's ability to continue to support the system.

References

16.1 Sandborn, P. (2008). Trapped on technology's trailing edge, *IEEE Spectrum*, 45(1), pp. 42-45.

16.2 Song, Y. and Lau, H. (2004). A periodic review inventory model with application to the continuous review obsolescence problem, *European Journal of Operations Research*, 159(1), pp. 110-120.

16.3 Fine, C. (1998). Clockspeed: Winning Industry Control in the Age of Temporary Advantage (Perseus Books, Reading, MA).

16.4 Pecht, M. and Tiku, S. (2006). Electronic manufacturing and consumers confront a rising tide of counterfeit electronics, *IEEE Spectrum*, 43(5), pp. 37-46.

16.5 Pecht, M. and Humphrey, D. (2006). Uprating of electronic parts to address obsolescence, *Microelectronics International*, 23(2), pp. 32-36.

16.6 Feng, D., Singh, P. and Sandborn, P. (2007). Optimizing lifetime buys to minimize lifecycle cost, *Proceedings of the 2007 Aging Aircraft Conference*.

16.7 Edgeworth, F. (1888). The mathematical theory of banking, *J. Royal Statistical Society*, 51, pp. 113-127.

16.8 Herald, T. E. (2000). Technology refreshment strategy and plan for application in military systems – A how-to systems development process and linkage with CAIV, *Proceedings of the National Aerospace and Electronics Conference (NAECON)*, pp. 729-736.

16.9 Porter, G. Z. (1998). An economic method for evaluating electronic component obsolescence solutions, Boeing Company White Paper.

16.10 Cattani, K. D. and Souza, G. C. (2003). Good buy? Delaying end-of-life purchases, *European Journal of Operational Research*, 146, pp. 216-228.

16.11 Singh, P. and Sandborn, P. (2006). Obsolescence driven design refresh planning for sustainment-dominated systems, *The Engineering Economist*, 51(2), pp. 115-139.

16.12 Romero Rojo, F. J., Roy, R., Shehab, E. and Cheruvu, K. (2010). A cost estimating framework for materials obsolescence in product-service systems, *Proceedings of the ISPA/SCEA Conference*.

16.13 McDermott, J., Shearer, J. and Tomczykowski, W. (1999). Resolution Cost Factors for Diminishing Manufacturing Sources and Material Shortages, ARINC.

16.14 Shaw, W., Speyerer, F. and Sandborn, P. (2010). *DMSMS Non-Recurring Engineering Cost Metric Update*, ARINC.

16.15 Sandborn, P. (2007). Software obsolescence - Complicating the part and technology obsolescence management problem, *IEEE Transactions on Components and Packaging Technologies*, 30(4), pp. 886-888.

16.16 De Grip, A. and Van Loo, J. (2002). The economics of skills obsolescence: A review, *Research in Labor Economics*, 21, ed. A. De Grip, J. Van Loo, and K. Mayhew, Elsevier, pp. 1-26.

16.17 Besanko, D. Doraszelski, U., Kryukov, Y. and Satterthwaite, M. (2010). Learning-by-doing, organizational forgetting and industry dynamics, *Econometrica*, 78(2), pp. 453-508.

16.18 Sandborn, P. A. and Prabhakar, V. J. (2015). The forecasting and impact of the loss of the critical human skills necessary for supporting legacy systems, *IEEE Transactions on Engineering Management*, 62(3), pp. 361-371.

Problems

16.1 Find an example on the web of a discontinued electronic part. What was the date on which the part was discontinued? Find an example of a part that is not discontinued yet, but for which the manufacturer has issued a last-time buy date.

16.2 Perform the discrete newsvendor problem calculations for the example in Section 16.2.1 for $Q = 8$ and for $Q = 23$. What are the expected total losses in this case?

16.3 For the demand distribution considered in Section 16.2.1, if $Q_{opt} = 18$, and $C_O = \$3$, what is C_U?

16.4 What does an expected total loss of zero imply in the newsvendor problem?

16.5 Assuming that holding cost is zero and that buying extra parts (if needed) from a broker happens during a short period of time at the end of the need for the part, why can't the cost of money simply be accounted for by modifying the penalty to account for the discount rate? What is wrong with this approach?

16.6 Why doesn't the example problem in Section 16.2.1 use a normally distributed demand?

16.7 Derive Equation (16.7). Note that the equation can be derived for either discrete demand (starting from Equation (16.5) with the second summation to ∞) or continuous demand (starting from Equation (16.6)).

16.8 Verify that Equation (16.11) works correctly by constructing a table of q_{ij}. Hint: Use $Q = 30$, $y = 10$, range i from 1 to 10 and D_i from 35 to 44, find q_{ij} for $j = 1$ to 10.

16.9 If the discount rate is r = 15%/year, what is the optimum Q for the lifetime buy problem considered in Section 16.2.2?

16.10 Derive a general holding cost to include in the Porter model. Assume that the demand quantity (Q) is the same in every year and that the demand is drawn at a constant rate throughout the year and that the holding cost per part per year is C_h.

16.11 A part becomes obsolete and there is no remaining manufacturing demand, but spare parts are needed to maintain the system. The reliability of the part is characterized by the Weibull distribution given in Equation (11.18) with $\beta = 4$, $\eta =$ 600 parts, and $\gamma = 0$. The parts can be purchased for \$2/part at the lifetime buy point and the cost of buying the part from a broker later is \$50, what is the optimum number of parts to buy? Ignore cost of money and holding costs. Calculate the exact solution.

16.12 Using the Porter model, what year should a design refresh be performed if C_{DR_0} = \$67,000, r = 22%/year, Q_i = 500 (for 15 years and zero thereafter), and P_0 = \$16?

16.13 Using a Porter model, if C_{DR_0} = \$100,000, r = 12%/year, Q_i = 500 (for all i from 0 to 20 and Q_i = 0 thereafter), P_0 = \$10 and an inflation rate of 3% is assumed, what is the optimum design refresh date? Assume that the inflation rate applies to both the part price and the cost of the design refresh.

16.14 Part "A14" is discontinued (becomes obsolete) at the beginning of year 1). The demand for the part is 2765 per year (constant, for all years) and the price of the part (in year 0 dollars) is \$2.34/part. A design refresh that will design out the part is scheduled to take place in year 9 (assume that the refresh is not finished and available until the end of year 9). Assume that the refresh, which will cost \$389,000 when it is performed and has to be paid for on completion (at the end of year 9). If the discount rate is 6%:

a) How much money should be budgeted at the beginning of year 1 for this management solution, assuming you need to get through the design refresh? Assume discrete compounding.
b) What is the cost avoidance (in year 0 dollars) of delaying the refresh 2 years (available at the end of year 11), assuming you need to get through the design refresh? Assume discrete compounding.
c) Assuming continuous compounding, what would the optimum year for the refresh be?
d) In the original case (year 9 refresh), how much should be budgeted if there is a holding cost for the parts that is 10% of the part price per year (paid on the last day of the year). For simplicity, assume that the entire year's part demand is drawn on the last day of the year (including year 9).

Chapter 17

Return on Investment (ROI)

When managers consider spending money they usually want to formulate a business case that not only describes the process they wish to follow, but also the value that they expect to gain through the investment. For electronic systems manufacturing and life-cycle support, business cases could be required for spending money to modify a manufacturing line, refresh the design of a system, add or expand product or system management activities, or adopt a new technology. One common way to quantify the value is to compute a return on investment (ROI) for a given use of money.

While the formulation of an ROI associated with investing money in a financial instrument is straightforward, the calculation of an ROI associated with the generation of an increase in the customer base, cost savings, or future cost avoidance is not as simple to perform. This chapter discusses the formulation and application of ROIs to activities relevant to electronic systems manufacturing and management.

17.1 Definition of ROI[1]

A rate of return is the benefit received from an investment over a period of time. Generally returns are ratios relating the amount of money that is gained or lost to the amount of money risked. Return on investment (ROI)

[1] The concept of ROI originated as part of what is known as the DuPont analysis (also known as the DuPont identity, DuPont equation, DuPont model or the DuPont method). The DuPont analysis was developed by an electrical engineer named F. Donaldson Brown and was first used in 1918, when DuPont purchased a substantial stake in General Motors, to examine the fundamental drivers of profitability at GM.

is the monetary benefit derived from having spent money on developing, changing, or managing a product or system. ROI is a common performance measure used to evaluate the efficiency of an investment or to compare the efficiency of a number of different investments. To calculate ROI, the benefit or gain associated with an investment is divided by the cost of the investment and the result is expressed as a percentage or a ratio:

$$ROI = \frac{Return - Investment}{Investment} = \frac{V_f - V_i}{V_i} \tag{17.1}$$

The second equality in Equation (17.1) is the form that the finance world uses to express ROI, where V_f and V_i are the final and initial values of an investment, respectively. The quantity behind the second equality is also known as a single-period arithmetic return. The quantity expressed in Equation (17.1) is the true rate of return on an investment that generates a single payoff after one period (where the period is the length of time over which the value is measured).[2]

A key to using Equation (17.1) is to realize that the return (or final value, V_f) includes the investment (or initial value, V_i) and that the difference of the two is the gain realized by making the investment. For the formulation in Equation (17.1), an ROI of 0 represents a break-even situation — that is, the value you get back exactly equals the value you invested. If the ROI is > 0, then there is a gain; if the ROI is < 0, there is a loss. Constructing a business case for a product does not necessarily require that the ROI be greater than zero; in some cases, the value of a product is not fully quantifiable in monetary terms, or the product is necessary in order to meet a system requirement that could not otherwise be attained, such as an availability requirement (discussed in Chapter 15). However, ROIs are still important parts of business cases, even if they are not > 0.

---

[2] Other forms of the rate of return that are used in finance include logarithmic (or continuously compounded) return; and arithmetic, geometric, and multiple periods as an average of single period returns (either arithmetically or geometrically determined), see [Ref. 17.1]. Note, the discount rate, r, defined in Equation (II.1) is also known as the internal rate of return.

The simplest application of ROI is the calculation of the return on a financial investment. If V_i = \$100 is invested in the stock market, and at some later time the value of the stock has increased to V_f = \$150, then the ROI associated with the investment is ROI = (150-100)/100 = 0.5, or 50% over the time period of the investment.

Keep in mind that the calculation for return on investment, and therefore the definition, can be modified to suit the situation, depending on what you include as returns and investments. The definition of ROI in the broadest sense attempts to measure the profitability of an investment and, as such, there is no single "right" calculation. For example, a marketing organization may compare two different products by dividing the revenue that each product generates by its respective marketing expenses. A financial organization, however, may compare the same two products using an entirely different ROI calculation, perhaps by dividing the net income of an investment by the total value of all resources that have been employed to make and sell the product. This flexibility has a downside because ROI calculations can be easily manipulated to suit the user's purposes and the result can be expressed in many different ways. When using this metric, make sure you understand what inputs are being used and use them consistently.

ROIs are easy to calculate but deceivingly difficult to get right. Financial investment ROIs are straightforward, but when evaluating the ROI of a cost savings, market share increase, or cost avoidance, the difference between costs that are investments and those that are returns is blurred.

17.2 Cost Reduction and Cost Savings ROIs

In this section we present several examples where an ROI associated with reducing or saving cost is desired. This type of ROI could be used to justify the investment (or use) of money when the return is a savings.

17.2.1 *ROI of a Manufacturing Equipment Replacement*

In this example, consider the ROI associated with replacing an old piece of manufacturing equipment with a newer piece of equipment. Consider the input data summarized in Table 17.1. The recurring cost per unit manufactured with the new machine is less, possibly due to the requirement for less labor oversight, or perhaps the new machine is more energy efficient. The new machine introduces fewer defects, as expressed in the increased yield. In addition, assume that defects introduced by the machines are non-repairable, that there is no salvage value in defective units, that the defects are detected immediately after the process step that uses the machine, and that there is no salvage value in the old machine. For simplicity, assume that the cost of maintenance is the same for both machines and that there is no depreciation schedule.

Table 17.1. Replacement Equipment Assumptions.

	Old Machine	New Machine
Purchase price	\$0 (owned)	\$100,000
Recurring cost (cost per unit manufactured with the machine)	\$0.50	\$0.40
Yield of manufactured parts	0.95	0.97

Two more key pieces of information are needed to perform an ROI calculation. First, we need to know where in the manufacturing process the machine resides. Why? Part of the value of the new machine is the increase in yield of the parts that are processed by it. In order to place a monetary value on the yield increase, we need to know how much money has been spent on a unit being manufactured when it arrives at the machine. For example purposes, let's assume \$1.35/unit has been spent prior to reaching the machine and that none of this is recoverable if defects are introduced to the unit by this machine. The other piece of input information needed is the volume of units that will be manufactured in the machine's lifetime. The ROI of the new machine purchase as a function of the volume (V) is

$$ROI = \frac{V[(\$0.50 - \$0.40) + (0.97 - 0.95)(\$1.35)] - \$100,000}{\$100,000} \tag{17.2}$$

We have assumed that both machines (new and old) would cost the same to maintain for whatever period of time is required to produce V units.

In this case, the return is a combination of reduced recurring cost per unit and increased yield per unit that results in lower scrap. The fact that the new machine is less expensive to operate (resulting in a lower recurring cost per unit manufactured) is not incorporated into the ROI calculation as a lower investment cost in the new machine but, rather, is a result of the investment in the new machine and therefore is included in the return.

Figure 17.1 shows the resulting ROI of the new machine as a function of the volume of units produced during the machine's lifetime. The conclusion from this example is that if more than 787,401 units are going to be manufactured during the machine's life, then there is a financial advantage to buying the new machine.

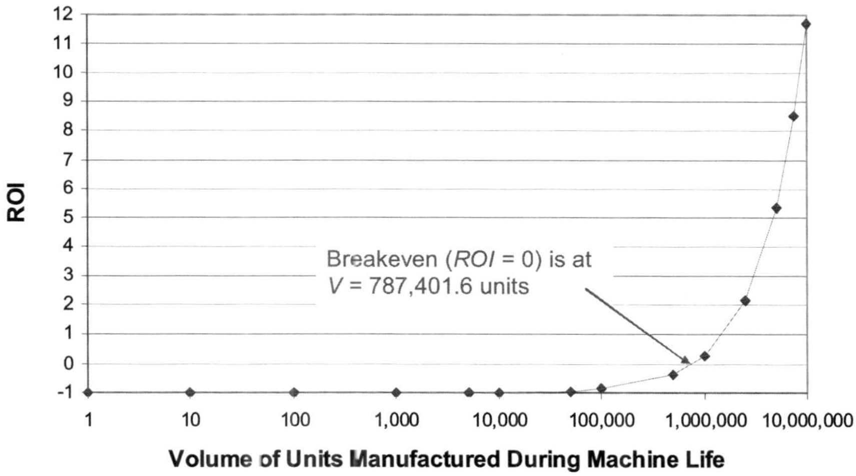

Fig. 17.1. ROI as a function of volume of units manufactured.

17.2.2 *Technology Adoption ROI*

In the early 1990s several companies invested in flip chip technology.[3] The investment was not cheap and the companies wanted to know, "how many years will it take for the investment to pay off?"

---

[3] Flip chip technology, originally known as controlled collapse chip connection (C4), was developed by IBM in the 1960s for ICs used in their mainframe computer systems. Although several other companies attempted to develop similar technologies, flip chip remained largely an IBM-only technology until the late 1980s and early 1990s, when IBM began to license the C4 technology to others. A history of flip chip technology appears in [Ref. 17.2].

Flip chip is a method for connecting semiconductor devices, such as integrated circuits, to the next level of the package (a single chip package or directly onto a board) with solder bumps that are deposited onto the die pads. The solder bumps are deposited on the die pads on the top side of the wafer during the final wafer processing step. In order to mount the die to external circuitry (e.g., a circuit board, the leadframe in a package or another die or wafer), it is flipped over so that its top side faces down, aligned so that its pads align with matching pads on the external circuit, and then the solder is reflowed to complete the interconnect. This is in contrast to wirebonding, in which the die is mounted facing up and wires are used to connect the chip pads to external circuitry as in Figure 17.2.

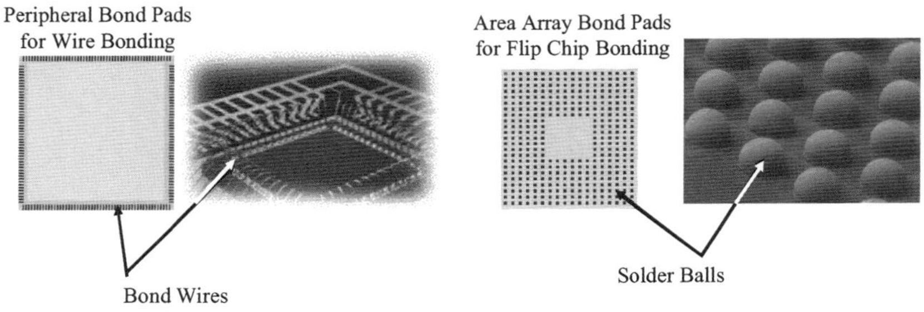

Fig. 17.2. Description of peripheral and flip chip bonding.

The possible contributions to ROI associated with technology transitions include: changes in engineering/design productivity, manufacturing cost, manufacturing productivity (throughput), product quality improvement, product reliability improvement (leading to warranty cost and/or other sustainment cost changes), product extensibility, and product performance.

For transitioning from peripherally bonded chips to area array (flip chip) bonded chips the investments to be considered include the following:

- purchasing a license to the technology
- hiring experts who know how to implement the technology
- buying new equipment
- implementing and characterizing the new processes
- training processing engineers and technicians (learning curves apply)

- performing qualification testing on the new parts
- ISO certification of the process
- purchasing new design software (plus training designers to use it)
- redesigning existing die
- redesigning package leadframes
- creating new user documentation and part datasheets.

The returns we will consider are:

- smaller die (more die up on the wafer) and associated cost and/or yield improvements
- higher electrical performance that may lead to an ability to maintain and improve the market share for the parts.

For simplicity, assume that the company is only going to use flip chip to replace wirebonded die inside of single chip packages (it is not going to sell bare flip chip die). The data shown in Table 17.2 is assumed for this example.

Unlike the example in Section 17.2.1, this problem is complicated because time is a factor — not everything happens at the same moment. For example, there is a significant amount of time between licensing the technology and producing the first article, during which there is no return on the investment; also, the cost of money must be considered. The type of analysis performed in this example is known as discounted cash flow ROI. Table 17.3 shows the investment costs as a function of time for this example. The cost values correspond to the end of each year.

All costs in Table 17.3 are in year 0 dollars and follow an end-of-year convention. For example, to determine the cost for the New Equipment category as a function of year we use Equation (II.1) and assume straight line depreciation to obtain

$$\text{New Equipment Cost in Year } i = \frac{\$1,200,000 / D_L}{(1+r)^i} \tag{17.3}$$

where $r = 0.07$ and $D_L = 5$ (depreciation life). Obviously, several of the costs appearing in Table 17.2 could be distributed differently among the years — the distribution assumed in Table 17.3 is only one possibility. After year 5 no more investment in the technology or process is assumed.

Table 17.2. Flip Chip Technology Adoption Assumptions.

License fee	\$5,000,000	One-time payment made at the end of year 0, i.e., beginning of year 1
Additional staff hired to adopt the new technology	10 people	Hired at the beginning of year 1 and only needed until production start
Burdened cost of additional staff per year	\$130,000	Per person
New equipment purchase price	\$1,200,000	Assume a D_L = 5 year straight-line depreciation, first charge made at the end of year 1
Number of process engineers that need to be trained	25	One-time training cost assumed to occur during year 1
Training cost per process engineer	\$3200	Per process engineer
ISO certification of process changes	\$50,000	One-time cost assumed to occur at the end of year 1
New design software	\$200,000	One-time cost assume at the end of year 0, i.e., beginning of year 1
Number of affected chips (N_c)	50	
Cost of redesign of a leadframe	\$5000	Per leadframe
Cost to redesign a die	\$10,000	Per unique die
Discount rate (r)	7%	Per year
Years to production	1.5	Number of years before the first flip chip product is produced
Average die shrink (D_s)	7%	Of die area
Average die cost	\$9	Per die
Profit per chip (P)	\$3.15	Per chip
Average sales volume per chip (S)	500,000	Per year
Market share increase (M_s)	3.5%	One time

Table 17.4 shows the return and ROI over the first 6 years of the flip chip technology adoption.

The return that is specific to the investment in flip chip technology at the beginning of each year (starting in year 3) is given by

$$\text{Return}_i = \left[\frac{PSN_c}{(1+r)^i}\right]\frac{D_s + M_s}{100} \tag{17.4}$$

where

P = the average profit per chip.

S = the average original sales volume per chip per year.

N_c = the number of chips effected.
D_s = the average die shrink (% area decrease).
M_s = the average market share one-time increase (%).

Table 17.3 Flip Chip Technology Adoption Investment as a Function of Time for the First Six Years (blank cells in the table have a value of zero), end of year convention

Year (*i*)	0	1	2	3	4	5	6
Licensing	$5,000,000						
Additional Staff		$1,214,953	$567,735				
New Equipment		$224,299	$209,625	$195,911	$183,095	171,117	
Training Process Engineers		$74,766					
ISO Process Certification		$46,729					
New Design Software	$200,000						
Redesign Leadframes		$233,645					
Redesign Die		$467,290					
Cumulative Investment	$5,200,000	$7,461,682	$8,239,043	$8,434,954	$8,618,049	$8,789,166	$8,789,166

Table 17.4 Flip Chip Technology Return for the First Six Years, end of year convention

Year (*i*)	0	1	2	3	4	5	6
Return	$0	$0	$3,611,123	$6,749,763	$6,308,190	$5,895,504	$5,509,817
Cumulative Return	$0	$0	$3,611,123	$10,360,886	$16,669,076	$22,564,581	$28,074,398
ROI	-1	-1	-0.56	0.23	0.93	1.57	2.19

The return at the end of year 1 is zero since the technology does not result in the sale of the first chip (with flip chip) until half-way through year 2 (note that in year 2, Equation (17.4) is multiplied by 0.5 to account for only half a year of sales of the new chips).

Using the cumulative investments from Table 17.3 and cumulative returns from Table 17.4, the cumulative ROI is computed in each year using Equation (17.1).[4] Figure 17.3 shows the ROI as a function of time for the first 12 years after technology licensing. Results of two different years to production and two different discount rates are shown. The break-even points (where the ROI is 0) range from 2.8 to 4.5 years. The discount rate reduces the value of money in the future, so when the discount rate is zero, the ROI becomes larger faster.

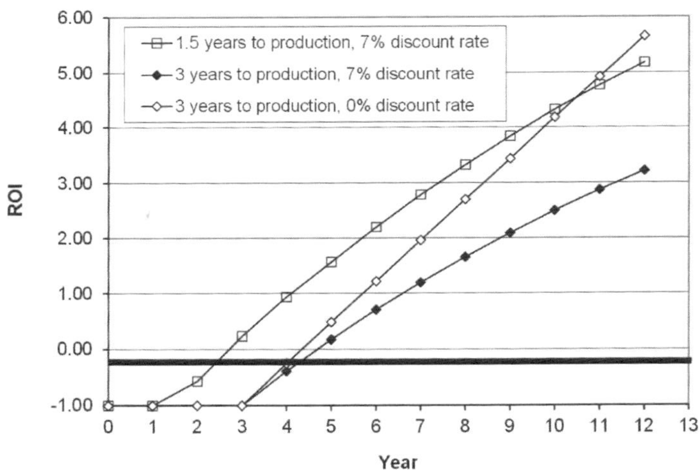

Fig. 17.3. Flip chip technology adoption ROI.

This example included many implicit assumptions that would need to be carefully evaluated if a true ROI for flip chip technology adoption was being determined. These assumptions included no inflation, no price erosion in the chips sold, no service contracts associated with the new equipment, no consumable or energy costs associated with the new

[4] Note, the cumulative ROI in year *i* is not computed from the ROI in year *i*-1. Rather, the cumulative ROI in year *i* is computed from the cumulative investment and return up to year *i*.

equipment, and no new permanent hires needed (the only new people are used to adopt the technology and then they are off the payroll). We have also not specified die areas in this model — we have simply assumed that a 7% die shrink will correspond to (on average) 7% more die being produced on a wafer. We have not assumed that there is any effect on the yield of the die, but the yield of the die would likely increase because the die are smaller (see Chapter 3).

17.3 Cost Avoidance ROI

Cost avoidance is a metric that results from a spend that is lower than what would have otherwise been required if the cost avoidance exercise had not been undertaken [Ref. 17.3]. Restated, cost avoidance is a reduction in costs that have to be paid in the future. Cost avoidance is commonly used as a metric by organizations that have to support and maintain systems to quantify the value of the services that they provide and the actions that they take.[5]

As an example of a cost avoidance ROI calculation, consider the determination of an ROI for performing condition-based maintenance (CBM) on a system. CBM uses real-time data from the system to observe the system's state (condition monitoring) and thus determine its health. CBM then allows action to be taken only when maintenance is necessary [Ref. 17.4]. The alternatives are to perform maintenance on a fixed schedule (whether it is actually needed or not) or to adopt an unscheduled maintenance policy in which maintenance is only performed when the system fails. CBM allows minimization of the remaining useful life of the system component that would be thrown away by implementing fixed scheduled maintenance policies and avoidance of failures that accompany unscheduled maintenance policies. CBM, however, is costly to implement and maintain. Is it worth it?

---

[5] These organizations do not like to use the term "cost savings" since a savings implies that there is unspent money, whereas in reality there is no unspent money, only less money that needs to be spent. Another way to put it is, if you told a customer that you saved $100, the customer could ask you for the $100 back; if you told a customer you avoided spending $100 there is no $100 to give back.

As an example, consider changing the oil in your car. A fixed scheduled maintenance approach is to change the oil every 3000 miles, but not every 3000-mile period is equivalent. During some 3000-mile periods the degradation of the oil may be minimal (due to the conditions under which the car was driven) and the oil could be left in the engine without any detrimental effects for 5000 miles. In this case the fixed scheduled interval of 3000 miles for the oil change results in throwing away significant remaining useful life in the oil (i.e., money lost). During another 3000-mile period the oil is significantly degraded and causes damage to the engine after 2000 miles, resulting in future maintenance costs on the engine. So, if it costs an extra \$500 per vehicle to implement an oil monitoring system that can sense when the oil needs to be changed, is it worth it?

Consider the maintenance of an electronic system. Electronics is almost always managed via an unscheduled maintenance policy — that is, electronics is only fixed when it breaks. The version of CBM applied to electronics is called prognostics and health management (PHM) [Ref. 17.5]. PHM is a broader concept than CBM, in addition to the current condition of the system, it also considers the expected future usage conditions for the system in order to provide advanced warning of system failures (it determines a remaining useful life – RUL) to avoid failure and/or optimize the maintenance of the system.

To formulate the ROI for adding PHM to an electronic system, we first have to decide what we are measuring the ROI relative to. In the case of electronics, we will measure the ROI of PHM relative to the unscheduled maintenance case, since this is the commonly used default maintenance policy. The ROI from Equation (17.1) becomes [Ref. 17.6]

$$ROI = \frac{V_f - V_i}{V_i} = \frac{C_u - C_{PHM}}{I_{PHM} - I_u} \tag{17.5}$$

where

C_u = the life-cycle cost of the system when managed using unscheduled maintenance.

C_{PHM} = the life-cycle cost of the system when managed using a PHM approach.

I_{PHM} = the investment in PHM when managing the system using a PHM approach.

I_u = the investment in PHM when managing the system using unscheduled maintenance.

To form Equation (17.5), replace V_f-V_i with C_u-C_{PHM} (which assumes $C_u > C_{PHM}$) and V_i with I_{PHM}-I_u. Note, C_u and C_{PHM} are total life-cycle costs that include their respective investment costs, I_u and I_{PHM}. The denominator is the investment (relative to the unscheduled maintenance case). By definition, $I_u = 0$ (contains no investment in PHM because there is no PHM). Therefore, Equation (17.5) simplifies to

$$ROI = \frac{C_u - C_{PHM}}{I_{PHM}} \tag{17.6}$$

In Equation (17.6) (C_u – C_{PHM}) excludes all the costs that are a "wash" (i.e., they are the same, independent of the maintenance approach). Formulation of the ROI in this manner solves the problem of splitting up the costs, because we never need to address which particular life-cycle costs are due to the maintenance policy. In Equation (17.6), if $C_u = C_{PHM}$, then ROI = 0, implying that the cost avoidance that results from PHM exactly equals the investment made (which is correct; again, note that C_{PHM} includes I_{PHM} within it).

In Equations (17.5) and (17.6) the investment cost is given by

$$I_{PHM} = C_{NRE} + C_{REC} + C_{INF} \tag{17.7}$$

where

C_{NRE} = PHM management non-recurring costs.

C_{REC} = PHM management recurring costs (cost of putting PHM hardware into each instance of the system).

C_{INF} = PHM management infrastructure costs.

The non-recurring engineering (NRE) costs associated with PHM management are the costs of designing hardware and software to perform the PHM. PHM infrastructure costs are the costs of acquiring and keeping PHM management resources in place (equipment, people, training, software, databases, plan development, etc.). One question that arises is, is I_{PHM} complete? Are there other investment costs that are not captured in Equation (17.7)? This is a difficult question to answer. Consider the following observations, for example:

- Since my PHM approach results in more maintenance actions (the need for more spare parts) than an unscheduled maintenance approach (since it will cause maintenance to be performed prior to failure), is the cost of the extra spare parts accounted for as part of the investment (I_{PHM})?
- What if (for simplicity) my PHM management approach resulted in buying exactly the same number of spare parts for exactly the same price per part as my unscheduled maintenance approach, but I buy them at different times. Due to the cost of money (non-zero discount rate), this does not end up costing the same. Is the cost of money part of I_{PHM}?

The costs in the examples above would not be included in the investment cost because they are the result of the PHM management approach (i.e., the result of the investment) and are reflected in the life-cycle cost C_{PHM}.

Performing the calculation in Equations (17.6) and (17.7) is not trivial and is beyond the scope of this chapter. However, it is useful to look qualitatively at a result (see [Ref. 17.6] for the details of the model that was used to generate this result). The ROI as a function of time for the application of a data-driven PHM approach to an electronic display unit in the cockpit of a Boeing 737 is shown in Figure 17.4. Unscheduled maintenance in this case means that the display unit will run until failure (no remaining useful life will be left) and then an unscheduled maintenance activity will take place. In the case of an airline, an unscheduled maintenance activity will generally be more costly to resolve than a scheduled maintenance activity because, depending on the time of the day that it occurs, it may involve delaying or canceling a flight. Alternatively, an impending failure that is detected by the PHM approach ahead of time will allow maintenance to be performed at a time and place of the airline's choosing, thus not disrupting flights and being less expensive to resolve. These effects can be seen qualitatively in Figure 17.4.

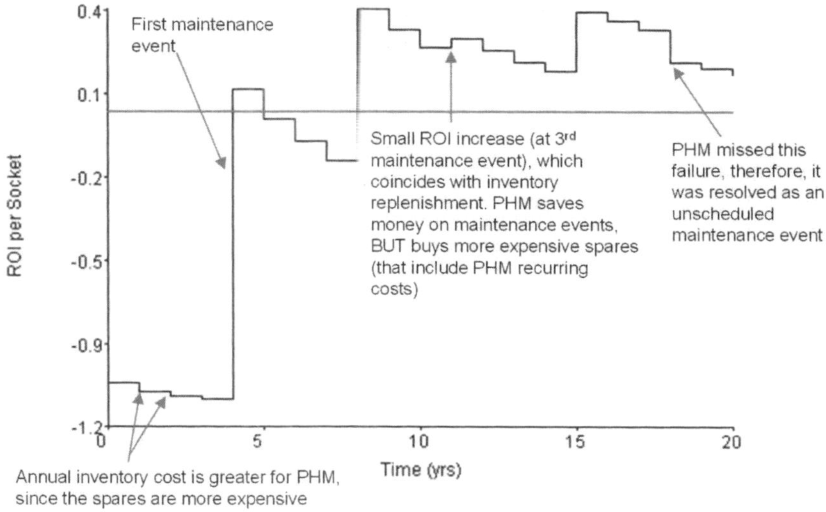

Fig. 17.4. ROI as a function of time for the application of a data-driven PHM approach to an electronic display unit in the cockpit of a Boeing 737.

Figure 17.4 was generated by simulating the life cycle of one instance of the socket that the display unit resides in, managed using unscheduled maintenance and the data-driven PHM[6] approach and applying Equations (17.6) and (17.7).[7] The ROI starts at a value of -1 at time 0, which represents the initial investment to put the PHM technology into the unit with no return ($C_u - C_{PHM} = -I_{PHM}$). After time 0, the ROI starts to step down.[8] In this analysis the inventory cost (the cost of holding spares in the inventory) is a percentage of the cost of the spares (10% of the spare purchase price per year, in this example). Since spares cost more for PHM due to PHM recurring costs in the display unit, inventory costs more. In

[6] Data-driven PHM means that you are directly observing the system and deciding that it looks unhealthy (e.g., monitoring for precursors to failure, use of canaries, or anomaly detection).

[7] A socket is the location in a system where a module or line replaceable unit (LRU) resides. Sockets are tracked instead of modules because a socket could be occupied by one or more modules during its lifetime and socket cost and availability are more relevant to systems than the cost and availability of the modules.

[8] In Figure 17.4 all the accounting is done on an annual basis, so the ROI is only recalculated once per year.

the period from years 0 to 4, C_{PHM} is increasing while C_u and I_{PHM} are constant (inventory costs are considered to be a result of the PHM investment, not part of the PHM investment). The step size decreases as time increases, in part due to a non-zero cost of money (the discount rate in this example is 7%). If there was no inventory charge, or if the inventory charge was not a function of the spare purchase price, then the ROI would be a constant −1 until the first maintenance event. The first maintenance event occurs in year 4 and is less expensive to resolve for PHM than for unscheduled maintenance, since PHM successfully caught the failure ahead of time. As a result, the ROI increases to above zero. During the period from years 4 to 8 the decreases in ROI are inventory charges and annual PHM infrastructure costs (even though PHM infrastructure costs are an investment, they still affect the ROI ratio). A second maintenance event that was successfully detected by PHM occurs at year 8. In year 11 a third maintenance event occurs and more spares are purchased. In year 18 there is a system failure that was missed by PHM.

Finally, the calculation of an ROI relative to an alternative PHM management approach (rather than unscheduled maintenance) can be found using

$$ROI = \frac{C_{PHM_1} - C_{PHM_2}}{I_{PHM_2} - I_{PHM_1}} \tag{17.8}$$

where PHM_1 and PHM_2 represent the two different PHM management approaches.

17.4 Stochastic ROI Calculations

Like every other cost calculation, the inputs to ROI analysis have associated uncertainties. How are these uncertainties accounted for in the process of assessing ROIs? Each instance in the population of products or systems potentially has a unique investment cost and unique return. The ROI is unique for each because each instance is slightly different and each instance is subjected to a different environmental stress history. The investment and return for the population can be expressed as a histogram (distribution), as shown in Figure 17.5.

The ROI can be compared using

$$ROI = \frac{\bar{R} - \bar{I}}{\bar{I}} \tag{17.9}$$

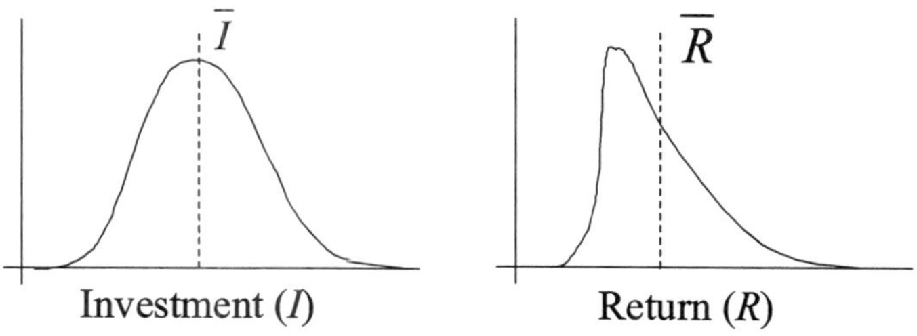

Fig. 17.5. Histograms of the investment and return associated with a population of products or systems. The mean investment and return are indicated.

Unfortunately, this calculation is static, <u>not</u> stochastic. It uses values that are averaged over the whole population. The problem is that a particular instance may be represented by the values shown in Figure 17.6.

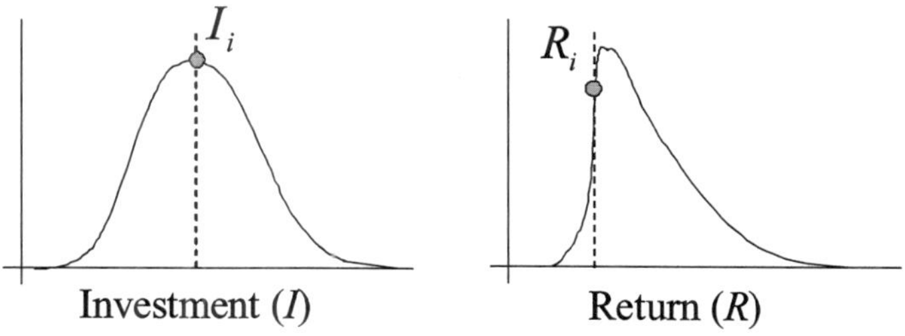

Fig. 17.6. Histograms of the investment and return associated with a population of products or systems. The investment and return from one instance from the population is indicated.

A separate ROI could be computed for each instance of the product or system using Monte Carlo analysis:

$$ROI_i = \frac{R_i - I_i}{I_i} \tag{17.10}$$

A histogram of the ROIs computed for each instance of the product or system can be formed as shown in Figure 17.7. Armed with an ROI distribution, the mean ROI, uncertainty, and confidence can be determined.

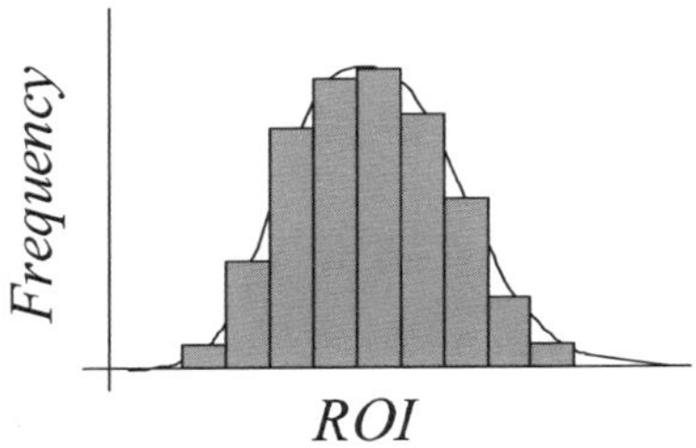

Fig. 17.7. Histograms of the ROIs for a population of products or systems.

17.5 Summary

ROI calculations are a key part of making business cases; however, they are often difficult to perform correctly and consistently. ROIs must be measured relative to something that is clearly defined, such as the current equipment, the current management approach, or doing nothing.

If there is no investment it does not make sense to calculate an ROI. For example, can we compute the ROI of switching the order of two process steps in a manufacturing process? If an investment is required in order to switch the steps — that is, if the line is shut down and labor (and possibly materials) are required to make the change — then there is an ROI. If, on the other hand, switching the order requires no disruption of production and no special labor or materials (maybe it just entails exchanging two people), then it does not make sense to compute an ROI.

The determination of whether costs are investments or the costs incurred as a result of the investment is at the discretion of the analyzer. However, consistency is important; define clearly what the investments are and stick with this definition when comparing the ROIs associated with various options. One of the major criticisms against ROI calculation is that it can easily be manipulated, which is true. For example, if a company invests in a new piece of manufacturing equipment, but does not include within the investment calculation the learning curve of the manufacturing personnel, then the ROI of the new equipment will be overestimated.

ROI is also dependent on the cost of money (discount rate). In the technology adoption example in Section 17.2, a constant discount rate was assumed, but discount rates are rarely constant over time (see Appendix B). Economies change, opportunities available to companies change, and markets change. However, the discount rate could be represented as a probability distribution and used within a Monte Carlo or other analyses that include uncertainties.

Several other types of ROI exist that have not been discussed in this chapter. These include: revenue enhancement in which the organization will increase its revenue as a result of an investment; profit enhancement in which revenue may not change, but profitability increases; and capital cost avoidance in which capital expenditures create future cost avoidances.

References

17.1 Groppelli, A. A. and Nikbakht, E. (2000). *Barron's Finance*, 4th Edition (Barron's Finance, New York, NY).

17.2 Gilleo, K. (March 2001). A brief history of flipped chips, http://flipchips.com/tutorial/other/a-brief-history-of-flipped-chips/. Accessed April 27, 2016.

17.3 Ashenbaum, B. (March 2006). *Defining Cost Reduction and Cost Avoidance*, CAPS Research.

17.4 Williams, J. H., Davies, A. and Drake, P. R. Editors (1994). *Condition-based Maintenance and Machine Diagnostics* (Chapman & Hall, London).

17.5 Pecht, M. G. (2008). *Prognostics and Health Management of Electronics* (John Wiley & Sons, Inc., Hoboken, NJ).

17.6 Feldman, K., Jazouli, T. and Sandborn, P. (2009). A methodology for determining the return on investment associated with prognostics and health management, *IEEE Transactions on Reliability*, 58(2), pp. 305-316.

Problems

ROI problems appear in other places in this book. See Problems 12.6c, 14.2, and 19.2d.

17.1 For what value of a new machine-manufactured part yield in Table 17.1 will the break-even point be 2,000,000 units?

17.2 If the old machine in Table 17.1 has a salvage value of $20,000, what is the break-even quantity of units?

17.3 In Problem 12.5, if spares cost $2000/spare and downtime is values at $80,000/month, what is the ROI associated with buying 9 spares? Ignore the cost of money (discount rate = 0). Hint, you do not need to solve Problem 12.5 to solve this problem.

17.4 If the cost of the new equipment to support flip chip bonding had to be depreciated over 10 years (instead of 5), recalculate the ROI as a function of time for the three cases shown in Figure 17.3.

17.5 How is the ROI changed if the technology licensing cost for flip chip technology considered in Table 17.2 is charged per chip sold at the rate of 0.2% of the chip sales price instead of as a lumped sum?

17.6 As described in Chapter 3, the yield of die is a function of the die area. The flip chip ROI example provided in this chapter ignored potential yield improvements due to the die shrink that accompanied the redesign of die using flip chip bonding. Include yield improvements into the flip chip ROI example in Section 17.2.2 assuming that the original die yield was 85%.

17.7 Show that the ROI of one PHM approach relative to another PHM approach is not the difference between their respective ROIs relative to unscheduled maintenance.

17.8 The application of the discount rate for computing the present value Equation (II.1) effectively results in the same multiplier on both the numerator and denominator in the ROI calculation. However, the cumulative ROI as a function of time is not independent of the discount rate. Why not? Hint: Create a simple example that includes investments and life-cycle cost changes over several years and compute ROI as a function of time, including the discount rate effects in each year.

17.9 Find examples in the engineering literature of incorrectly (or inconsistently) performed ROI analyses.

17.10 Prognostics and health management (PHM) is to be included within a system that your company has to support. In order to make a business case for the inclusion of PHM into the system, its ROI has to be assessed. Assume the following:

- The system will fail 3 times per year
- Without PHM, all 3 failures will result in unscheduled maintenance actions
- With PHM, 2 out of the 3 failures per year can be converted from unscheduled to scheduled maintenance actions (the third will still result in an unscheduled maintenance action)
- The cost of an unscheduled maintenance action is $200,000 (downtime = 12 hours)
- The cost of a scheduled maintenance action is $20,000 (downtime = 4 hours)
- The effective cost (per system instance) of putting PHM into the system is $1,200,000 (assume that this is all charged at the end of the first year)
- In addition you have to pay $50,000 per year (per system instance) to maintain the infrastructure necessary to support the PHM in the systems

- The system has to be supported for 25 years
- There is a non-zero after tax discount rate that can vary from 0 to 20% - assume that it is a constant over the whole 25 years.

Assume that all the costs above are year 0 costs and that all the charges for maintenance are charged at the end of each year. Assume all the maintenance actions are field repairs (no spares are used).

a) Calculate the ROI of the investment in PHM relative to all unscheduled maintenance as a function of the after tax discount rate.

b) For a discount rate of 5%, how much can you afford to spend to put PHM into each system instance and still break even?

c) Assuming that the system needs to be operational 100% of the time, what is the increase in availability when PHM is used? (give your answer to 4 significant figures).

17.11 You are having a new home built and have the option of installing conventional toilets or "low-flush" toilets, and you want to understand the return on investment of this decision. Assume the following data:

Number of people living in the house = 5
Number of toilet usages per day per person = 4
Number of toilets in the house (assume all are used equally) = 3
Water/sewer cost per 1000 gallons = $6.13
Plumber cost per call = $200

The toilets are characterized by the following:

	Conventional	Low-Flush	Ultra Low-Flush
Average number of re-flushes per day per person	0	1	1.5
Liters of water per flush	19	13.2	5.7
Purchase price of the toilet	$200	$300	$400
Average number of plumber calls per year per toilet	0.2	0.22	0.25
Lifetime of the toilet (years)	25	23	22

Calculate and plot the total life-cycle cost of each toilet for 100 years. Calculate and plot the return on investment (relative to the conventional toilet) for 100 years for the low-flush and ultra low-flush toilets. Hint: Consider the investment cost to be only the year zero cost to purchase the toilet.

Chapter 18

The Cost of Service

X. X. Huang[1], M. Kreye[1], G. Parry[2], Y. M. Goh[3] and L. B. Newnes[1]

[1]*University of Bath, Bath, UK*
[2]*University of the West of England, Bristol, UK*
[3]*University of Loughborough, Loughborough, UK*

Sustainable production and consumption have become increasingly important internationally, which has led to the transformation of market structures and competitive situations in the direction of servitization. To adapt to these changes, many manufacturers have had to move towards primarily providing a service (capability and availability) rather than a product with support as a subsidiary activity. This trend toward product service systems (PSS) focuses on creating value from an asset throughout the life cycle. For example, Rolls Royce estimates the value of their after-sales service market at $280 billion, while their engine sales are worth only $170 billion [Ref. 18.1]. This means that the supply of services offers important business opportunities; however, one of the challenges industry faces is how to estimate the cost of providing this service.

Manufacturing is defined as creating value and delivering a service through life [Ref. 18.2]. Estimating the through-life (or life-cycle) cost of a product service system can be characterized as a stream of events through various life-cycle phases — concept, assessment, development, manufacturing, in-service, and disposal, as depicted in Figure 18.1. This chapter discusses how to estimate the cost of the in-service phase, which includes the utilization and support of the product service system, which we will consider an engineering service. It addresses the following four questions and how they influence the service cost estimate:

- Can product cost estimation techniques be used to estimate the cost of a service?
- How can uncertainty be taken into consideration in the estimation process?

- How can the cost estimate be used to inform the bidding process?
- How can uncertainties be accounted for in the bidding process?

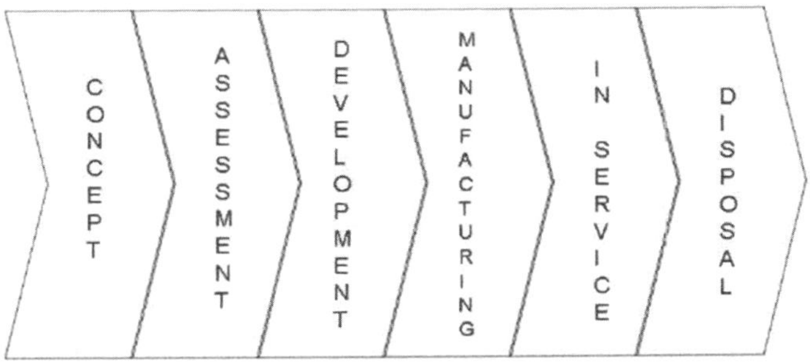

Fig. 18.1. Life-cycle phases of product service systems (PSS).

The aim of this chapter is to illustrate a process that could be used to ascertain the cost of providing an engineering service. To illustrate this, an example of an original equipment manufacturer is presented and an approach that could be used to estimate the cost of providing a service for the equipment is shown. To illustrate how this estimate could be adopted within a commercial environment the implications for the pricing decisions in the contracting and bidding process are discussed.

18.1 Why Estimate the Cost of a Service?

The importance of estimating service costs has been highlighted in various engineering industries such as defense, aerospace, manufacturing and construction sectors. For example, the in-service costs of military equipment can account for up to 75% of the total expenditure through the product's life [Ref. 18.3]. One of the key challenges is the uncertainty connected to the process; which can, for example, impact schedules, creating delays that cause budgets to be exceeded [Ref. 18.4]. Examples include the Deh Cho Bridge in Canada, which was planned to be opened in fall 2010. The costs associated with the redesign and a delay of over a year increased the budget by at least $15 million over the estimated cost of $182 million [Ref. 18.5].

The delivery of a service is usually embedded in a contract that is a legally binding agreement between the parties concerning the technical details of the service. When competing for these service contracts, particularly during bidding, decision makers face various uncertainties that influence their decisions. One of the main uncertainty factors is the cost forecast. How can we estimate the costs of providing such a service, especially when cost modeling methods and software are primarily product-oriented [Ref. 18.6]? Currently, there are very few cost estimation tools that model the provision of an engineering service.

18.2 An Engineering Service Example

The following example shows what an engineering service is and why it is important for today's business. Consider the following challenge: You have been asked to travel from California to New York by car. How do you get a car? Here are two possible scenarios:

(a) You buy a car.
You purchase a car from a car dealer and the ownership of the car is transferred from the dealer to you. You can keep the car for as long as you wish. However, the drawback is that you are responsible for maintaining and servicing the car. This exposes you to continuous expenses, such as the costs of fuel, licensing, car insurance, repairs, and breakdown coverage. Sadly, you have a breakdown on the way to New York. You may need to wait for maintenance staff to come to the scene, attempt to fix the problem, or transport your car to the nearest garage. If you did not purchase breakdown coverage, you face the additional expense to tow and/or repair your car. Either way, your journey is disrupted, delayed, or even cancelled. The cost to you is probably greater than you first planned.

(b) You rent a car.
In this scenario, you rent a car for the trip from California to New York. You do not own the car but have use of it during the rental period. Hence, you are only responsible for keeping the car in reasonable condition during the trip and the cost of insurance and fuel. You do not need to worry about the car once the lease period ends and the rental company has made sure the car is in good condition and is safe to drive. Hence, the likelihood of a breakdown during the trip may be smaller than if you bought the car. Even if a breakdown occurs on the road, you do not have to fix it yourself, as

this is usually covered by the rental car company. However, the consequence of the breakdown might still have an adverse impact on you, as the trip plan and schedule are interrupted.

To improve the rental, you could alternatively purchase the *availability* of a car. This means that you pay the car owner for not only obtaining the use of the car but also for being guaranteed an acceptable level of performance and reliability. This could include repairing the car more efficiently or providing you with a replacement car to reach New York without affecting your schedule in the event of a breakdown. Any failures to provide the availability of a car leads to a penalty charge for the car owner. Hence, you have been provided a service contract where the service is to get you to New York within a timeline you dictate.

Many customers may feel that the second scenario, with availability coverage, delivers the better value because the car is guaranteed to be available for you to travel from California to New York. Hence, in some cases customers have shifted from purchasing a physical product to demanding service-added products or service solutions. Interest in this type of service contract has been observed in other sectors, such as aerospace, defense, manufacturing and construction sectors [Ref. 18.7]. This type of service is now being offered by many companies, such as BAE Systems, Rolls Royce, and ABB. They offer long-term engineering service support solutions, or PSS, which tend to concentrate on performance outcomes rather than individual parts and repair actions.[1]

Engineering service focuses on the maintenance, repair and training of staff within the in-service phase of a PSS.

18.3 How to Estimate the Cost of an Engineering Service

Literature on estimating service costs is scarce, since most approaches focus on the estimation of the costs of products. Quantitative cost estimation techniques are categorized as parametric cost estimation (Chapter 6) and analytical cost estimation (Chapters 2, 4 and 5). The

---

[1] A specific example of servitization in the form of availability-based contracting, previously described in Section 15.6, is an availability contract through which customers buy the availability of a product rather than the product itself.

parametric approach focuses on the characteristics of the product, identifying the cost estimating relationship (CER) between costs and cost-related factors. Further details on parametric cost modeling and how to generate CERs are discussed in Chapter 6. The principle behind the parametric approaches is to identify any trends or rules between costs and cost-related drivers during the product's life cycle. It is preferable to do this with sufficient estimating time and when clear relationships between different cost variables can be identified.

Parametric cost modeling is used to demonstrate the process of estimating the costs of an engineering service using the following steps:

1. Identify cost variables, such as labor costs, machine breakdown, training programs, and stock levels.
2. Construct a hypothesis for each cost-related variable. For example, it can be assumed that a good training program provided to machine operators will assist in the proper operation of the machine, reducing the number of failures and consequently costs.
3. Collect cost-related data from the company's database, complemented by an internal survey of the appropriate staff.
4. Test and analyze each hypothesis using historical data and/or maintenance staff's and customers' questionnaires.
5. Generate relationships for different cost-related factors. Key cost-related factors, such as machine breakdown, must be identified before establishing these relationships.
6. Develop cost estimation relationships (CERs) for different cost drivers. Key cost drivers should be identified by analyzing and choosing the relationship that best predicts the dependent cost variable.

18.4 Application of the Service Costing Approach within an Industrial Company

To illustrate how the approach described in Section 18.3 can be used to estimate the cost of a service, consider the following case. A company in China, with annual revenues of £5-6 million provides extrusion laminating machines. The machines are sold in China and other countries, including

India, Japan and Russia. The company focuses on designing and selling these machines, as well as providing after-sales services, such as maintenance and training. The company is seeking a model to estimate the costs of providing their after-sales services in order to achieve a more profitable service contract at the purchasing stage of a machine. It is considering offering and delivering a service contract guaranteeing the machine will be maintained for a specified length of time when it is in-service, and wants to estimate the cost of providing such a service.

Available data includes billing and service charges generated from 2003 to 2010. An internal survey also collected information from the employees in the after-sales service department. During this period, the service operations of maintenance staff were examined, and customers were visited to observe how the repairs were carried out on-site.

Let's consider an example that describes how to establish the relationships for service-related cost drivers for this machine. First, a relationship between machine breakdown (failure rate) and number of years in service is established.

Step 1: Identify the in-service cost variable

The in-service cost variables to derive the CERs are the rate of machine breakdowns (failure rate) and the total service costs per failure.

Step 2: Construct hypotheses

In order to establish CERs for estimating the cost of providing an engineering service, the following hypothesis is tested: *The longer the machine has been in service, the less likely it is to fail.*

Step 3: Collect cost data

The model is created based on the five assumptions below and the in-service cost data covering seven years (2003-2009) of data from 71 extrusion laminating machines. Several assumptions need to be made in order to gather a consistent and usable data set. These include:

1. All machines are identical in terms of components and are sourced from the same supplier.

2. All machines have the same operating conditions, despite being introduced into service in different years.
3. All failures are repairable.
4. Total overhead cost in 7 years is 5% of the total service cost.
5. Total training costs in 7 years (C_{TR}) are included in the total labor costs (C_L).

The following quantities are defined for use in the analysis that follows:

$C_{average}$ = the average service cost per failure for machines.
C_j = the total cost for a population of machines in their j^{th} year of service.
i = years in service.
I_s = the year the machine is sold (and enters service).
j = year of service.
λ_j = the failure rate for machines in their j^{th} year of service.
N_{f_j} = the total number of failures for machines in their j^{th} year of service.
N_i = the total number of machines in service for at least i years.
N_{ij} = the number of failures of machines in service for at least i years during their j^{th} year of service.
N_s = the total number of machines sold in year I_s.
T = the service contract length (in years).

When machine breakdowns occurred, the operator recorded the failure time and called the service provider to repair the machine. If the problem could not be resolved over the telephone, the service provider sent maintenance staff to the customer's site to fix it. A single machine could breakdown numerous times during the service period, and repair service is provided throughout the service contract life of each machine.

In total there were 71 machines from the same production line sold during 2003-2009. These new machines were purchased by customers during different years, so they have different numbers of in-service years during the seven-year period studied. The number of machines sold and the number of machine failures occurring during the 2003-2009 period are summarized in Table 18.1.

Table 18.1. The Number of Machines Sold and the Number of Failures Recorded.

I_s	i	N_{s_i}	*Number of Machine Failures* (N_{ij})						
			1st year in service ($j = 1$)	2nd year in service ($j = 2$)	3rd year in service ($j = 3$)	4th year in service ($j = 4$)	5th year in service ($j = 5$)	6th year in service ($j = 6$)	7th year in service ($j = 7$)
2003	1	8	4	1	1	3	1	0	0
2004	2	15	17	14	2	2	1	1	-
2005	3	9	15	11	3	9	1	-	-
2006	4	10	17	9	25	13	-	-	-
2007	5	7	28	4	6	-	-	-	-
2008	6	12	27	8	-	-	-	-	-
2009	7	10	27	-	-	-	-	-	-

Table 18.1 shows that eight machines were sold in 2003 that as of 2009 had been in service for seven years, fifteen machines were sold in 2004 that had six in-service years as of 2009 and so on. The eight machines sold in 2003 had a total of four failures during their first year in-service, and this reduced to one failure in their second and third year of service. The number of failures increased to three in the fourth year, and so on.

Based on Table 18.1, the total number of machines and total number of machine failures are presented in Table 18.2. It shows that there are 71 machines in-service for at least one year, 61 out of 71 were in service for at least two years, and so on. Furthermore, the 71 machines had 135 failures in their first service year, 61 machines had 47 failures in the second service year, and so on. Each machine could fail more than once or not at all during a service year. Of the machines that failed during the first year, the repaired machines could fail again in subsequent years. For example, the 47 failures occurring during the second year includes machines that failed in the first year, were repaired, and failed again in their second year in service.

The costs incurred at the in-service stage (years one to seven) of providing service includes the costs for labor, training, travel, accommodation, spare parts, telephone services, subsidies for travel, bonus for providing a good service and overheads. The cost data collected from the industrial company is tabulated in Table 18.3. The total service provided includes service provided both by telephone and on-site.

Table 18.2. Number of Machines in Service for at Least *i* Years; Number of Failures and Failure Rate in the jth Year of Service.

i	N_i	j	N_{f_j}	$\lambda_j = N_{f_j}/N_i$
1	$\sum_{i=1}^{7} N_{s_i} = 71$	1	$\sum_{i=1}^{7} N_{i1} = 135$	1.9014
2	$\sum_{i=1}^{6} N_{s_i} = 61$	2	$\sum_{i=1}^{6} N_{i2} = 47$	0.7705
3	$\sum_{i=1}^{5} N_{s_i} = 49$	3	$\sum_{i=1}^{5} N_{i3} = 37$	0.7551
4	$\sum_{i=1}^{4} N_{s_i} = 42$	4	$\sum_{i=1}^{4} N_{i4} = 27$	0.6429
5	$\sum_{i=1}^{3} N_{s_i} = 32$	5	$\sum_{i=1}^{3} N_{i5} = 3$	0.0938
6	$\sum_{i=1}^{2} N_{s_i} = 23$	6	$\sum_{i=1}^{2} N_{i6} = 1$	0.0435
7	$\sum_{i=1}^{1} N_{s_i} = 8$	7	$\sum_{i=1}^{1} N_{i7} = 0$	0.00

Table 18.3. Total Service Cost Variables (years 2003-2009).

Total Costs in 7 Years	Seven years (year 2003-2009)
$C_L + C_{TR}$	\$198,277
C_{TP} = Total transportation costs	\$300,580
C_A = Total accommodation costs	\$116,184
C_{SP} = Total costs for spare parts	\$461,183
C_P = Total telephone service costs	\$19,870
C_S = Total subsidies for travelling	\$101,198
C_{BO} = Total bonus for providing a good service	\$226,035

Steps 4 and 5: Test hypothesis and establish relationships for cost-related factors

The hypothesis is tested based on the historical data listed in Table 18.1. The machine failure rate is calculated as a ratio of the total number of failures divided by the total number of machines in service for at least *i* years. The relationship calculated between the machine failure rate and the number of years in service is shown in the last column of Table 18.2.

The failure rates are shown in Figure 18.2. A 190.14% failure rate occurred on 71 machines during their first year in-service, which means that on average, every machine had almost two failures during its first year in service. However, in year two this reduced significantly to ~77% based on a sample of 61 machines. During the third and fourth in-service years, the machines failed less frequently. After machines had been in service for more than four years, the failure rates reduced significantly to less than 10%. In general, within the seven in-service years, the longer the machine had been in service, the less likely it was to fail.

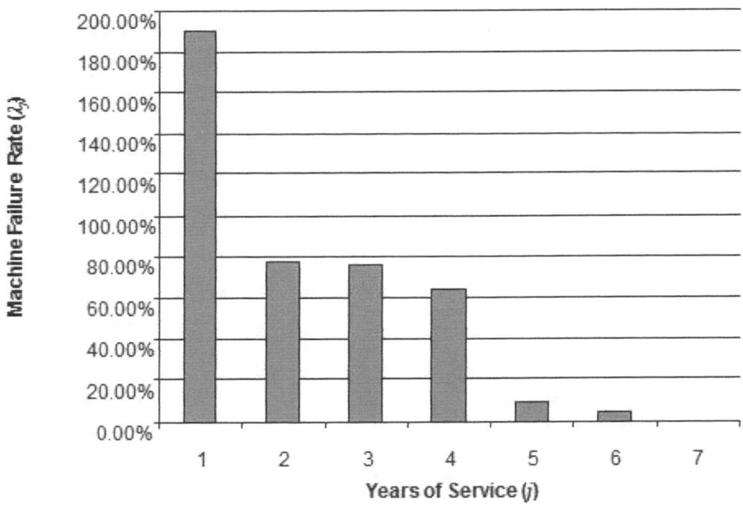

Fig. 18.2. The relationship between machine failure rate and years in-service.

Step 6: Establish a CER

The average service cost per failure is calculated as a quotient of the total service costs (which includes 5% overhead, $b = 0.05$) divided by the total number of failures during seven in-service years. Based on Table 18.2, the average service cost per failure, $C_{average}$, is

$$C_{average} = \frac{(C_L + C_{TR} + C_{TP} + C_A + C_{SP} + C_P + C_S + C_{BO})(1+b)}{\sum_{j=1}^{7} N_{f_j}}$$

$$= \frac{(198{,}277 + 300{,}580 + 116{,}184 + 461{,}183 + 19{,}870 + 101{,}198 + 226{,}035)(1 + 0.05)}{(135 + 47 + 37 + 27 + 3 + 1 + 0)}$$

$$= \$5977.97$$

The relationship between the total service cost and the number of years in-service was determined from Tables 18.2 and 18.3. The service costs are estimated by multiplying the average cost per failure ($C_{average}$) by the number of failures that occurred in the year, as shown in Table 18.4.

Table 18.4. Total Service Cost in the jth Year of Service.

j	N_{f_j}	$C_j = N_{f_j} C_{average}$
1	135	\$807,026
2	47	\$280,965
3	37	\$221,185
4	27	\$161,405
5	3	\$17,934
6	1	\$5978
7	0	\$0

where C_j is the total service cost for machines in their j^{th} year of service.

Table 18.4 shows the cost of providing different lengths of a service contract for the original 71 machines. The longer the machine stayed in service, the smaller the costs of servicing the machines was. The average service cost per failure, $C_{average}$, can be used to estimate the service costs for different numbers of machines in the first seven service years.

Application of the Model

We wish to sell 100 machines to a customer and they are requesting that we enter into an engineering service contract with them. The options are different contract lengths — one, three, five or seven years. What are the costs for providing such a service?

Table 18.5 calculates the total costs for servicing 100 machines for from one to seven in-service years.

Table 18.5. Total Service Costs for 100 Machines in the jth Year of Service.

j	λ_j	$N_{f_j} = \lceil 100\lambda_j \rceil$	$C_j = N_{f_j} C_{average}$
1	190.14%	191	\$1,141,793
2	77.05%	78	\$466,282
3	75.51%	76	\$454,326
4	64.29%	65	\$388,568
5	9.38%	10	\$59,780
6	4.35%	5	\$29,890
7	0.00%	0	\$0

Based on these cost estimates for each service year we can determine the costs of providing different lengths of service contracts (T). This is calculated as a yearly average over the contract period with contract periods of one, three, five or seven years, as depicted in Table 18.6.

Table 18.6. The Per Year Cost of Servicing 100 Machines for a One, Three, Five and Seven Year Contract.

T	Mathematical relationship	Per-year cost of servicing 100 machines for T years
1	C_1	\$1,141,793
3	$\frac{C_1 + C_2 + C_3}{3} = \frac{1}{T}\sum_{j=1}^{T} C_j$	\$687,467
5	$= \frac{1}{T}\sum_{j=1}^{T} C_j$	\$502,150
7	$= \frac{1}{T}\sum_{j=1}^{T} C_j$	\$362,948

The cost for a one-year service contract for the 100 machines was estimated at \$1,141,793, whereas the per-year cost reduced approximately by half for a three-year contract. Further cost reductions were calculated for a five-year contract (\$502,150) and a seven-year contract (\$362,948). In general, the longer the service contract, the less expensive it is to provide an engineering service per year.

18.5 Bidding for the Service Contract

Cost estimates like those developed in this chapter can be used as input for the decision process when bidding for a service contract. The bids offered to the customer should cover the estimated costs calculated in Section 18.4 and also yield a suitable profit. Thus, the prices for the different service contract lengths may differ significantly based on the costing information. In the bidding process, the decision is reached through a strategic evaluation of the uncertainty factors.

The most important factor is the uncertainty influencing the cost forecast (the accuracy of the cost estimate). The calculation presented in Section 18.4 offers different cost values for the different contract periods, based on the assumption that the behavior of the 100 machines investigated is accurately described by the serviced machines. However, this assumption may not hold true — the current set of 100 machines may show a higher or lower failure rate than estimated by λ_j and may realize different service costs per failure in comparison to the estimation in $C_{average}$. These uncertainties have to be considered in the pricing decision process.

Furthermore, uncertainties that are connected to the cost model itself have to be considered. For example, the number of machine breakdowns can be influenced by the level of training of the operator, the capacity utilization, or the environmental circumstances, such as temperature and humidity. Including a training program for machine operators within the service contract may increase the short-term costs but decrease the number of machine breakdowns in later years, and thus decrease the service costs per failure later.

In addition, the strategic evaluation process must include the customer, who may accept or reject the price bid. A price must be established that can convince the customer to buy the service contract. Uncertainty arises from a lack of knowledge about the customer's buying strategy, budget constraints, or evaluation criteria and processes [Ref. 18.8]. For example, the customer may be willing to pay a higher price for an availability guarantee. These uncertainties can be addressed through modeling and management techniques (such as Monte Carlo, subjective probabilities, or interval analysis). This can form the basis for an informed decision at the bidding stage to secure a profitable service contract and realize the business opportunities connected with servitization.

References

18.1 Rolls Royce (2015). http://www.rolls-royce.com/~/media/Files/R/Rolls-Royce/documents/investors/annual-reports/2015-annual-report-v1.pdf. Accessed April 27, 2016.

18.2 Foresight (2013). *The Future of Manufacturing: A new era of opportunity and challenge for the UK*, Project Report (The Government Office for Science, London).

18.3 Mathaisel, D. F. X., Manary, J. M. and Comm, C. L. (2009). *Enterprise Sustainability: Enhancing the Military's Ability to Perform its Mission* (CRC Press, Boca Raton, FL).

18.4 Gray, B. (2009). Review of acquisition for the secretary of state for defence. https://www.bipsolutions.com/docstore/ReviewAcquisitionGrayreport.pdf Accessed April 27, 2016.

18.5 Northern News Services (2010). http://www.nnsl.com/frames/newspapers/2010-03/mar26_10dc.html, Accessed April 27, 2016.

18.6 Newnes, L. B, Mileham, A. R. and Hosseini-Nisab, H. (2007). On-screen real-time cost estimating, *International Journal of Production Research*, 45(7), pp.1577-1594.

18.7 Brax, S. (2005). A manufacturer becoming service provider – challenges and a paradox, *Managing Service Quality*, 15(2), pp. 142-155.

18.8 Kreye, M. E., Newnes, L. B. and Goh, Y. M. (2011). Uncertainty Analysis and its Application to Service Contracts. *Proceedings of the IDETC/CIE 2011: International Design Engineering Technical Conferences & Computers and Information in Engineering Conference*, Washington, DC, USA.

Problems

18.1 What is an engineering service? Can you provide an example of an engineering service?

18.2 What is the service costing process?

18.3 How do you use parametric costing to estimate the costs for an engineering service (refer to Chapter 6)?

18.4 Using the data in Section 18.4, we wish to sell 200 machines to a customer who is requesting a three-year engineering service contract. What is the total service cost for providing such a contract?

Chapter 19

Software Development and Support Costs

Software is the most expensive component in many types of electronic systems. Software costs are comprised of development, which includes specifying, designing, and developing software, and maintenance, which is the process of optimizing and enhancing deployed software (software release), as well as remedying defects (fixing bugs).

Estimating software costs is critical to both developers and customers. Cost estimates are important in order to prioritize projects, forecast necessary resources, forecast the impacts of changes, and budget.

Software cost estimation involves determining one or more of the following metrics: human effort, project duration, and/or cost [Ref. 19.1]. The majority of software cost estimation models calculate an effort estimate that is then used to determine the project duration and cost. Table 19.1 summarizes the basic elements that are used in most software costing models.

Different approaches to *a priori* software cost estimation include the calculation of, lines of code, functions, and objects. All of these methods approach cost estimation through estimating the human effort necessary to complete a project.

Definitions

Several definitions are useful for the discussion that follows:

- Source lines of code (SLOC) = the sum of all the data declaration statements and executable statements that are delivered in a software program (does not include comments).

- Delivered source instructions (DSI) = similar to SLOC but counts physical lines of code in the SLOC (e.g., an "if-then-else" statement would be counted as several DSI but only one SLOC).

Table 19.1. Factors Affecting Software Cost Estimation [Ref. 19.2].

Group	Factor
Size Attributes	Source instructions Number of routines Number of output formats Quantity of personnel Number of functions Number of objects
Program Attributes	Type Complexity Language Required reliability
Personnel Attributes	Personnel capability Personnel continuity Hardware experience Application experience Language experience
Project Attributes	Tools and techniques Customer interface Requirements definition

19.1 Software Development Costs

In traditional software cost models, costs are derived based on the required effort (measured in person-months). Empirical estimation models provide formulae for determining the effort based on statistical information about similar projects. The precise software development situation is taken into account using complexity factors. Complexity factors are empirically derived coefficients that model possible deviations from the nominal case. Models usually require calibration to the actual software development process used by an organization.

Fundamentally the traditional models (called "algorithmic models") are parametric models (see Chapter 6). Algorithmic models are constructed by analyzing the attributes and costs of many completed software development projects. The attributes that are cataloged typically include a count of either size (number of SLOC) or points (function, feature, or object). The models discussed in this chapter have the same pros and cons as the parametric models discussed in Chapter 6.

Most algorithmic estimation models use a model of the form:

$$Effort = b + ca^x \tag{19.1}$$

where

a = the product metric variable, e.g., size.
b, c, and x = parameters chosen to best fit the observed data.

The exponent in algorithmic estimation models (x in Equation (19.1)) is associated with the size estimate. This models the fact that costs do not generally increase linearly with project size (a). Normally as the size of a software project increases, additional costs are incurred due to the management overhead associated with a number of factors including: larger teams of developers, more complex configuration management, increased complexity of system integration, etc. As a result, larger size systems, have larger exponent values.

The challenge using Equation (19.1) is that it is difficult to estimate a (the size) at the start of a project, and c and x are subjective — that is, they vary depending on the type of software being developed and the experience of the software developers.

19.1.1 *The COCOMO Model*

COCOMO [Ref. 19.3] is the best known algorithmic software costing model (COCOMO = COnstructive COst MOdel). The COCOMO model is an empirical model constructed by collecting data from a large number of completed software projects. The data was analyzed to construct parametric models (see Chapter 6) that represent the best fit to the observations. The parametric models in COCOMO provide a quantitative linkage between the size of the system and the project and development team characteristics, and the effort required to develop (and maintain) the system.

The original version of COCOMO defined three software development models:

1. *Organic (Simple) Model* – relatively simple, well understood projects in which small teams work to satisfy an informal set of requirements (e.g., an electric field simulation program for an electronics design group).
2. *Semi-Detached (Moderate) Model* – an intermediate complexity project that requires mixed teams to satisfy a set of requirements. In this case not all team members necessarily have a view of the whole system, i.e., they may have limited experience and/or knowledge about the portions of the system they are not working on.
3. *Embedded Model* – a project that operates within a tightly defined set of regulations, constraints and operational procedures (e.g., control software for a safety-critical system).

The basic COCOMO effort calculation is

$$PM = Ec(KDSI)^x \tag{19.2}$$

where

PM = person-months.
E = 1.0 (effort adjustment factor).
c = 2.4 (organic), 3.0 (semi-detached), 3.6 (embedded).
x = 1.05(organic), 1.12 (semi-detached), 1.20 (embedded).
$KDSI$ = thousands of delivered source instructions.

The software development time (in months) in basic COCOMO is given by

$$TDEV = 2.5(PM)^{0.38} \tag{19.3}$$

For example, if the estimated size of an organic software development project is 50,000 delivered source instructions (DSI). Using basic COCOMO we can estimate the following project attributes:

$$PM = 2.4(50)^{1.05} = 146 \text{ person months} \tag{19.4a}$$

$$Productivity = \frac{50{,}000 \text{ DSI}}{146 \text{ person months}} = 342 \text{ DSI/person month} \tag{19.4b}$$

$$TDEV = 2.5(146)^{0.38} = 16.6 \text{ months} \tag{19.4c}$$

$$Average\ Staffing = \frac{146 \text{ person months}}{16.6 \text{ months}} = 8.8 \text{ people} \tag{19.4d}$$

In Intermediate COCOMO [Ref. 19.3], c = 3.2 (organic), 3.0 (semi-detached), 2.8 (embedded), and the effort adjustment factor (E) is calculated using fifteen cost drivers. The cost drivers are grouped into the following four categories: product, computer, personnel, and project, as shown in Table 19.2. Each of the cost drivers are rated on an ordinal scale ranging from low to high importance. Using the rating, an effort multiplier is determined.

Table 19.2. Intermediate COCOMO Cost Drivers [Ref. 19.3].

Cost Driver	Description	Rating					
		Very Low	Low	Nominal	High	Very High	Extra High
Product							
RELY	Required software reliability	0.75	0.88	1.00	1.15	1.40	-
DATA	Database size	-	0.94	1.00	1.08	1.16	-
CPLX	Product complexity	0.70	0.85	1.00	1.15	1.30	1.65
Computer							
TIME	Execution time constraint	-	-	1.00	1.11	1.30	1.66
STOR	Main storage constraint	-	-	1.00	1.06	1.21	1.56
VIRT	Virtual machine volatility	-	0.87	1.00	1.15	1.30	-
TURN	Computer turnaround time	-	0.87	1.00	1.07	1.15	-
Personnel							
ACAP	Analyst capability	1.46	1.19	1.00	0.86	0.71	-
AEXP	Applications experience	1.29	1.13	1.00	0.91	0.82	-
PCAP	Programmer capability	1.42	1.17	1.00	0.86	0.70	-
VEXP	Virtual machine experience	1.21	1.10	1.00	0.90	-	-
LEXP	Language experience	1.14	1.07	1.00	0.95	-	-
Project							
MODP	Modern programming practices	1.24	1.10	1.00	0.91	0.82	-
TOOL	Software Tools	1.24	1.10	1.00	0.91	0.83	-
SCED	Development Schedule	1.23	1.08	1.00	1.04	1.10	-

For example, if your product is rated very high for complexity (CPLX effort multiplier of 1.30), and low for language experience (LEXP effort multiplier of 1.07), and all of the other cost drivers are assumed to have a nominal-effort multiplier of 1.00. The effort adjustment factor is E = (1.30)(1.07) = 1.39. For the example given previously, the calculated effort becomes

$$PM = (1.39)(3.2)(50)^{1.05} = 270 \text{ person months}$$

In the original versions of COCOMO it was assumed that software development follows a "waterfall" process. However, much of today's software is developed by "gluing" reusable components and off-the-shelf systems together — that is, by re-engineering existing software to create new software. The COCOMO II model accommodates various software development approaches including: prototyping, development by component composition, and the use of database programming. As an alternative to the original versions of COCOMO, COCOMO II supports a "spiral" development process.[1]

19.1.2 *Function-Point Analysis*

Instead of using size (e.g., lines of code) as the estimated attribute, the functionality of the code can be used. The basic tenant of function-point analysis is that functionality is independent of implementation language. There are several function-based measures of software development effort. The best known of these measures is function-point counting. Function-point analysis sizes a software application from an end-user perspective instead of using the technical details of the specific coding language.

---

[1] Waterfall development is a sequential design process in which progress is seen as flowing steadily downwards through conception, initiation, analysis, design, construction, testing, production/implementation and maintenance. Alternatively, spiral development combines elements of both design and prototyping-in-stages in an effort to combine the respective advantages of top-down and bottom-up concepts. The spiral development process is most often used for large, expensive, complicated projects.

Software development cost estimation based on user functionality (function points) was proposed by Albrecht in 1979 [Ref. 19.4]. Feature points, which is an extension of function points, developed by Jones in 1986 [Ref. 19.5] are used to estimate effort for real-time systems, embedded systems, operating systems and communications software.

Function and feature points are constant regardless of the programming language used and can also be used to measure the effort associated with non-coding activities, such as management. Function and feature points can be converted into code statements for many languages.

This cost estimation requires counts of the following five unique function points to be made [19.6]:

- External inputs – items provided by the user that describe distinct application-oriented data (such as file names and menu selections)
- External outputs – items provided to the user that generate distinct application-oriented data (such as reports and messages, rather than the individual components of these)
- External inquiries (or queries) – interactive inputs requiring a response
- External files (or external interfaces) – machine-readable interfaces to other systems
- Internal files – logical master files in the system.

Each count is multiplied by a complexity weight and the results are summed to determine the unadjusted function-point count (*UFC*), using

$$UFC = \sum_{\text{Function Point}} \left[\sum_{\text{Complexity}} (CW)(Count) \right] \qquad (19.5)$$

where *Count* is the raw function-point count and *CW* is the complexity weight given in Table 19.3.

Function-point metrics account for the fact that some inputs and outputs (and other interactions) are more complex than others by multiplying the unadjusted function-point estimate by a technical complexity-weighting factor (*TCF*). The adjusted function-point count (*FP*) is given by

$$FP = (UFC)(TCF) \qquad (19.6)$$

Table 19.3. Function Point Complexity Weights (*CW*s).

	Simple	Average	Complex
External inputs	3	4	6
External outputs	4	5	7
External inquiries	3	4	6
External files	7	10	15
Internal files	5	7	10

The components of the *TCF* (F_i) are given in Table 19.4.

Table 19.4 Technical Complexity Factors (*TCF*s)

F_1	Reliable back-up and recovery	F_2	Data communications
F_3	Distributed functions	F_4	Performance
F_5	Heavily used configuration	F_6	Online data entry
F_7	Operational ease	F_8	Online update
F_9	Complex interface	F_{10}	Complex processing
F_{11}	Reusability	F_{12}	Installation ease
F_{13}	Multiple sites	F_{14}	Facilitate change

Each factor in Table 19.4 is rated from 0 to 5, where 0 means the component has no influence on the system and 5 means the component is essential. The *TCF* is then formed using

$$TCF = 0.65 + 0.01\sum_{i=1}^{14} F_i \tag{19.7}$$

Finally, function points can be converted to *Effort* using an appropriate form of Equation (19.1), without ever estimating lines of code.

As an example of function-point analysis, assume that a planned software application has the following attributes:

15 simple external inputs
2 complex external outputs
12 simple external inquiries
1 simple external file
2 complex external files
3 complex internal files.

Let's estimate how much effort will be required to implement this software application. Figure 19.1 shows the determination of the *UFC* using Equation (19.5) for this example. The factors of the *TCF* for this example are

$$
\begin{array}{lll}
F_1 = 0 & F_2 = 2 & F_3 = 0 \\
F_4 = 3 & F_5 = 3 & F_6 = 5 \\
F_7 = 5 & F_8 = 0 & F_9 = 3 \\
F_{10} = 2 & F_{11} = 1 & F_{12} = 0 \\
F_{13} = 0 & F_{14} = 0 &
\end{array}
$$

and the *TCF* is found using Equation (19.7) as

$$TCF = 0.65 + 0.01(0+2+0+3+3+5+5+0+3+2+1+0+0+0) = 0.89 \tag{19.8}$$

Using the *UFC* from Figure 19.1 and the *TCF* from Equation (19.8) we obtain the number of function points, $FP = (162)(0.89) = 144.18$ function points.

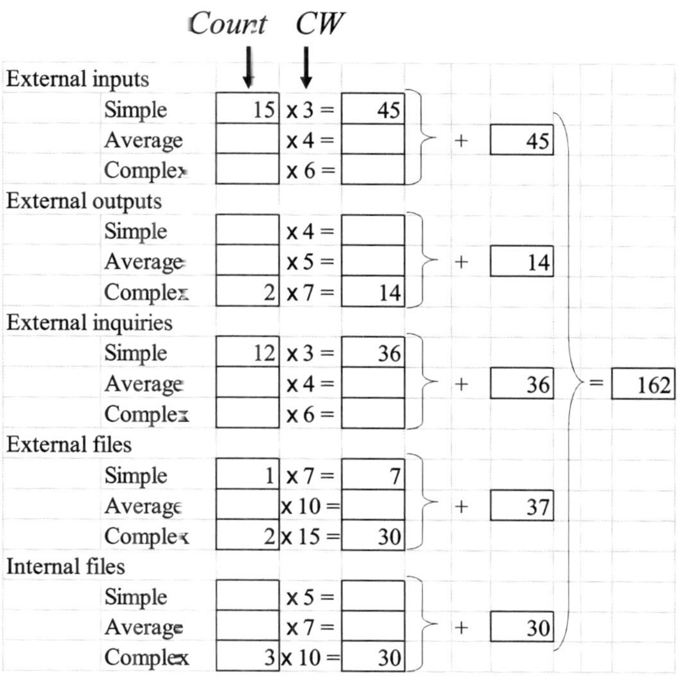

Fig. 19.1. Example function point counting process, the *UFC* for this example is 162.

Now we need to use a parametric model to determine the effort from the function points. Many models have been developed; the following model is attributed to Kemerer [Ref. 19.7]:

$$Effort = (60.62)(7.728 \times 10^{-8})FP^3 \tag{19.9}$$

where *Effort* is in person-months. Application of Equation (19.9) to our example problem predicts *Effort* = 14 person-months.

Function points can be mapped to programming language-specific source lines of code. A list of conversions for over 600 languages is given in [Ref. 19.8]. Note that using function points to find *SLOC* and then estimating cost based on *SLOC* is not considered a technically correct approach — function points are the only independent variable needed to calculate costs.

The number of function points that are implemented per person-month is a measure of the productivity.

In 1986, Software Productivity Research, Inc. (SPR) developed an experimental method for applying function-point logic to system software such as operating systems, or telephone switching systems. The resulting SPR Feature-Point metric is a superset of the function-point metric that introduces a new parameter — number of algorithms — in addition to the five standard function point parameters. Algorithms are defined as a set of rules that must be completely expressed to solve a computational problem.

Overall, the function-point models appear to more accurately predict the effort needed for a specific project than models based on lines of code. However, because complexity estimates are subjective, function-point count depends on the estimator — that is, different estimators measure complexity differently. Therefore, accurate counting requires certified function-point specialists; function-point counting can be time-consuming and expensive, and function-point counts are erratic for applications or systems below fifteen function points in size [Ref. 19.9].

19.1.3 *Object-Point Analysis*

Given the popularity of object-oriented programming (OOP) and object-oriented CASE tools, software cost estimation methods based on objects are also used. Object-point analysis [Ref. 19.10] is similar to function-point-based cost estimation, but it counts objects instead of functions. Object points avoid the subjectivity of function points by more clearly defining the complexity adjustment factor. Object points also take into account the fraction of code reuse. In COCOMO II object points are referred to as application points.

The number of object points in a program is a weighted estimate of the following [Ref. 19.11]:

- The number of separate screens that are displayed – simple, moderately complex, and very complex screens count as different numbers of object points.
- The number of reports that are produced – simple, moderately complex, and difficult-to-produce reports count as different numbers of object points.
- The number of third-generation (3-GL) components that will be used by each object that makes up the application. A third-generation component is a software module written in a third-generation language such as COBOL, C, C++, VB.NET, or Java.

In this process, the number of objects are estimated, the complexity of each of those objects is estimated, and finally the weighted total (the object-point count) is computed.

Object points are easier to estimate for a high-level software specification than function points. The advantage is that object points are only concerned with screens, reports, and modules in conventional programming languages — they are not concerned with implementation details, and the complexity factor estimation is much simpler than for function-point counting.

19.2 Software Support Costs

Software maintenance is the process of modifying and maintaining existing software while leaving its primary functions intact. This includes minor enhancements, bug fixes, addition of new drivers, and additions to or corrections of documentation.

Generally software maintenance tasks are classified as the following:

- *Corrective tasks* – corrections related to the diagnosis, localization, and fixing of errors in the software. Often correction-type tasks are the easiest and thus the least costly maintenance tasks; however, they usually have to be performed on a tight schedule.
- *Adaptive tasks* – interfacing existing software into a changing (technical) environment
- *Perfective tasks* – additions, enhancements and modifications made to the code based on changing user needs
- *Preventive maintenance tasks* – enhancement of the future maintainability of the system. Preventive maintenance should be considered when the software has a long lifetime.

Maintenance also generally includes configuration management, change control, and a number of other code management tasks.

Software maintenance is often characterized using a metric called annual change traffic (ACT), which is the fraction of the source code that undergoes change:

$$ACT = \frac{DSI_{maint}}{DSI_{develop}} \tag{19.10}$$

Consider the maintenance of the example case described at the end of Section 19.1.1. In this case there are 50,000 DSI. If we assume maintenance adds 5000 DSI and modifies 3000 of the existing DSI in a particular year, then the resulting *ACT* is given by

$$ACT = \frac{5000 + 3000}{50{,}000} = 0.16 \tag{19.11}$$

The number of person-months necessary for this maintenance is given by

$$PM = (ACT^{1.05})(PM_{development}) = (0.16^{1.05})(146) = 21.3 \text{ person months} \tag{19.12}$$

which gives

$$Maintenance\ Staffing = \frac{21.3 \text{ person - months}}{12 \text{ months/year}} = 1.8 \text{ person years} \tag{19.13}$$

19.3 Discussion

There is no simple way to make an accurate estimate of the effort required to develop software [Ref. 19.11]. Initial estimates may have to be based on the requirements definition, and be made for software developers whose skills are unknown and whose productivity is a function of numerous factors that cannot be easily estimated.

Before a project is implemented, there is always uncertainty about the project's attributes. Any cost estimate produced at this stage is guaranteed to be inaccurate. Most software cost models produce exact results with little regard for these uncertainties.

Like parametric modeling, software cost models should be calibrated to the particular organization developing and/or supporting the software.

References

19.1 Leung, H. and Fan, Z. (2002). Software cost estimation. *Handbook of Software Engineering & Knowledge Engineering, Volume 2 – Emerging Technologies* (World Scientific Publishing Co. Singapore).

19.2 Taylor, R. (1996). Project management, cost estimation, and team organizations. ICS 125 Lecture Notes (University of California, Irvine, CA) http://www.ics.uci.edu/~taylor/ics125_fq99/management.pdf. Accessed April 28, 2016.

19.3 Boehm, B. W. (1981). *Software Engineering Economics* (Prentice Hall, Englewood Cliffs, NJ).

19.4 Albrecht, A. J. and Gaffney, J. E. (1983). Software function, source lines of code, and development effort prediction: a software science validation, *IEEE Transactions on Software Engineering*, SE-9(6), pp. 639-648.

19.5 Jones, C. (1986). *Applied Software Measurement – Assuring Productivity and Quality*, 2nd Edition. (McGraw-Hill, New York, NY).

19.6 Fenton, N. E. and Pfleeger, S. L. (1997). *Software Metrics: A Rigorous and Practical Approach* (International Thomson Computer Press, Boston, MA).

19.7 Pressman, R. S. (2001). Software Engineering – A Practitioner's Approach, 5th Edition (McGraw-Hill, Boston, MA).

19.8 Jones, T. C. (2001). *Table of Programming Languages and Levels – Version 8.2* (Software Productivity Research, Burlington, MA).

19.9 Jones, T. C. (2005). *Strengths and Weaknesses of Software Metrics (SMM01051)* (Software Productivity Research, Burlington, MA).

19.10 Banker, R. D., Kauffman, R. J., Wright, C. and Zweig, D. (1994). Automating output size and reuse metrics in a repository-based computer aided software engineering (CASE) environment, *IEEE Transactions on Software Engineering*, 20(3), pp. 169-187.

19.11 Sommerville, I. (2007). Chapter 26 – Software cost estimation, *Software Engineering*, 7th Edition (Addison-Wesley, Harlow, England).

Bibliography

In addition to the references, several good books on software cost estimation include:

Boehm, B. W., Abts, C., Brown, A. W., Chulani, S., Clark, B. K., Horowitz, E., Madachy, R., Reifer, D. J. and Steece, B. (2000). *Software Cost Estimation with COCOMO II* (Prentice Hall, Upper Saddle River NJ).

Jones, T. C. (1998). *Estimating Software Costs* (McGraw-Hill, Inc., New York, NY).

Problems

19.1 A particular software functionality needs to be implemented. Two different groups within your organization could perform the work and have provided you with details of their proposed approaches. The proposals from the two groups are as follows:

$F_1 = 2$	$F_2 = 2$	$F_3 = 0$	$F_4 = 3$	$F_5 = 3$
$F_6 = 4$	$F_7 = 5$	$F_8 = 0$	$F_9 = 3$	$F_{10} = 2$
$F_{11} = 1$	$F_{12} = 5$	$F_{13} = 4$	$F_{14} = 0$	

a) Assuming the Kemerer model (Equation (19.9)), and based only on burdened labor costs for the original development of the software, which group should you use to develop your software? Assume 52 weeks per year and 40 hours a week from each software developer.

b) How many source lines of code need to be developed for each group in part (a)?

c) Assuming an annual change traffic of 0.23, how many people do you need to commit to software maintenance for the group chosen in part (a)?

Property	Group A	Group B
Implementation language	C++	SMALLTALK
External inputs	4 simple 10 average 1 complex	4 simple 10 average 1 complex
External outputs	1 simple 3 average 10 complex	1 simple 3 average 10 complex
External inquiries	0 simple 5 average 0 complex	1 simple 4 average 0 complex
External files	0 simple 3 average 2 complex	0 simple 5 average 0 complex
Internal files	4 simple 10 average 0 complex	10 simple 5 average 0 complex
Labor rate for the proposed software developers	\$18.50/hour	\$20/hour
Effective overhead multiplier	3.5	4.1

19.2 You are the owner of a small company that develops software applications. The software engineers in your group want to switch from C to COBOL because COBOL will make external files easier to handle.

a) A software development job in C has the following: adjusted function points = 300; technical complexity factor = 0.716; simple external files = 2, average external files = 5, complex external files = 7. Assuming that switching to COBOL <u>only</u> changes the external files to: simple external files = 3, average external files = 6, complex external files = 5 (everything else remains the same). How many adjusted function points (*FP*) will the COBOL implementation require?

b) Assuming a profit of 35%, a labor rate of \$20/hr, a labor burden (overhead rate) of 0.6, and 160 working hours per month, how much will switching from C to COBOL save the customer on the software development job described above, using the Kemerer model — Equation (19.9)?

c) Unfortunately, your software developers are C programmers who do not have much experience programming in COBOL, so there is a learning curve associated with each engineer's effort. Assuming that the numbers in part (b) represent how the developers are expected to perform on the fifth development job and with a 90% learning curve, calculate the expected developer effort in months as a function of the job number. How many months of effort will it take to do the <u>first</u> job in COBOL?

d) Suppose that you can avoid the learning curve by sending each software engineer to a one-week class (40 hours) on COBOL. After the class, the developers can perform at the level described in part (a). If the class costs \$5000 per person, what is the return on investment (ROI) of the training after the first job, assuming that the job needs to be done in exactly 12.788 months? Hint: You need to use information from parts (b) and (c) to solve this problem.

Chapter 20

Total Cost of Ownership Examples

From a customer's viewpoint, understanding the total cost of owning a product is the most important aspect of the product's cost. In many cases, the cost of purchasing a product may be insignificant compared to the cost of operating and maintaining it. Figure II.2 in the Part II introduction summarizes the elements that are included in a total cost of ownership analysis.

This chapter presents three examples of total cost of ownership. The first is an estimation of the total cost of ownership of color printers; the second looks at an electronic part selection decision. In the final example we introduce the levelized cost of energy.

20.1 The Total Cost of Ownership of Color Printers

In this example, we determine the total cost of ownership of three printers that are used to print black and white and color pages, and demonstrate that the purchase price of printers is not always the best way to assess their real cost to the customer.

Table 20.1 lists the assumptions associated with three different printers that are all manufactured and marketed by the same company.

To determine the total cost of ownership (C_{TCO}), consider the costs of the printer, paper, and ink or toner:

$$C_{TCO} = C_{printer} + C_{paper} + C_{ink/toner} \tag{20.1}$$

The cost of the printers is determined from

$$C_{printer} = N_{printers} P_{printer} \tag{20.2}$$

where

$N_{printers}$ = the number of printers needed —equals $\lceil N_{pages}/L_{printer} \rceil$.

N_{pages} = the total number of pages printed.

$L_{printer}$ = the lifetime of the printer measured in the number of printed pages.

$P_{printer}$ = the purchase price of the printer.

Table 20.1. Comparison Data for Three Color Printers [Ref. 20.1].

Description	Inkjet printer	Home laser color printer	Business laser color printer
Printer purchase price (including 6% sales tax), $P_{printer}$	$67.18	$210.94	$952.94
Printer lifetime (pages/warranty period). This is the manufacturer's maximum suggested pages/month multiplied by the warranty length in months, $L_{printer}$	12,000	12,000	90,000
Ink/toner cartridge cost per set*, $I_{ink/toner}$	$76.32	$297.82	$934.88
Cartridge set life (pages printed), Z	500	2,200	7,500
Number of pages printed with cartridges included with printer when purchased. Cartridge life is based on standard pages as defined in ISO/IEC 19798, $N_{withprinter}$	125	550	7,500
Paper cost (including 6% sales tax)	$3/500 sheets	$3/500 sheets	$3/500 sheets

*A cartridge set includes black, cyan, yellow, and magenta; the price includes 6% sales tax.

Assume that each printer is disposed of (has zero salvage value) after $L_{printer}$ pages have been printed and that printers do not malfunction during the printing of these pages.

The cost of the paper per printed page is $3/500 = $0.006/page; therefore, $C_{paper} = \$0.006 N_{pages}$.

The cost of the ink/toner is given by

$$C_{ink/toner} = N_{refill} I_{ink/toner} \tag{20.3}$$

where N_{refill} is the number of ink refills needed — that is,

$$N_{refill} = \left\lceil \left(N_{pages} - N_{printers} N_{withprinter} \right) / Z \right\rceil \tag{20.4}$$

where N_{refill} is constrained to be ≥ 0, and

$I_{ink/toner}$ = the cost of an inkjet cartridge set or toner cartridge set.

Z = the number of pages that can be printed with one ink/toner cartridge set.

$N_{withprinter}$ = the number of pages that can be printed with the ink/toner cartridge set that comes with the original printer purchase.

The quantity N_{refill} gives the number of ink or toner cartridge sets that need to be purchased and accounts for the amount of ink or toner included with each printer when it is purchased.

Using the data in Table 20.1, Table 20.2 summarizes the cost calculations corresponding to printing 15,000 pages on each of the three printers. Figure 20.1 shows the total cost of ownership as a function of the total number of pages printed. From this figure it can be seen that the inkjet printer is the least expensive solution up to approximately 5000 pages, at which point the total cost of ownership of all the printers becomes comparable. The steps that appear in Figure 20.1 represent the purchases of ink/toner cartridge sets.

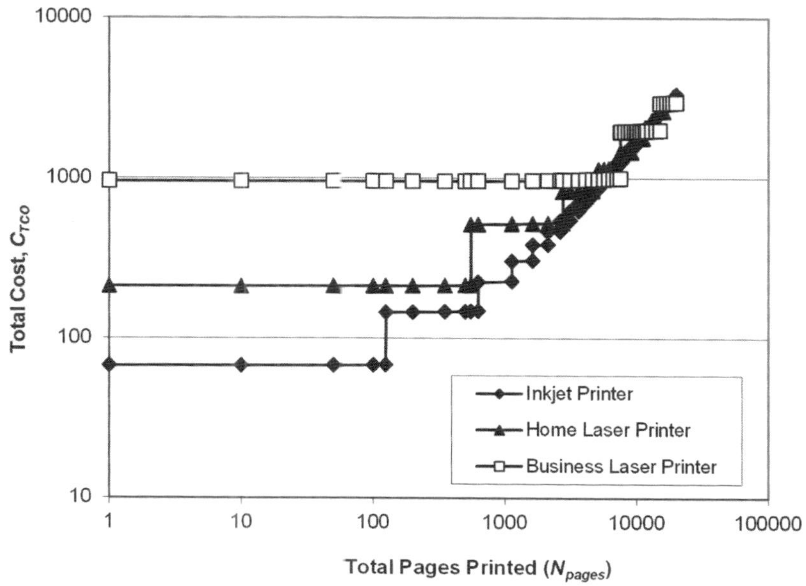

Fig. 20.1. Total cost of ownership as a function of the number of pages printed.

Table 20.2. Example Cost Calculations for Three Color Printers (N_{pages} = 15,000).

	Inkjet printer	Home laser color printer	Business laser color printer
$N_{printer}$	$\left\lceil \frac{15,000}{12,999} \right\rceil = 2$	$\left\lceil \frac{15,000}{12,999} \right\rceil = 2$	$\left\lceil \frac{15,000}{90,000} \right\rceil = 1$
$P_{printer}$ Table 20.1	\$67.18	\$210.94	\$952.94
$C_{printer}$ Equation (20.2)	2(\$67.18) = \$134.36	2(\$210.94) = \$421.88	1(\$952.94) = \$952.94
N_{refill}	$\left\lceil \frac{15,000 - 2(125)}{500} \right\rceil = 30$	$\left\lceil \frac{15,000 - 2(550)}{2200} \right\rceil = 7$	$\left\lceil \frac{15,000 - 1(7500)}{7500} \right\rceil = 1$
$C_{ink/toner}$ Equation (20.3)	30(\$76.32) = \$2,289.60	7(\$297.82) = \$2,084.74	1(\$934.88) = \$934.88
C_{paper}	15,000(\$0.006) = \$90.00	15,000(\$0.006) = \$90.00	15,000(\$0.006) = \$90.00
C_{TCO} Equation (20.1)	\$134.36 + \$90.00 + \$2,289.60 = \$2,514.26	\$421.88 + \$90.00 + \$2,084.74 = \$2,596.62	\$952.94 + \$90.00 + \$934.88 = \$1,977.82

An alternative measure of the total cost of ownership that might be useful to organizations that provide printing services is the cumulative average cost per page. This value is obtained from C_{TCO}/N_{pages} and is shown in Figure 20.2 for the three printers considered.

Several important effects have not been considered in this analysis. We have assumed that the quality of the printed page and speed of printing are not issues. We have also not considered what is being printed; for example, it takes more ink/toner to print photos than text. In this example, we have used the page counts cited by the printer manufacturers on their ink/toner cartridges. We have also not considered the option of refilling the ink cartridges rather than purchasing new ones. Refilling is an option that may reduce ink costs at the risk of decreasing the lifetime of the printer. Lastly, Equations (20.2) and (20.3) do not assume that there is any credit provided for unused printer life or unused ink/toner after the specified number of pages is printed.

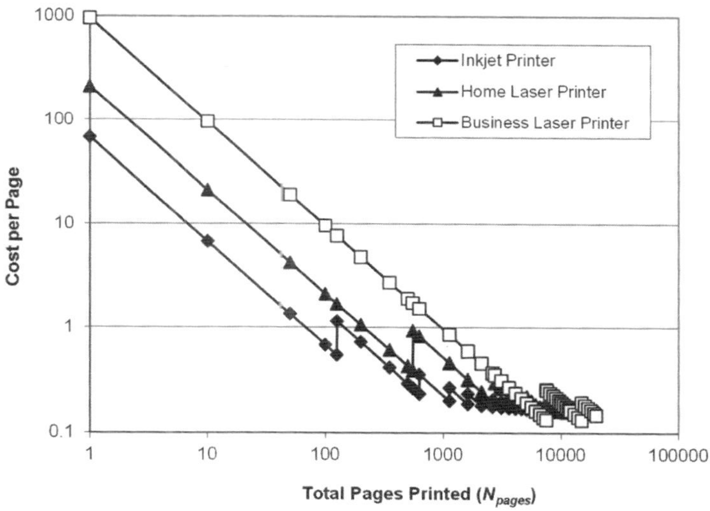

Fig. 20.2. Effective cost per page as a function of the total number of pages printed.

20.2 Total Cost of Ownership for Electronic Parts [Ref. 20.2]

Electronic part selection for products is often driven or significantly influenced by procurement management processes that have little or no understanding of the effective cost of ownership or through-life cost of the part. Procurement organizations are often motivated by minimizing procurement cost or selecting suppliers that offer parts at lower prices, and may not take into account life-cycle costs.

This section describes the formulation and use of an electronic part total cost of ownership model that allows part selection and management organizations to predict the total cost of ownership of a part in order to enable better informed fundamental part selection decisions. This model focuses on optimal part management from a part selection and management organization's viewpoint, as opposed to optimum part management from a product group's perspective. These perspectives differ because the part selection and management group has a more holistic view of a part's cost of ownership than a product group, and because a part (especially an electronic part) may be concurrently used in many different products within the same organization. This approach requires a cost model that comprehends long-term supply chain constraints associated with specific parts and their effects downstream at the product level. Therefore the cost that we wish to predict and minimize is the effective total cost of ownership (TCO) of the part as used across multiple products. Assessing the total cost incurred over the life cycle of the part as an effective total cost of ownership will allow part management organizations to quantify the cost spent (inclusive of procurement) per part.

20.2.1 *Part Total Cost of Ownership Model*

The part total cost of ownership model is composed of the following three sub-models: part support model, assembly model, and a field failure model. This model contains both assembly costs (including procurement) and life-cycle costs associated with using the part in products.

Part Support Model

The part support model captures all non-recurring costs associated with selecting, qualifying, purchasing, and sustaining the part (these costs may recur annually, but do not recur for each part instance). The total support cost in year i (in year 0 dollars) is given by

$$C_{support_i} = \frac{\left(C_{ia_i} + C_{pa_i} + C_{as_i} + C_{ps_i} + C_{ap_i} + C_{or_i} + C_{nonPSL_i} + C_{design_i}\right)}{(1+r)^i} \quad (20.5)$$

where

C_{ia_i} = the initial part approval and adoption cost — all costs associated with qualifying and approving a part for use (i.e., setting up the initial part approval). This could include reliability and quality analyses, supplier qualification, database registration, added NRE for part approval, etc. The approval cost occurs only in year 1 (i = 1) for each new part.

C_{pa_i} = product-specific approval and adoption — all costs associated with qualifying and approving a part for use in a particular product. This approval cost occurs exactly one time for each product that the part is used in and is a function of the type of part and the approval level of the part within the organization when the part is selected. This cost depends on the number of products introduced in year i that use the part.

C_{as_i} = the annual cost of supporting the part within the organization — all costs associated with part support activities that occur for every year that the part must be maintained in the organization's part database, including database management, product change notice (PCN) management, reclassification of parts, and services provided to the product sustainment organization. This cost depends on the part's qualification level.

C_{ps_i} = all costs associated with production support and part management activities that occur every year that the part is in a manufacturing (assembly) process, for one or more products; this includes volume purchase agreements, services provided to the manufacturing organization, reliability and quality monitoring, and availability (supplier addition or subtraction).

C_{ap_i} = the purchase order generation cost, which depends on the number of purchase orders in year i.

C_{or_i} = the obsolescence case resolution costs, which are only charged in the year that a part becomes obsolete.

C_{nonPSL_i} = setup and support for all non-PSL (preferred supplier list) part suppliers, which depends on the number of non-PSL sources used.

C_{design_i} = the non-recurring design-in costs associated with the part, which are only charged in years of introduction of new products using the part; this includes the cost of a new CAD footprint and symbol generation, if needed.

r = the after-tax discount rate on money.

i = the year.

C_{ia_i}, C_{pa_i}, C_{as_i}, and C_{ps_i} are determined from an activity-based cost model in which cost activity rates can be calculated by part type.

Assembly Model

The assembly model captures all the recurring costs associated with the part: purchase price, system assembly cost (part assembly into the system), and recurring functional test/diagnosis/rework costs. The total assembly cost (for all products) in year *i*, assuming exactly one part site[1] per product, is given by

$$C_{assembly_i} = \frac{N_i C_{out_i}}{(1+r)^i} \tag{20.6}$$

where

N_i = the total number of products assembled in year *i*.

C_{out_i} = the output cost/part from the model shown in Figure 8.4. C_{out} is a function of C_{in} as shown in Figure 8.4.

C_{in_i} = the incoming cost/part = $P_i + C_{a_i}$.

P_i = the purchase price of one instance of the part in year *i*.

C_{a_i} = the assembly cost of one instance of the part in year *i*.

This model uses the test/diagnosis/rework model for the assembly process of electronic systems described in Section 8.3.2. The approach includes a model of functional test operations characterized by fault coverage, false

[1] A "part site" is defined as the location of a single instance of a part in a single instance of a product.

positives, and defects introduced in testing, in addition to rework and diagnosis (diagnostic test) operations that have variable success rates and their own defect introduction mechanisms. The model accommodates multiple rework attempts on any given product instance and enables optimization of the fault coverage and rework investment during assembly tradeoff analyses.

The model discussed in this section contains inputs to the test/diagnosis/rework model that are specific to the part type and how the part is assembled — automatic, semi-automatic, manual, pre-mount, lead finish, extra visual inspection, special electrostatic discharge (ESD) handling. The output of the model is the effective procurement and assembly cost per part site. For simplicity, the application of this test/diagnosis/rework model assumes that all functional and assembly-introduced part-level defects are resolved in a single rework attempt — that is, $Y_{rew} = 1$ when there are no defects introduced by the testing process, that $Y_{beforetest} = Y_{aftertest} = 1$, and that there are no false positives in testing ($f_p = 0$). These yield assumptions guarantee that Y_{out} will always be 1.

Field Failure Model

The field failure model captures the costs of warranty repair and replacement due to product failures caused by the part. Equation (20.7) gives the field failure cost in year *i*.

$$C_{field\ use_i} = \frac{N_{f_i}\left(1-f\right)C_{repair} + N_{f_i} f C_{replace} + N_{f_i} C_{proc_i}}{(1+r)^i} \tag{20.7}$$

where

N_{f_i} = the number of failures under warranty in year *i*. This is calculated using 0-6, 6-18 and > 18 month FIT rates[2] for the part; the warranty period length (an ordinary free replacement warranty is assumed with the assumption that no single product instance fails more than once during the warranty period); and the number of parts sites that exist during the year.

---

[2] FIT (failure in time) rate – Number of part failures in 10^9 device-hours of operation.

f = the fraction of failures requiring replacement (as opposed to repair) of the product.
C_{repair} = the cost of repair per product instance.
$C_{replace}$ = the cost of replacing the product per product instance.
C_{proc_i} = the cost of processing the warranty returns in year i.

Total Cost of Ownership

Traditionally, the term "part" is used to describe one or more items with a common part number from a parts-management perspective. For example, if the product uses two instances of a particular part (two part sites), and one million instances of the product are manufactured, then a total of two million part sites for the particular part exist. The reason part sites are counted (instead of just parts) is that each part site could be occupied by one or more parts during its lifetime (e.g., if the original part fails and is replaced, then two or more parts occupy the part site during the part site's life). For consistency, all cost calculations are presented in terms of either annual or cumulative cost per part site.

The total cost of ownership expressed as an effective cumulative cost per part site is given in Equation (20.8) up to year i:

$$C_i = \frac{\sum_{j=1}^{i}\left(C_{support_j} + C_{assembly_j} + C_{field\ use_j}\right)}{\sum_{j=1}^{i} N_j} \tag{20.8}$$

where N_j is the number of part sites assembled in a particular year j.

In this model we focus on the effective cost per part site rather than the cost per part, because when product repair and replacement are considered there is effectively more than one part consumed per part site. All computed costs in the model are indexed to year 1 for reference, where year 1 refers to the period between time 0 and the end of 1 year.

20.2.2 *Example Analyses*

This section includes example analyses performed using the model described in Section 20.2.1. The model was populated with data from Ericsson AB for a generic surface-mount capacitor. Figure 20.3 shows a summary of inputs to the model that correspond to all of the example analyses presented. A part site usage profile indicating the number of part sites used for each product annually is provided as an input to the model. The profile in Figure 20.3 also describes the number of unique products using the part each year and the total quantity of part sites assembled each year (N_i). In all cases, inflation or deflation in cost input parameters can be defined (electronic part prices generally decrease as a function of time).

As an example of the part total cost of ownership model, consider the part data shown in Figure 20.3. For this part used in the products given in Figure 20.3 (for which a resultant total annual part site usage is also shown), the results in Figure 20.4 are obtained. The plots on the left side of Figure 20.4 show that initially, all the costs for the part are support costs — that is, initial selection and approval of the part. Manufacturing and procurement costs approximately follow the production schedule shown in Figure 20.3. This example part becomes obsolete in year 17 (*YTO* is 16.7 years at year 0) and a lifetime buy of 4,000 parts is made at that time, indicated by the small increase in procurement and inventory costs in year 17. Year 18 is the last year of manufacturing after which field use costs dominate.

For the case shown in Figure 20.4, the initial procurement price per part ($0.015/part) is only 11% of the cumulative effective cost per part site ($0.14/part site) during a 20-year usage life. The results in Figure 20.4 show that, at high volumes, the procurement and inventory cost after 20 years is 7% of the total effective cost per part site. Assembly and support costs contribute to a combined share of the total cost of ownership of 93% (88% and 5% of the total effective cost per part site, respectively). The organization dedicates an annual average of $1.85 per operational hour of support cost over 20 years per part site in this high-volume case.

PART-SPECIFIC INPUTS:

Parameter	Value
Part name	SMT Capacitor
Existing part or new part?	New
Type	Type 1
Approval/Support Level	PPL
Procurement Life (*YTO* at beginning of year 1)	16.7 years
Number of suppliers of part	7
How many of the suppliers are not PSL but approved?	5
How many of the suppliers are not PSL AND not approved?	0
Part-specific NRE costs	0
Product-specific NRE costs (design-in cost)	0
Number of I/O	2
Item part price (in base year money)	$0.015
Are order handling, storage and incoming inspection included in the part price?	Yes
Handling, storage and incoming inspection (% of part price)	10.00%
Defect rate per part (pre electrical test)	5 ppm
Surface mounting details	Automatic
Odd shape?	No
Part FIT rate in months 0-6 (failures/billion hours)	0.05
Part FIT rate in months 7-18 (failures/billion hours)	0.04
Part FIT rate after month 18 (failures/billion hours)	0.03

GENERAL NON-PART-SPECIFIC INPUTS:

Parameter	Value
Part price change profile (change with time)	Monotonic
Part price change per year	-2.0% per year
Part price change inflection point (year)	5
Manuf. (assembly) cost change per year	-3.00%
Manuf. (test, diagnosis, rework) cost change per year	-3.00%
Admin. cost change per year	0.00%
Effective after-tax discount rate (%)	10.00%
Base year for money	1
Additional material burden (% of price)	0.00%
% of part price for LTB storage/inventory cost (per part per year)	66.67%
LTB overbuy size (buffer)	10%
Expected obsolescence resolution	LTB
Fielded product retirement rate (%/year)	5.00%
Operational hours per year	8760 hours
Product warranty length	18 months
% of supplier setup cost charged to non-PSL, approved suppliers	0.00%

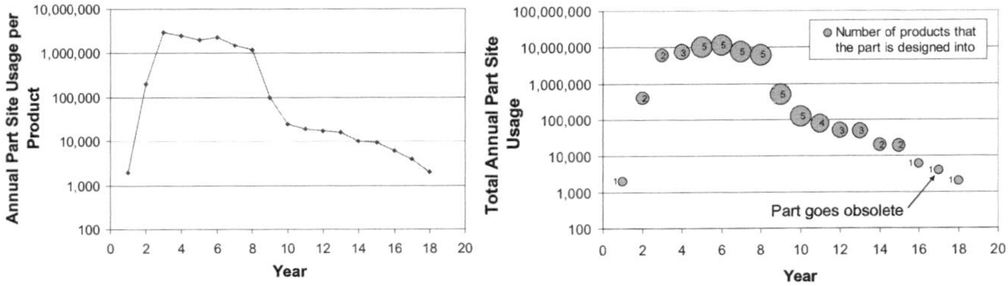

Fig. 20.3. Inputs used in the part total cost of ownership cost model for the examples provided in this section. (© 2011 Taylor & Francis)

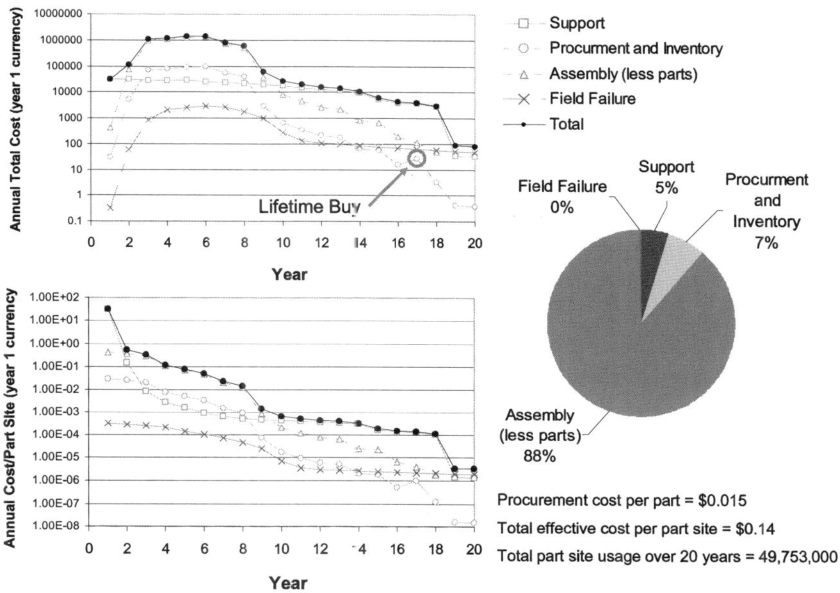

Fig. 20.4. Example part total cost of ownership modeling results (high-volume case). (© 2011 Taylor & Francis)

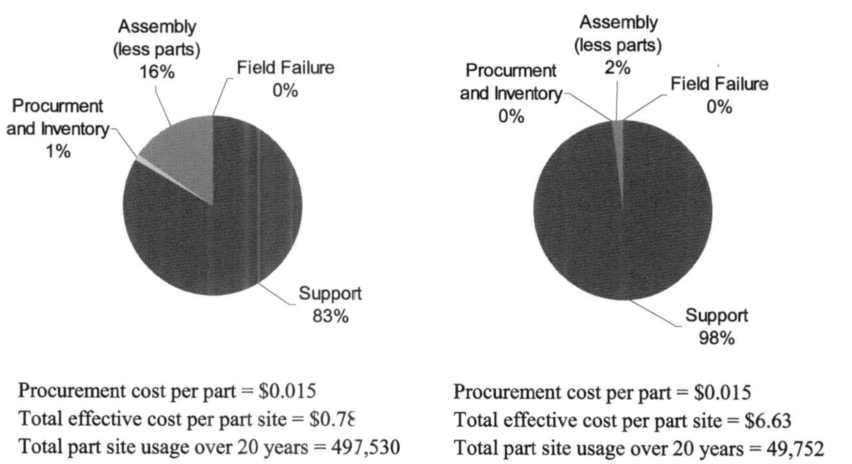

Fig. 20.5. Part total cost of ownership results for different part volumes (lower-volume cases). (© 2011 Taylor & Francis)

At lower volumes, support costs dominate, with significant contributions from fixed and variable costs that may be a hundred times larger (for example, in the case of production support costs) than costs incurred by field failures, procurement and inventory. The effect of economy of scale, a benefit of high-volume production, is demonstrated in Figure 20.5, which compares two lower-volume cases as variations of the SMT capacitor considered in Figure 20.4. Support costs make up 83% of the $0.78 spent per part site (shown on the left side of Figure 20.5) when a total of 497,530 parts are consumed over 20 years. When the volume consumed is further reduced to 49,752 parts over 20 years (shown on the right side of Figure 20.5), support costs contribute to 98% of the $6.63 spent per part site.

20.3 Levelized Cost of Energy (LCOE)

The levelized cost of energy, also known as the levelized cost of electricity, or the levelized energy cost (LEC), is an economic assessment of the average total cost to build and operate a power-generating asset over its lifetime divided by the total power output of the asset over that lifetime. LCOE is often taken as a proxy for the average price that the generating asset must receive in a market to break even over its lifetime. It is a first-order economic assessment of the cost competitiveness of an electricity-generating system that incorporates all costs over its lifetime: initial investment, operations and maintenance, cost of fuel, and cost of capital.

In the following, we derive the most common form of the LCOE. The LCOE is the cost that, if assigned to every unit of energy produced (or saved) by the system over the analysis period, will equal the TLCC (total life-cycle cost) when discounted back to the base year [Ref. 20.3]. This definition of LCOE is represented by,

$$\sum_{i=1}^{n} \frac{E_i LCOE}{(1+r)^i} = TLCC \qquad (20.9)$$

where

r = discount rate (per year).
i = Year.
n = number of years over which the LCOE applies.
E_i = quantity of energy produced in year *i*.
$TLCC$ = total life-cycle cost.

Discrete compounding has been assumed in Equation (20.9). Since *LCOE* is by definition constant (not dependent on *i*), we can factor it out of the summation and rewrite Equation (20.9) as,

$$LCOE = \frac{TLCC}{\sum_{i=1}^{n} \frac{E_i}{(1+r)^i}} \tag{20.10}$$

Equation (20.10) is the most common form of LCOE used. Note, the denominator of Equation (20.10) appears to be discounting the energy, however, only costs can be discounted; the apparent discounting is actually a result of the algebra carried through from the previous formula in which revenues were discounted.

The total life-cycle cost (*TLCC*) can include several contributions depending on the application. Commonly it is formulated as,

$$TLCC = \sum_{i=1}^{n} I_i + M_i + F_i \tag{20.11}$$

where

I_i = investment expenditure in year *i*.
M_i = operations and maintenance expenditures in year *i*.
F_i = fuel expenditures in year *i*.

Typically the LCOE is calculated over the lifetime of an asset, which is usually 20 to 40 years. However, care should be taken in comparing different LCOE studies and the sources of the information as the LCOE for a given energy source is highly dependent on the assumptions, financing terms and technological deployment analyzed. In particular, the assumption of capacity factor has a significant impact on the calculation of LCOE.

References

20.1 Magrab, E. B., Gupta, S. K., McCluskey, F. P. and Sandborn, P. A. (2009). *Integrated Product and Process Design and Development: The Product Realization Process*, 2nd Edition (CRC Press, Boca Raton).

20.2 Prabhakar V. and Sandborn, P. (2011). A part total ownership cost model for long-life cycle electronic systems, *International Journal of Computer Integrated Manufacturing*, 24.

20.3 Short, W. Packey, D. J. and Holt, T. (1995). A Manual for the Economic Evaluation of Energy Efficiency and Renewable Energy Technologies, *NREL/TP-462-5173*, March. http://www.nrel.gov/docs/legosti/old/5173.pdf. Accessed April 28, 2016.

Chapter 21

Cost, Benefit and Risk Tradeoffs

Analyzing costs is usually only a portion of the challenge when one needs to make critical decisions. Another important part of the decision process is the value of the benefit gained or the risk reduced. The evaluation of the benefit and risk is often less straightforward than cost.

As an example, many materials that are hazardous to humans and the environment are widely used in technology and commerce today, why? Very simply, these materials are used because they provide benefits that are considered substantial enough to warrant use (i.e., benefits that outweigh their risks). For example, pesticides are used worldwide to manage agricultural pests. Pesticides are used widely because they increase food production, increase profits for farmers, and control disease (e.g., Malaria, Typhus, Bubonic plague, etc.). However, pesticides have also been shown to disrupt the balance of an ecosystem by killing non-pest organisms. In addition to causing harm to wildlife, human exposure to pesticides has caused poisonings, the development of cancer and deaths. Despite the negative consequences, without pesticides a large fraction of the world would starve to death and would be at a considerably higher risk from serious diseases.

So, how do we appropriately weigh costs against risks and non-monetary benefits? This chapter attempts to shed some light on this problem by looking at cost-benefit analysis, cost of risk, and rare event modeling.

21.1 Cost-Benefit Analysis (CBA)

For many public enterprises, it is difficult to justify spending money based solely on return on investment arguments. In these cases the decision to spend money has to be based on more than just economics. CBA provides a framework to assess the combination of costs and benefits associated with a particular decision or course of action.

Ideally cost-benefit analyses take the broadest possible view of costs and benefits, including indirect and long-term effects, reflecting the interests of all stakeholders affected by the program. If all relevant benefits are simply increases in revenue or cost savings, then CBA is not necessary — a simple cash flow analysis or ROI will suffice. CBA is used when the benefits are not monetary, but can be monetized. It is precisely the process of monetizing the non-monetary benefits that makes CBA challenging.

The idea of CBA is usually attributed to Jules Dupuit, a French engineer in the mid-1800s [Ref. 21.1]. The practical development of CBA came as a result of the Federal Navigation Act of 1936 [Ref. 21.2]. This act required that the U.S. Corps of Engineers carry out projects for the improvement of the waterway system when the total benefits of a project exceed the costs of that project making it necessary for the Corps of Engineers to develop systematic methods that enabled the concurrent measurement of benefits and costs.

21.1.1 *What is a Benefit?*

Simply defined, a benefit is something that promotes well-being or value received. It is a positive effect on a relevant stakeholder that results from the successful implementation of a project. There are several different types of benefits including:

- Monetary (pecuniary)
- Personal or national security
- Environmental improvement, restoration or impact minimization
- Aesthetic improvement
- Safety
- Elimination and/or reduction of future damages and losses.

Benefits can be organized into the following general categories:

- Intangible: political, prestige, satisfaction, social

- Direct Tangible: improvements in cost, capability, availability, risk, productivity
- Indirect Tangible: fallout benefits that result from the direct benefits (e.g., job creation, property value increase, etc.).

21.1.2 *Performing CBA*

CBA involves determining the effective monetary value of initial and ongoing expenses and all expected returns. For comparison purposes, CBA must put all relevant costs and benefits on a common temporal footing (usually present value) using the applicable discount rates.

An Electronic Signaling System CBA Example

Consider a commuter rail system in a large city. Today the system averages 800,000 passenger trips per day. The system was originally opened for operation in the mid 1970s. The rail lines in the system consist of two parallel tracks, one dedicated to each direction of travel. The tracks that the trains use have an automatic electronic signaling system that indicates the presence of other trains on the same track. Due to the age of the signaling system, it has a high incidence of failure. When the signaling system fails in a section of track, it becomes necessary to route all trains (going both directions) onto a single track to move trains around an issue until the issue can be resolved. This process, known as "single tracking," can create delays as trains must wait for the affected area to be clear of trains traveling in the opposite direction.

The city in which this commuter rail system is operating has proposed an upgrade to the signaling system that will alleviate the need to single track trains during regular operation periods (non-scheduled maintenance periods). The upgrade will cost $200 million and take 2 years to implement. The alternative to upgrading the signaling system is to continue operation with the present system. We wish to perform a CBA to assess the value of the proposed signaling system upgrade.

Table 21.1 provides the assumed data for this analysis. Notice that when the system reliability improves (the elimination of failure and thus the elimination of single tracking), the public responds by increasing the number of trips taken. Also note that the trip delay due to single tracking is smaller during non-rush hour times because there are fewer trains running in the system.

Table 21.1. Data for Commuter Rail System Upgrade.

	Current System	Upgraded System
Frequency of failures causing single tracking during operational hours	1 per day	0
Rush Hour Passenger trips per hour	75,000	76,000
Rush Hour Average trip delay when single tracking	7 min	-
Rush Hour Value of passenger time ($/min)	0.10	0.10
Non-Rush Hour Passenger trips per hour	25,000	25,400
Non-Rush Hour Average trip delay when single tracking	4.5 min	-
Non-Rush Hour Value of passenger time ($/min)	0.08	0.08

In addition to Table 21.1, the following data applies:

- 20 hours per day of operation (5 am to 1 am)
- 6 hours of rush hour per day (14 hours of non-rush hour per day)
- 5 days a week have rush hours, 2 days a week have no rush hours
- Average fare (per passenger trip) = $5.50
- Effective discount rate = 2%/year (includes cost of money and inflation)
- Installation takes 2 years (no benefits are accrued prior to completion of installation)
- 20 total years of support (2 years of installation + 18 more years).

First let's calculate the value of removing the single-tracking delays. For the rush hour trips that would be taken anyway (u) and the trips generated by the improvement (g), the value per day is,[1]

$$Ru = 75{,}000\left(\frac{6}{20}\right)(1)(7)(0.10) = \$15{,}750/\text{day} \qquad (21.1\text{a})$$

$$Rg = (76{,}000 - 75{,}000)\left(\frac{6}{20}\right)(1)(7)(0.10) = \$210/\text{day} \qquad (21.1\text{b})$$

where the 6/20 ratio is the fraction of the 1 single-tracking event occurring per day occurring during rush hour. Similarly for non-rush hour trips,

$$Nu = 25{,}000\left(\frac{14}{20}\right)(1)(4.5)(0.08) = \$6300/\text{day} \qquad (21.2\text{a})$$

$$Ng = (25{,}400 - 25{,}000)\left(\frac{14}{20}\right)(1)(4.5)(0.08) = \$100.8/\text{day} \qquad (21.2\text{b})$$

Since there are 260 weekdays a year (days with rush hours), and we will assume 364 total days a year,[2] the per year values become,

$Ru = (260)(15750) = \$4{,}095{,}000/\text{year}$
$Rg = (260)(210) = \$54{,}600/\text{year}$
$Nu = (260)(6300) + (364\text{-}260)(20/14)(6300) = \$2{,}574{,}000/\text{year}$
$Ng = (260)(100.8) - (364\text{-}260)(20/14)(100.8) = \$41{,}184/\text{year}.$

The *Nu* and *Ng* calculations account for weekends during which all travel is non-rush hour. There is an additional benefit, which is the increased fare collected due to the increased number of trips taken by the public,

---

[1] There are a host of assumptions buried in Equations (21.1) and (21.2). We are assuming that the average delay per rider in the system is 7 or 4.5 min depending on whether the delay is during a rush hour or not. This does not mean that the single-tracking event is necessarily in the path of each rider, it is simply somewhere in the system. We also assume that there is an equal probability of the delay happening in every operational hour of the day.

[2] 364 days per year is 52 weeks multiplied by 7 days per week. 364 was chosen for convenience, if 365 days/year is used then on average 5/7 of the additional day would fall on a weekday and 2/7 of the additional day would fall on a weekend.

$$F_I = (5.50)\begin{bmatrix}(76000-75000)(6)(260)\\ +(25400-25000)(14)(260)\\ +(25400-25000)(20)(364-260)\end{bmatrix} = \$21{,}164{,}000/\text{year} \tag{21.3}$$

In order to properly accumulate costs and benefits we must discount everything to present dollars. The required discounting factors (assuming discrete compounding) needed are:[3]

$$(P/A,r,n_t) = \frac{(1+r)^{n_t}-1}{r(1+r)^{n_t}} \tag{21.4a}$$

$$(P/F,r,n_t) = \frac{1}{(1+r)^{n_t}} \tag{21.4b}$$

When the discount rate $r = 0.02$, $(P/A,r,18) = 14.992$ and $(P/F,r,2) = 0.961$. Using these discounting factors, the present value of the total rider benefit over 20 years becomes,

$$(Ru+Rg+Nu+Ng)(P/A,r,18)\,(P/F,r,2) = \$97{,}462{,}354$$

This assumes that the rider benefit is in years 3 through 20 only (no benefit before the system upgrades are completed). Similarly, the total increased fare collection discounted back to year 0 is,

$$(F_I)(P/A,r,18)(P/F,r,2) = \$304{,}916{,}351$$

So the total benefit is, $402,378,705 in year 0 dollars.

Now we must consider the costs. The costs associated with the system are:

- System upgrade = $200,000,000 (we assume half of this is paid at the end of the first year and half is paid at the end of the second year)
- Maintenance cost of the improved system = $2,000,000/year
- Maintenance cost of the unimproved system = $3,400,000/year.

[3] These discounting factors can be found in any engineering economics book.

The present value of the system upgrade is,

$$\frac{100,000,000}{(1+r)^1}+\frac{100,000,000}{(1+r)^2}=\$194,156,094$$

The present value of the total (20 years) maintenance when the system is improved is,

$$(3,400,000)(P/A,r,2) + (2,000,000)(P/A,r,18)(P/F,r,2) = \$35,410,624$$

The total cost of the improved system in year 0 dollars is 194,156,094 + 35,410,624 = \$229,566,718. The present value of maintaining the current system for 20 years (without the improvement) is,

$$(3,400,000)(P/A,r,20) = \$55,594,873$$

The net improved system cost is 229,566,718 - 55,594,873 = \$173,971,845.

In summary, for comparison purposes, everything is computed at the same point in time (present value is convenient). If this is not done then the various benefits and costs cannot be accumulated and the comparison will not be "apples to apples". This is a basic tenant of engineering economics analysis. Secondly, we monetized everything, i.e., all the benefits were mapped to costs — this is a fundamental attribute of CBA.

Benefit-Cost Ratio (BCR)

A BCR is a measure of the value of a project or proposal. BCR is the ratio of the benefits of a project, to the costs of a project. Both the benefits and the costs must have the same units (i.e., monetary units) to be compared using a ratio.

Generally if the BCR is greater than one, the project or proposal is worthwhile and if it is less than one it is not. Aside from the BCR's size relative to one, the magnitude of the BCR is somewhat arbitrary. The magnitude is arbitrary because some costs (e.g., the operating costs) may or may not be "netted out". This means that the operating costs (more precisely the present value of the operating costs) is either subtracted from the present value of the benefits and divided by the initial cost (netting out) or the present value of the benefit is divided by the initial cost plus the present value of the operating cost (not netted out). The present value of the operating cost is either subtracted from the numerator or the denominator, but not both. Netting out may be done for some projects and not for others. Under no circumstances will netting out raise a BCR that is less than one to greater than one.

The benefit-cost ratio for the electronic signaling example is given by:

$$\frac{\$402{,}378{,}705}{\$173{,}971{,}845} = 2.31$$

The Cost of the Status Quo

In many CBAs one of the choices may be to continue with the status quo.[4] When we are talking about sustaining systems, the cost of the status quo often escalates over time as the system ages. For example it is common knowledge that the maintenance cost of cars increases as they get older and things start to wear out.[5]

For the signaling system example presented in this section, while we have included annual maintenance costs for continuing the use of the current system (the non-upgraded system), we have assumed that the annual maintenance costs stay constant. In reality, it is quite possible that the annual maintenance costs will increase over time for this system (see Problem 21.2).

21.1.3 *Determining the Value of Human Life*

Many CBAs must place a value on human life. Although there is a deep aversion amongst many people to the idea of placing a monetary value on human life, some rational basis is needed to compare projects when human life is a factor.

The most commonly used monetary value of life is called the value of a statistical life (VSL). Most of the analyses that have been performed to determine this value focus on the following premise: "the VSL should

---

[4] The status quo is not the same as the cost of doing nothing. The cost of doing nothing literally means doing nothing, whereas the status quo means continuing to do the same thing you have been doing.

[5] This is not to say that the cost of ownership increases as cars get older, it may not. This statement is purely about the cost of maintenance.

roughly correspond to the value that people place on their lives in their private decisions" [Ref. 21.3]. If asked, most people would say that they will spare no expense to avoid death, however, economists know that the public's actual behavior (job choice, spending patterns, lifestyle choices) don't agree with this statement. Given choices, people will often choose style, convenience, or low cost over safety. Consider the simple task of commuting to work via an automobile. In many places one could drive on "surface streets" to work. Driving the surface streets, where the speed limit is relatively low has nearly no risk of death but may represent a very long and arduous commute. Alternatively, using a high-speed highway reduces the commuting time significantly, but carries a much higher risk of accidental death. Similarly, there are many occupations in which people accept increased risks in return for higher pay — transmission line workers, oil field workers, miners, construction workers, etc.[6] Using the choices that people make, the value that people place on increased risk (and thus the value of reduced risk) can be determined.

The VSL is the value that an individual person places on a marginal[7] change in their likelihood of death. Note, the VSL is NOT the value of an actual life. It is the value placed on changes in the likelihood of death, NOT the price someone would pay to avoid death.

Economists use several methods to estimate the VSL (a review of VSL is provided in [Ref. 21.4]). *Stated preference* methods are based on surveys of the willingness of people to pay to avoid a risk.[8] *Revealed preference* methods study wage-risk relationships associated with actual jobs.

Hedonic valuation is a revealed preference method used to estimate economic values for ecosystem or environmental services that directly affect market prices. Hedonic valuation can be used to analyze the risks

---

[6] In fact many occupations define hazard pay to mean additional pay for performing hazardous work, which includes work that carries an increased risk of injury and death.

[7] In the context of this discussion, marginal refers to a specific change in a quantity as opposed to some notion of the over-all significance of the quantity.

[8] Asking people how much they would be willing to pay for a reduction in the likelihood of dying suffers from a problem called "hypothetical bias", where people tend to overstate their valuation of goods and services.

that people voluntarily take and how much they must be paid for taking them. The most common source of data for these studies is the labor market, where jobs with a greater risk of death can be correlated with higher wages.

Consider the following example: suppose that a revealed preference study estimates that when the annual risk of death associated with a particular job increases by 0.0001 (1 in 10,000), workers receive \$750 more per year for the additional risk. The VSL is given by,

$$VSL = \frac{W_p}{P_i} \tag{21.5}$$

where W_p is the wage premium (\$750 per year in this case) and P_i is the increased probability of death (0.0001 per year). In this case VSL = \$7,500,000.[9]

VSL calculation is obviously controversial, after all, how can we assign a monetary value to a human life? Unfortunately, without the ability to assign a monetary value to life, we have no quantitative basis for economic damages due to wrongful death. Alternatively, if we do assign a monetary value to human life, it implies that high wage-earners' lives are more valuable than low wage-earners' lives.[10] While the whole idea of VSL may be ethically troubling, simply ignoring the value of life (and economic cost of death) and leaving it out of CBA results in a substantial underestimation of the value of the benefits associated with many types of projects.

The Value of Human Life in the Electronic Signaling System Example Case

Let's introduce another benefit into the example case presented in Section 21.1.2. Suppose that the new switching system also avoids some head-on collisions of trains (that would otherwise occur). The current incidence of fatalities due to head-on collisions is 1 fatality per year (this is an average

---

[9] This simple result assumes that workers are fully informed of (and understand) the risks and that the labor market is competitive.

[10] One problem is that the value of a statistical life varies from country to country. As a result the logic of CBA suggests locating hazardous jobs in poorer regions of the world where the VSL is smaller.

that could, for example, represent one train crash every 7 years that kills 7 people). Using the VSL value calculated using Equation (21.5) of \$7,500,000, we obtain an additional benefit over 20 years of,

$$(7{,}500{,}000)(P/A,r,20) = \$122{,}635{,}750$$

and the resulting overall benefit-cost ratio increases to 3.02.

Note, there is no assumption here about lawsuits that result from fatalities, which are brought against the transit authority that runs the commuter rail system. This is purely the value to the public of the avoided fatalities.

21.1.4 *Comments on CBA*

In CBA, every benefit is monetized and every cost and benefit is discounted to the same point in time so that a valid accumulation and comparison can be done.

Special care must be taken to insure that the benefit of a project is not double counted, this is easier said than done and often requires some careful thought. For example, consider an improved highway that reduces travel time and the risk of injury, as a result property values increase in areas served by the highway. The increase in property value is a good way to measure the benefit in this case. However, if the property value increase is used then one cannot include the value of the time and lives saved by the highway project. In this case the property value went up because of the time savings and risk reduction, not in addition to it. Including both the property value increase and the time and risk reduction would be double counting.

CBA is not without its detractors. The argument has been made that CBA is flawed due to [Ref. 21.5]:

- In a framework in which money is all that matters, some benefits will be valued at zero.
- CBA makes an implicit assumption that everything can be traded for everything else (some things can't be traded for other things).
- Costs and benefits of public policies do not always occur simultaneously. For example, the benefits to health and the

environment are usually realized over much longer timeframes than other benefits. When the time span is so great that different generations are involved, the analogy to an individual investment decision breaks down (e.g., what discount rate should be used?).

- CBA is often constrained by the range of alternatives it considers. Biases may enter into the choice of alternatives for analysis, and in the interpretation of complex, technical data.

Several other types of analyses are also available and may be confused with CBA. A good question is what is the difference between a Business Case Analysis (BCA), Return On Investment (ROI) and a Cost Benefit Analysis (CBA)? All three are tools for enabling fact-based project decisions. Briefly, CBA focuses on evaluation (comparison) of alternatives, ROI's focus is on the valuation of the investment in a particular alternative, and BCA (the business case) communicates the argument for making an investment in a particular alternative.

There are several other types of analyses that are similar to CBA. Whereas CBA monetizes of all effects, Cost Effectiveness Analysis (CEA) does not require the monetization of either the benefits or the costs. Unlike CBA, CEA determines which alternative has the lowest costs (with the same benefit level). CEA is particularly applicable to situation where a specific safety level is required. Lastly, Multicriteria Analysis (MCA) compares alternatives based on multiple criteria (CBA uses only cost). MCE results in a ranking of alternatives.

21.2 Modeling the Cost of Risk

Sometimes the benefit received as the result of a money spent is the mitigation of a risk and often the risk of interest for electronics is product or system failure. One way to define the cost of reliability is the cost of activities that are performed to keep the system free from failure. Risk in this case is defined as the product of the severity of a failure and the probability of the failure's occurrence, [Ref. 21.6].

The remainder of this section presents one approach to modeling the cost of risk and its application to technology insertion.

21.2.1 *A Multiple Severity Model for Technology Insertion*[11]

To assess the cost of risk associated with technology insertion we will determine the difference in the cost of failure consequence between a system with and without the technology insertion. It is important to note that the method discussed in this section does not calculate the actual life-cycle cost of the system, but rather the cost difference between the system with and without the insertion. This is referred to as a "relative accuracy" cost model, see footnote 1 in Chapter 1. In this case we may only have to include costs that differ between the two cases, as all other costs are a "wash" that subtract out of the difference.

Systems and products fail in many different ways, and each way that a system can fail has a unique financial consequence.[12] For example, a failure that requires a maintenance (repair) action is probably less costly than a failure that requires replacement of the system or product. The owner/operator of the system needs to be able to predict the cost (resolution and consequences) of the failure events that could occur over the service life of the system (or population of systems). This prediction must take into account that each system instance can fail more than once where each failure is due to the same or different reasons that have different financial consequences.

Taubel [Ref. 21.7] determines a total mishap cost. As shown in Figure 21.1, each severity level has a distinct cost and an associated probability of occurring and the area under the curve is the expected total mishap cost. For the model described in this section, we will not use the term "mishap" since mishap implies accident, which in turn implies safety.

---

[11] In the context of this section, technology insertion can be broadly defined as any change to a product or system. This could include a manufacturing process change, a material substitution, a part change, etc.

[12] Consequence refers to the economic impacts of the unavailability of the system (due to failure) and the restoration of the system to operation. This may include: diagnosis, maintenance, testing, documentation, and various unavailability penalties. The consequences of a reduction in the safety of the system are not addressed in this model, i.e., the modeling assumption is that safety is always preserved.

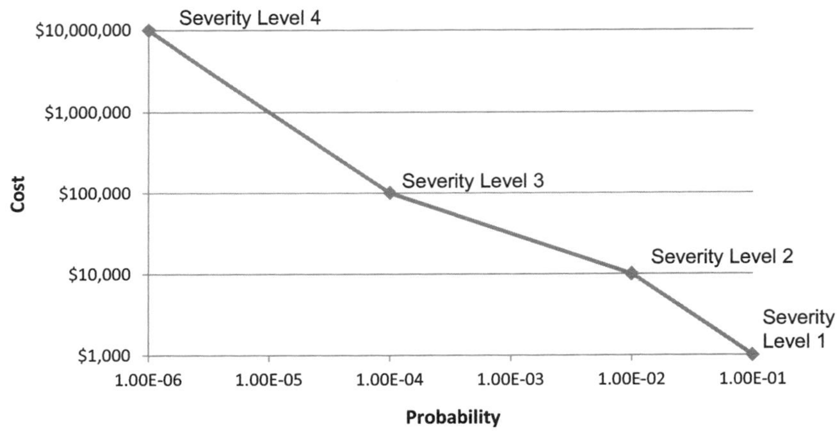

Fig. 21.1. Multiple severity model. Reprinted from [Ref. 21.8], © 2015 with permission from Elsevier.

Rather than calculating the probability of failure at each severity level (as in the Taubel model), the model described from this point forward in this section determines the expected number of failures at each severity level. This approach is used because some failures occur more than once during the life of the system and the cost of these multiple failures is accounted for. The product of the cost of individual failures and the number of times those failures occur is referred to as the Projected Cost of Failure Consequences (PCFC) for a population of products.[13]

Using an expected number of failure occurrences for each failure severity, and the cost associated with each failure occurrence, the PCFC for the system can be determined. Figure 21.2 shows the expected number of failures and associated cost for a five severity level. Note, Figure 21.2 is not the same model as shown in Figure 21.1. In Figure 21.2 the vertical axis is the expected number of failures occurring per product per service life, where the service life is the required system lifetime expressed in time or another applicable usage parameter (e.g., miles, thermal cycles, etc.)

---

[13] The model described in this section is a continuous risk model that assumes that probabilities are continuous. In the continuous model the PCFC is the area under the curve. In discrete risk models (which are also valid) the cost of failure is the sum of the probability of failure at each discrete severity level multiplied by the cost of failure resolution at that severity level.

associated with the relevant failure mechanism(s). The horizontal axis in Figure 21.2 is the cost per failure event.[14]

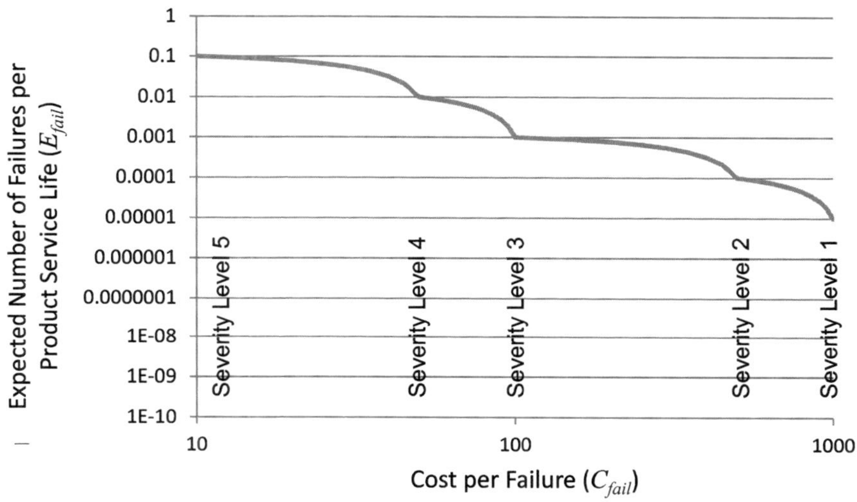

Fig. 21.2. Expected number of failures vs. cost per failure. Reprinted from [Ref. 21.8], © 2015 with permission from Elsevier.

In practice the PCFC is the area under the curve in Figure 21.2, which is the total area of a set of discrete trapezoids (they actually are trapezoids, their tops only appear curved in Figure 21.2 because they are plotted on a log-log plot, see footnote 14). The area formed by the points under the curve is determined using,

$$PCFC = \sum_{i=1}^{m} \left[E_{fail}(i-1) + 0.5E_{fail}(i)\right]\left[C_{fail}(i+1) - C_{fail}(i)\right] \quad (21.6)$$

where $E_{fail}(x)$ is the expected number of failures per product per unit lifetime of point x (a particular severity level) on the curve, $C_{fail}(x)$ is the cost of failure at point x, and m is the number of severity levels.

[14] The model described in this section, and in Equation (21.6) assumes that the cost of failure changes linearly between severity levels. When graphed on a log-log plot, this linear change appears as shown in Figures 21.2 and 21.3.

Evaluating Risk Mitigation Activities

As defined in [Ref. 21.8], "an activity is a sub-process, process, or group of processes that when performed (or applied) changes the expected number of failures over the service life of the product or system." Mitigation activities are not free, so the tradeoff here is the cost of mitigation versus the change in the PCFC ($\Delta PCFC$).

In general, activities affect specific failure mechanisms.[15] When a mitigation activity is performed it may reduce the number of expected failure occurrences. In general, several mitigation activities that may or may not be independent will be performed resulting in a modified PCFC for the system. The difference between the initial PCFC and the modified PCFC, is the reduction in failure cost.

For example, the reduction in failure cost is the difference in the areas under the curves in Figure 21.3. The top curve is the expected number of failures without mitigation activities, and the bottom curve is the expected numbers of failures after mitigation activities are included.

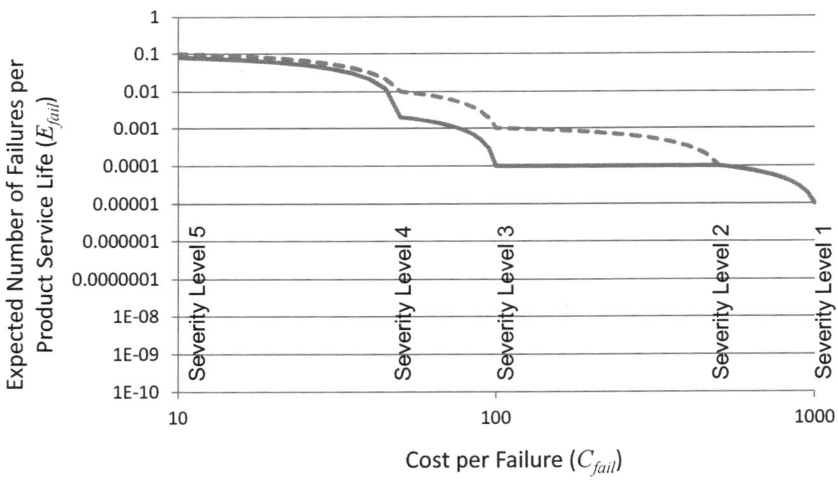

Fig. 21.3. The dashed curve represents the number of failures per product per unit lifetime at each severity level before activities are considered, and the solid line represents the expected number of failures with the activities performed. Reprinted from [Ref. 21.8], © 2015 with permission from Elsevier.

---

[15] See [Ref. 21.8] for a specific example of this.

In this case the Return on Investment (Chapter 17) is defined as,

$$ROI = \frac{\Delta PCFC - C_{Risk\ Total}}{C_{Risk\ Total}} \qquad (21.7)$$

where $C_{Risk\ Total}$ is the money spent on the risk mitigation activities and the $\Delta PCFC$ is the reduction in the projected cost of failure consequence due to the risk mitigation activities.

Comments on Cost-Based FMEA Methods

The example in this section is a cost-based failure modes and effects analysis (FMEA) approach. In general these methods measure the cost of risk and apply it to the selection of design alternatives. Other variations on this type of modeling exist, e.g., [Ref. 21.9]. Scenario-based FMEA predicts failure costs in order to make investment tradeoff decisions between reliability improvement and maintenance, [Ref. 21.10]. Similarly one can calculate the total "mishap cost" by relating the known costs associated with mishaps to the probability of mishap for different mishap severities, where mishap is defined by the Department of Defense's Military Standard 882C [Ref. 21.11]: "an unplanned event or series of events resulting in death, injury, occupational illness, or damage to or loss of equipment or property, or damage to the environment."

21.3 Rare Events

There are two different classes of risk. The first is the risk of volatility or fluctuations. If a particular event is common-place, then it is likely that we know with some certainty what the resulting frequency and cost of the event is. In this case a CBA analysis is a viable solution to determine the effective costs of resolution. The other type of risk is different; it is rare, but its consequences may be catastrophic. In the case of rare events, the costs of the events may be impossible to determine (i.e., there is no viable historical basis for them).[16]

[16] "Infrequent events," e.g., [Ref. 21.12], refers to events that are relatively rare, but not disastrous.

21.3.1 *What is a Rare Event?*

A rare event is an event whose probability of occurrence is low. This means that the probability of occurrence is low enough that it is difficult to observe in the real world.[17] In this chapter we define a "forecast" as a probability of occurrence assigned to an event or class of events. A "rare event" is an event that occurs with a very low spatial or temporal frequency when measured relative to the parent population or reference class.

It is important to very accurately assess the probability of a rare event because prediction errors may have significant consequences (in this case underestimation must be completely avoided).

Examples of rare events include: airplane and train accidents, satellite and space debris collisions, and tornado damage. Some of the events that we read about every day in the newspaper may also be classified as rare events because they are statistically negligible, e.g., credit card fraud, terrorist attacks, earthquakes, etc.

21.3.2 *Unbalanced Misclassification Costs*

In many real-world situations the probability of a bad outcome may be very small, but the consequence of that bad outcome may be very large. Examples include: medical screening, terrorist attack, and fraud detection. For example, when one trades off the cost and value of reliability or safety improvements, the cost of the consequence of a failure must be considered. This situation is generally referred to as "unbalanced classes".[18] For unbalanced classes, the cost of misclassifying a minority-class event is substantially greater than the cost of misclassifying a majority-class event.

A data set is unbalanced if the classes are unequally distributed. In this case the minority class is often much smaller (or rarer) than the majority

[17] It may also be difficult to observe via simulation, i.e., it cannot be easily estimated with Monte Carlo simulation. Note, there is an area of study that focuses on performing accelerated simulations. The most widely used methods for improving the efficiency of estimating small probabilities are "importance sampling" and "particle splitting".

[18] A "class" is a set of instances or observations that share a common attribute of interest.

class, i.e., there is much less data for the minority class than the majority class. When classes are unbalanced, classifiers can have good accuracy on the majority class but very poor accuracy on the minority class(es). So the problem becomes one of minimizing misclassification errors, and understanding that all misclassification errors do not have the same cost.

Consider the medical diagnosis of a patient with cancer, if the cancer is regarded as the positive class, and non-cancer (healthy) as negative, then missing a cancer (the patient actually is positive but is classified as negative; i.e., a false negative) is much more serious (and expensive) than a false-positive error (diagnosing the patient as positive when they are actually negative, i.e., healthy). In the false negative case, the patient could lose his/her life because of a delay in treatment. Similarly, if a passenger on an airplane carrying a bomb is positive (no bomb is the negative class), then it is much more serious and expensive to miss (false negative) a passenger who carries a bomb on a flight than to search an innocent person (a false positive).

The bottom line is the cost of missing a minority class is typically much higher that missing a majority class.

ROC (Receiver Operating Characteristic) Curves

A common way to visualize the performance of a binary classifier (a classifier with only two possible output classes) is to use ROC curves [Ref. 21.13].[19] A ROC curve represents the tradeoff between the true positive rate and the false positive rate. This allows an observer to see how well a classifier performs.

Consider a binary classification problem where the two classes are Positive and Negative. The classification model (or classifier, also called a learning algorithm) maps from particular instances to classes. Given a classification model (and an instance), there are four possible outcomes: 1) the instance is positive and classified as positive; 2) the instance is positive and classified as negative; 3) the instance is negative and classified as negative; or 4) the instance is negative and classified as

[19] ROC curves originated from the radio signal analysis and were later adopted by the machine learning and data mining communities.

positive. The true positive (*tp rate*) and false positive (*fp rate*) rates are given by,

$$tp\ rate = \frac{Positives\ correctly\ classified}{All\ positives} = \frac{TP}{TP+FN} \quad (21.8)$$

$$fp\ rate = \frac{Negatives\ incorrectly\ classified}{All\ negatives} = \frac{FP}{FP+TN} \quad (21.9)$$

In Equations (21.8) and (21.9), TP = number of true positives, FN = number of false negatives, FP = number of false positives, and TN = number of true negatives, where "number" refers to the number of instances in the test set.

The true positive rate answers the question: when the actual classification is positive, how often does the classifier predict positive? The false positive rate answers the question: when the actual classification is negative, how often does the classifier incorrectly predict positive?

A ROC curve (Figure 21.4) can be created by varying the probability threshold for predicting positive examples from 0 to 1.[20] Three cases are shown on Figures 21.4 and 21.5. In the first case (Case 1), there is no overlap between the positive and negative instances. In the second case (Case 2) there is an overlap between the instances, and in the third case (Case 3) the distributions of positive and negative instances exactly match indicating a random prediction. A ROC curve is considered to be better if it is closer to the top left corner. The ROC curve allows one to visualize the regions in which one model (classifier) is superior to another. A ROC curve implicitly conveys information about a classifier's performance across all possible combinations of misclassification costs and class distributions.

The Area Under Curve (AUC) is a way to summarize a classifier's performance (the larger the AUC, the better). The AUC measures the probability that a classifier will rank positive instances higher than negative instances. AUC measures the classifier's skill in ranking a set of patterns according to the degree to which they belong to the positive class. The overall accuracy of a classifier does not only depend on its ability to

---

[20] The "threshold" is the value of the classifier that defines the boundary between the first class and the second class.

rank patterns, but also on its ability to select a threshold in the ranking used to assign patterns to the positive class. If one classifier ranks patterns well, but selects the threshold badly, it can have a high AUC but a poor overall accuracy.

ROC curves are, however, insensitive to class balance. This is demonstrated by the fact that the rates in Equations (21.8) and (21.9) are independent of the actual positive/negative balance in the test set. For example, increasing the number of positive samples in the test set by a factor of two would increase both *TP* and *FN* by a factor of two, which would not change the true positive rate at any threshold. Similarly, increasing the number of negative samples in the test set by a factor of two would increase both *TN* and *FP* by a factor of two, which would not change the false positive rate at any threshold. Thus, both the shape of the ROC curve and AUC are insensitive to the class distribution.

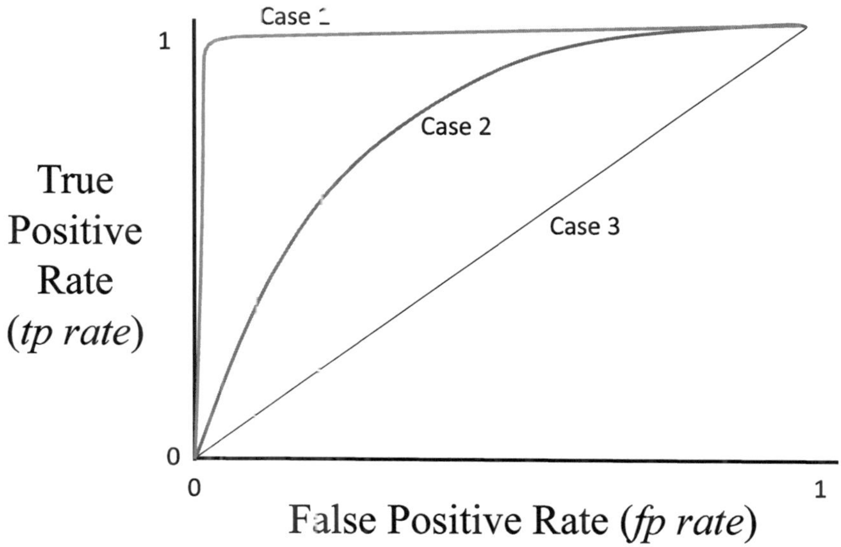

Fig. 21.4. ROC curve.

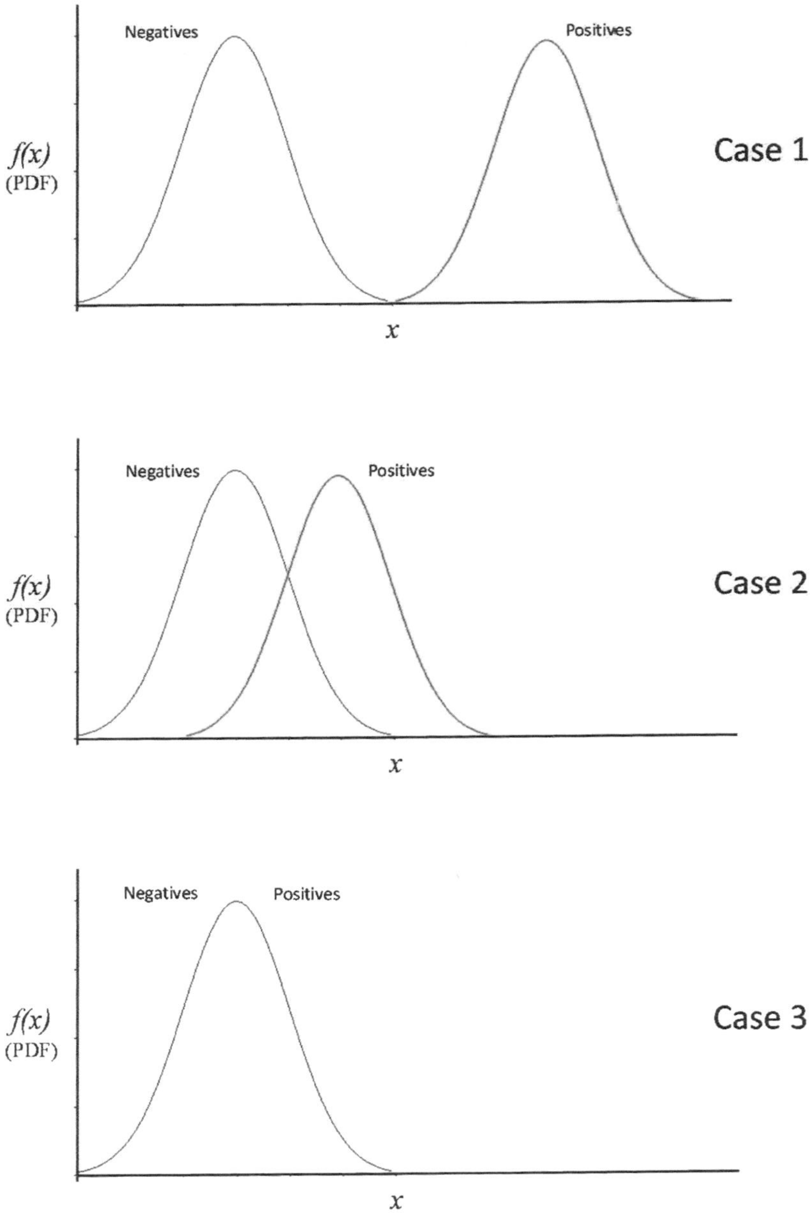

Fig. 21.5. Positive and negative instance distributions.

21.3.3 *The False Positive Paradox*

The false positive paradox is a statistical result where false positive tests are more probable than true positive tests when the overall population has a low incidence of a condition and the incidence rate is lower than the false positive rate. This paradox is common when trying to detect very low incidence infections (e.g., rare diseases) and very rare situations (e.g., terrorists in general populations). It can also present itself in testing high yield products for rare defects.

Consider the following example of a printed circuit board test. Assume that you have manufactured n = 100,000 boards. Let's consider the case where the boards are 60% yield (Y = 0.6) with respect to the defect of interest. Assume that a test has a false positive rate of 5% (*fp rate* = 0.05) giving a test accuracy of,[21]

$$T_A = 1 - fp\ rate \tag{21.10}$$

The test accuracy is 95% in this case. If the test produces no false negatives (FN = 0) then the number of true positives from the test is,

$$TP = n(1-Y) \tag{21.11}$$

which is 40,000 in this case, i.e., the test says that 40,000 defective boards are in fact defective. The number of false positives from the test is,

$$FP = nY(fp\ rate) \tag{21.12}$$

which is 3000 in this case, i.e., the test says 3000 non-defective boards are defective. So the number of true negatives (boards that are not defective and are passed by the test is),

$$TN = n - TP - FP \tag{21.13}$$

which is 57,000 in this case, i.e., the test correctly determines that 57,000 boards are not defective. The confidence that if the test says a board is defective, it actual is defective is given by,

---

[21] In this case "positive" means that the defect is present and "negative" means that the defect is not present. So a false positive means that the test says the defect is present and it is not present, while a true positive means that the test says the defect is present and it is present. Similarly, a false negative means that the test says the defect is not present and it is present, while a true negative means that the test says the defect is not present and it is not present.

$$\frac{TP}{TP+FP} = \frac{40{,}000}{40{,}000+3000} = 0.9302 \qquad (21.14)$$

Note, the confidence calculated in Equation (21.14) is not the *tp rate*. The *tp rate* is the fraction of true positives in the population of all the boards that have the defect (all positives), whether the test successfully found the defective boards or not. Alternatively, the confidence calculated in Equation (21.14) is the fraction of true positives (defective) in the population of everything the test claims is positive (defective). The important conclusion here is that when the yield is low, the test accuracy (0.95) and the confidence (0.9302) are about the same. A graphical representation of the board testing case is shown in Figure 21.6.

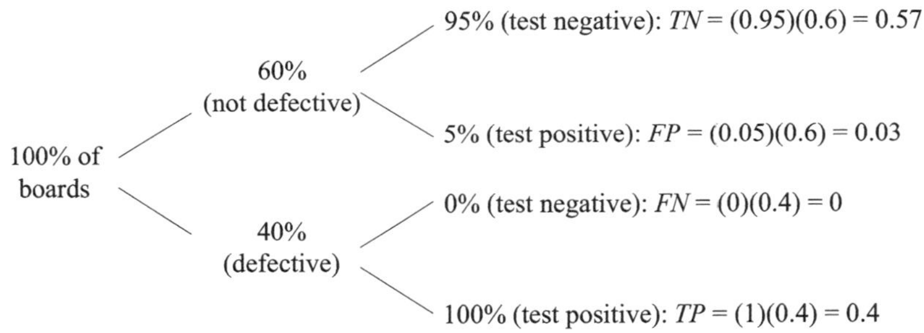

Fig. 21.6. Board testing (1 board).

Now consider the same problem, but for boards that have a high yield with respect to the defect of interest, $Y = 0.98$. Assuming that n and *fp rate* are the same as in the first case, and that there are no false negatives, we get,

T_A = 0.95 (same as before).
TP = 2000.
FP = 4900.
TN = 93,100.

Now the confidence that if the test says a board is defective, it actual is defective is $2000/(TP+FP) = 28.99\%$.

The second case presented here demonstrates the false positive paradox. If you have a low-incidence population, even with a high test accuracy the probability of a false positive (4900/100,000 = 0.049) is larger than the probability of a true positive (2000/100,000 = 0.02).

The lesson here is that the probability of a positive test result is not only determined by the accuracy of the test (which was high in the example provided), but also by the characteristics of the sampled population. When the incidence within the population of having a given condition (0.02) is lower than the test's false positive rate (0.05), even tests that have a very low probability of giving a false positive *in an individual test* will give more false than true positives *overall*. This means that if you are trying to test for something really rare, your test accuracy has to match the rarity of the thing you're looking for. Not adjusting for the scarcity of the condition (the defect in our case) in the population, and concluding that a positive test result probably indicates a positive condition (the defect is present), even though the incidence of the condition (the defect) in the population is below the false positive rate is a fallacy.[22]

References

21.1 Ekelund, R. B. and Hébert, R. F. (1999). *Secret Origins of Modern Microeconomics: Dupuit and the Engineers* (University of Chicago Press, Chicago).

21.2 Fuguitt, D. and Wilcox, S. J. (1999). *Cost-Benefit Analysis for Public Sector Decision Makers* (Quorum Books, Connecticut).

21.3 Brannon, I. (2004-2005). What is a life worth? *Regulation*, Winter, pp. 60-63.

21.4 Viscusi, W. K. and Aldy, J. E. (2003). The value of a statistical life: A critical review of market estimates throughout the world, *Journal of Risk and Uncertainty*, 27(1), pp. 5–76.

21.5 Ackerman, F. (2008). *Critique of Cost-Benefit Analysis, and Alternative Approaches to Decision-Making*. Global Development and Environment Institute, Tufts University. http://www.ase.tufts.edu/gdae/pubs/rp/ack_uk_cbacritique.pdf.

21.6 Hauge, B. S. and Johnston, D. C. (2001). Reliability centered maintenance and risk assessment, *Proceedings of the Reliability and Maintainability Symposium*, pp. 36-40.

[22] This is called a "base rate fallacy". If presented with related base rate information (i.e., generic, general information) and specific information (information only pertaining to a certain case), the mind tends to ignore the former and focus on the latter.

21.7 Taubel, J. (2011). Use of the multiple severity method to determine mishap costs and life cycle cost savings, *Proceedings of the International System Safety Conference*.
21.8 Lillie, E., Sandborn, P. and Humphrey, D. (2015). Assessing the value of a lead-free solder control plan using cost-based FMEA, *Microelectronics Reliability*, 55(6), pp. 969-979.
21.9 Rhee, S. and Ishii, K. (2003). Using cost based FMEA to enhance reliability and serviceability, *Advanced Engineering Informatics*, 17, pp. 179–188.
21.10 Kmenta, S. and Ishii, K. (2004). Scenario-based failure modes and effects analysis using expected cost, *ASME Journal of Mechanical Design*, 126, pp. 1027-1035.
21.11 *MIL-STD-882C* (1993). U.S. Department of Defense.
21.12 Sherman, G., Menachof, D., Aickelin, U. and Siebers, P.-O. (2010). Towards modelling cost and risks of infrequent events in the cargo screening process, *Proceedings of the Operational Research Society Simulation Workshop*.
21.13 Fawcett, T. (2006). An introduction to ROC analysis, *Pattern Recognition Letters*, 27, pp. 861-875.

Bibliography

In addition to the sources referenced in this chapter, there are many books and other good sources of information on cost-benefit analysis and risk tradeoffs including:

Taleb, N. N. (2010). *The Black Swan: The Impact of the Highly Improbable,* 2nd edition (Penguin, London).
Rubino, G. and Tuffin, B. (2009). *Rare Event Simulation Using Monte Carlo Methods* (Wiley, West Sussex UK).

Problems

21.1 In the example presented in Section 21.1.2, what would the cost-benefit ratio be if there was no value in the rider's delay time? Ignore the VSL.

21.2 For the example in Section 21.1.2, assume that the annual maintenance costs (A_m) do not remain constant over time, but rather escalate according the following functional relationships,

Non-upgraded system: $A_m = 3{,}400{,}000[1{+}0.1(y{-}1)]$

Upgraded system: $A_m = 2{,}000{,}000[1{+}0.05(y{-}3)]$

where y is the year (e.g., y = 1 represents year 1). Assume an end-of-year convention. Assume 20 total years of support and that the A_m for the upgraded system is the same as the A_m for the non-upgraded system in years 1 and 2 (because the upgrade is not in place in years 1 and 2). The A_m equation for the upgraded system does not apply until you get to year 3. Calculate the new benefit-cost ratio.

21.3 When the U.S. Environmental Protection Agency lowered its Arsenic standard for drinking water the annual cost to public utilities to meet the new standards was estimated to be $210 per household. Assuming that there are 100 million households in the U.S. and that the new standard saves 60 lives per year. If each human life is valued at $4 million, what is the benefit-cost ratio of the regulation?

21.4 If one wants to show the benefits of inserting a new technology into the Supply Chain, would it be better to conduct a Business Case Analysis or a Cost-Benefit Analysis?

21.5 A particular disease afflicts 1% of the population. Doctors have a test that correctly determines someone is healthy (determines that they do not have the disease) 98% of the time. Conversely the test correctly determines that someone has the disease 97% of the time. Your test results come back positive (claiming you have the disease). What is the probability (confidence) that you actually have the disease?

21.6 In the example in Section 21.3.3 what is the *tp rate*? Where would the example plot on a ROC curve?

21.7 If the *tp rate* in the example in Section 21.3.3 is changed to 0.8, what is the confidence that if the test says a board is not defective, it actual is not defective?

21.8 In Problem 21.7, what is the confidence that if the test says the board is defective, it actually is defective?

21.9 Mammography data is as follows: of all women with breast cancer, 86% will test positive. Of all women without breast cancer 9% will test positive. If only 1% of women between the ages of 55 and 70 have breast cancer,

a) What is the probability that that a women between the ages of 55 and 70 has breast cancer if she tests positive?
b) What is the probability that that a women between the ages of 55 and 70 has breast cancer if she tests negative?
c) If the incidence of breast cancer was only 0.1%, does the answer to part a) go up or down?

Chapter 22

Real Options Analysis

Cash flow analysis is the analysis of cash inflows and outflows over time representing a particular investment or project, such as the life-cycle cost of supporting a system. Conventionally in engineering economics, cash flow analysis is performed using discounted cash flow analysis (DCF).[1] DCF captures the time value of money and the uncertainties in the cash flow, but it does not reflect the flexibility that projects may have to change their actions during their life. By flexibility we mean the ability of decision makers to change what they do or how they do it as a result of things that have happened in the time that has passed since the start of the project. For example, a system development project that takes several years might be cancelled due to a change in the price of oil or a change in world economics.

Before discussing real options analysis, we briefly describe traditional cost flow analyses in order to illuminate the difference between classical engineering economics analyses and real options.

22.1 Discounted Cash Flow (DCF) and Decision Tree Analyses (DTA)

Consider the simple cash flow described in Figure 22.1. In this case the expected present value of the payoff from the investment or project is given by (see Section II.4),

$$PV_{investment} = \frac{1800}{(1+0.13)^1} = \$1593 \tag{22.1}$$

[1] Discrete-event simulation (DES), described in Appendix C, is simply an implementation of DCF.

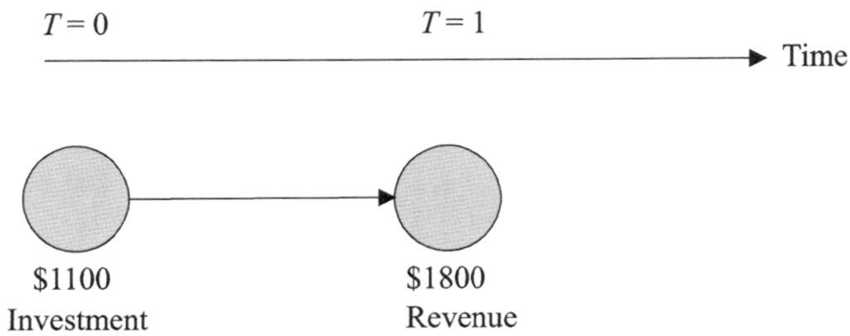

Fig. 22.1. Simple cash flow example.

where the cash flow is discounted using 13% per period and discrete compounding is assumed.[2] The $1593 is in $T = 0$ dollars, i.e., it is present value (PV). The net present value (NPV), i.e., the gain through investing, is: $NPV = 1593\text{-}1100 = \$493$.

What if there is uncertainty in the outcome of the investment (Figure 22.2)? In this case the expected value of the investment becomes,

$$PV_{investment} = \frac{(0.5)(1800)}{(1+0.13)^1} + \frac{(0.5)(675)}{(1+0.13)^1} = \$1095 \tag{22.2}$$

and the $NPV = 1095\text{-}1100 = \$\text{-}5$. A negative NPV may suggest that one should not make this investment.

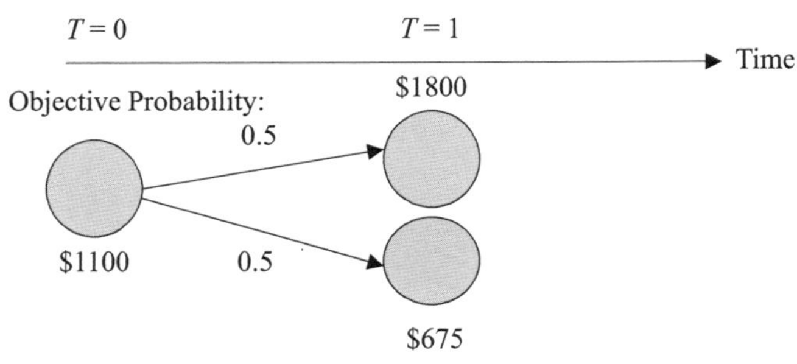

Fig. 22.2. Simple cash flow example with uncertainty.

[2] In this chapter we will refer to the rate at which compounding occurs (the 13% in this case) as the "risk adjusted discount rate". This term is consistent with the real options literature. WACC (Appendix B) is a risk-adjusted discount rate that reflects the risk perceived by the sources of the money used for the project.

Now, what if there is uncertainty and an option? The option in this case is that you can pay an additional \$320 at $T = 1$ to get an increase in return at $T = 1$ of 25% (later we will call the \$320 the "strike" price). In this case, the expected value of the investment is the same as in Equation (22.2) if the option is not exercised. If the option is exercised, you get,

$$PV_{investment} = \frac{(0.5)[(1.25)(1800)-320]}{(1+0.13)^1} + \frac{(0.5)[(1.25)(675)-320]}{(1+0.13)^1} = \$1086 \tag{22.3}$$

In this case, the objective probability of the two states is 0.5, i.e., there is a 50/50 chance of ending in the up or down states. The $NPV = 1086\text{-}1100 = \$\text{-}14$.

The analysis in Equations (22.1) through (22.3) is a simple discounted cash flow (DCF) analysis. DCF implicitly assumes that management commits to a particular course of action at the time the investment is made (or the project is launched), i.e., either the option will not be exercised as in Equation (22.2) or it will be exercised as in Equation (22.3).

Alternatively, consider a decision tree analysis (DTA). DTA can model managerial flexibility. DTA allows some of the limitations of simple DCF to be overcome. In this case the option is only exercised for the "up" side of the investment and the value of the investment is,

$$PV_{investment} = \frac{(0.5)[(1.25)(1800)-320]}{(1+0.13)^1} + \frac{(0.5)(675)}{(1+0.13)^1} = \$1153 \tag{22.4}$$

and the $NPV = 1153\text{-}1100 = \$53$. If the choice of whether to invest in the option can be delayed to $T = 1$, after you know whether you have an upside or downside situation, then Equation (22.4) is a more accurate model of the value of this investment. In this case you only exercise the option for the upside, which is referred to as "in the money" .

So, what would you be willing to pay for the option, i.e., what is it worth to you at $T = 0$ to have the opportunity to pay the extra \$320 at $T = 1$ assuming you can wait until $T = 1$ to decide to exercise the option or not? Considering the value of the option alone (as opposed to the whole investment), the present value of the option is,

$$PV_{option} = \frac{(0.5)Max[(0.25)(1800) - 320{,}0]}{(1+0.13)^1} + \frac{(0.5)Max[(0.25)(675) - 320{,}0]}{(1+0.13)^1}$$
$$= \$58 \qquad (22.5)$$

Equation (22.5) assumes that half the time the option is not exercised and no additional money is spent and half the time the option is exercised and the higher payoff would occur.[3]

DTA assumes that management makes the optimal decision at all future states — best possible case. The problem is that the DTA assumes that the risk is constant, allowing decisions to be made during the project changes how risky the project is and therefore, changes the effective discount rate required by investors. So in this case, it may be valid to discount the base project at 13% (one that has equally likely payoffs of \$1800 or \$675), but the risk changes for an option that results in \$130 or \$0.[4] In the case of a \$130 or \$0, we have no information on which to choose 13% or any other value for the WACC.

22.2 Introduction to Real Options

Real options[5] takes a different perspective than DCF on the valuation of cash flows. Real options is able to account for the additional project value that real projects have due to the presence of management flexibility that DCF cannot. However, DCF and real options are also fundamentally different in their approach to risk discounting. Real option valuation applies a risk-adjustment to the source of the uncertainty in the cash flow, alternatively DCF only adjusts for risk at an aggregate net cash flow level. Because of this, real options differentiates between projects based on each project's unique risk characteristics, whereas DCF does not, [Ref. 22.2].

---

[3] Note that using the NPV from Equation (22.4) and the *NPV* from Equation (22.2), 53 - (-5) = \$58.

[4] \$130 = *Max*[(0.25)(1800) - 320,0] and \$0 = *Max*[(0.25)(675) -320,0].

[5] The term "real options" was originated by Stewart Myers at MIT in 1977 [Ref. 22.1]. Myers used financial option pricing theory to value non-financial or "real" investments in physical assets and intellectual property.

In the financial world, options are derivative financial instruments that specify a contract between two parties for a future transaction on an asset at a reference price. For financial options the buyer of the option obtains the right, but not the obligation, to engage in the specified transaction at a specified future date. The seller incurs the corresponding obligation to fulfill the specified transaction at that future date. An option that conveys the right to buy something at a specific price is called a "call" option; an option that conveys the right to sell something at a specific price is called a "put" option.[6]

As an example of a call option, consider the following: Company XYZ's stock price is $100/share today. I (the option buyer) believe that XYZ's stock will go up in the next 6 months. I offer you (the option seller) the following deal: I will pay you $5/share today for the option to buy (from you) the share for $120, 6 months from today. If XYZ's stock is selling for >$120/share 6 months from today, I will exercise the option. In this case, I gain the difference between the current price of the stock and $120 (less the $5 I paid you for the option); you lose the difference between the current price of the stock and $120 (less the $5 you got paid for the option). If XYZ's stock is selling for < $120/share 6 months from today I will let the option expire. In this case, I lose the $5/share I paid you for the option and you gain $5/share for doing nothing.

Real options are based on financial options but are applied to real assets (e.g., real estate, products, intellectual property) rather than tradeable securities. Real options represent the flexibility to alter the course of action in a real assets decision, depending on future developments. Financial options represent a "side bet" that is not issued by the company whose stock is involved, but by some other entity that has no influence or connection with the company on which the bet is placed. In the case of real options, the bet is placed by the company that controls the underlying asset.

As an example of a real option, assume that company XYZ pays $20M for patent rights on a new technology. They estimate that it will cost

---

[6] A futures contract differs from an option in that it is a commitment to buy or sell (not the option to buy or sell) at a future date, i.e., it must be exercised, whereas the owner of an option has the right to choose not to exercise the option.

another $100M to develop and commercialize the technology. The payoff for developing and commercializing the technology is uncertain. Buying the patent is equivalent to buying an option. Company XYZ may never invest the additional $100M, in which case the patent rights (the "option") expires. Company XYZ can wait before investing more (for the uncertainty in the payoff to reduce).

There are many different types of options, but the most general two types are: European options that can only be exercised on a predetermined future date; and American options that can be exercised on any date up to a predetermined future date.

22.3 Valuation

The DCF value calculation is predicated on selecting an appropriate risk and time discount rate (e.g., 13% for the examples in Section 22.1). Often, WACC (Appendix B) is used as a proxy for risk, but this can lead to problems.[7]

For financial options, the goal is to determine the right price to pay for the option. This may also be the valuation goal for real options, however, real options may also want to determine the value obtained from a particular option and/or the optimum date on which to exercise the option. In either case, valuation is all driven by uncertainty. If there was no uncertainty in the future outcomes, then the valuation of an option would be trivial. However, everything is uncertain and in many cases, the outcomes are highly asymmetric (the magnitude of the upside and downside are not equal).

---

[7] Risk-return models of the financial markets explicitly value individual assets based on each asset's unique risk profile. Therefore, the use of a company's WACC to discount the risk of individual projects is often misleading.

22.3.1 *Replicating Portfolio Theory*

Most real option value calculations use no-arbitrage arguments to derive risk-neutral probabilities.[8] These probabilities are then used to adjust uncertain future one-period project value and cash flow outcomes for risk. Replicating portfolio theory is based on the "no-arbitrage principle", i.e., assets providing identical payoffs in the future must have the same present value.

In reality arbitrage can (and does) exist in the market, but, the assumption is that it cannot persist. If arbitrage opportunities arise, they are assumed to be eliminated by price adjustments. Options valuation makes the theoretical assumptions that financial markets are both efficient and complete. Market efficiency refers to the ability of markets to incorporate all available information about an asset into its market price. When markets are efficient, individuals with inside information cannot make excess returns because this information is already accounted for in the asset price. Complete markets allow investors to protect themselves (hedge) against any future outcome through market transactions. Note, DCF also requires these assumptions to maintain its validity.

Replicating portfolio theory identifies a "portfolio" that exactly mimics ("replicates") the option's state-contingent payoffs. For the example in Section 22.1, this means we construct a portfolio of existing assets (for which current values are known) and a riskless asset that provides the same payoff. Referring to Figure 22.2, if I invest in the option the upside and downside values are: S_u = \$1800 and S_d = \$675 (both in T = 1 dollars). The option values for the upside and downside are given by,

$$C_u = Max[(0.25)S_u - X, 0] = \$130 \tag{22.6a}$$

$$C_d = Max[(0.25)S_d - X, 0] = \$0 \tag{22.6b}$$

[8] Arbitrage refers to "the simultaneous purchase and sale of an asset in order to profit from a difference in the price. It is a trade that profits by exploiting price differences of identical or similar financial instruments, on different markets or in different forms. Arbitrage exists as a result of market inefficiencies." [Ref. 22.3]

where X is the strike price,[9] which is \$320 in this case. In Equation (22.6), the first term in the brackets is the exercised value and the second term (0) is the unexercised value. Note, C_u and C_d are the value of the option (not the payoff or the value of the project or investment). The portfolio we are going to consider has a fraction (m) of the base project and b_{rb} dollar holdings of a riskless bond. If V is the value of the portfolio, then,

$$\text{at } T = 0: \; V_0 = S_0 m + b_{rb} \tag{22.7a}$$

$$\text{at } T = 1: \; V_1 = S_1 m + (1 + R_f)^1 b_{rb} \tag{22.7b}$$

where

V_0 = value of the portfolio at $T = 0$.
V_1 = value of the portfolio at $T = 1$ ($V_1 = C_u$ or C_d).
R_f = interest rate paid by the riskless bond (riskless or risk-free rate).
m = fraction of the base project in the portfolio.
b_{rb} = dollar holdings of the riskless bond in the portfolio.
S_0 = project value at $T = 0$.
S_1 = project value at $T = 1$.

The goal is to find V_0, which is the value of the portfolio at $T = 0$ and therefore, since the portfolio replicates the option, V_0 is the value of the option at $T = 0$. Assuming $R_f = 0.04$ and using Equation (22.7), the value of the portfolio created at $T = 1$ is:

$$\text{upside}: \; V_1 = 1800m + (1 + 0.04) b_{rb} \tag{22.8a}$$

$$\text{downside}: \; V_1 = 675m + (1 + 0.04) b_{rb} \tag{22.8b}$$

where the upside V_1 is C_u and the downside V_1 is C_d. Equation (22.8) is two equations and two unknowns that when solved give $m = 0.1156$ and $b_{rb} = -\$75$.[10] The replicating portfolio has now been created, so we can use it to solve for the value of the portfolio at $T = 0$,

$$V_0 = 1100(0.1156) + (-75) = \$52.1 \tag{22.9}$$

---

[9] The strike price is the price to exercise the option. It is a fixed price at which the owner of the option can purchase (call option) the underlying commodity. The value of the option is different, it is the price for buying the option (or having the option available). It is not the price of the commodity.

[10] The fact that b_{rb} is negative implies that we have to borrow the \$75 at the riskless rate.

Note, the value of b_{rb} is not discounted in Equation (22.9) because this is the value at $T = 0$. The \$52.1 in Equation (22.9), is the value of the option. \$52.1 is less than the \$58 from DTA in Equation (22.5), which it should be, DTA represents the best case situation, i.e., management always makes the optimal decision.

22.3.2 *Binomial Lattices*

Replicating portfolio theory can be continued through additional time steps, however, a generalization of replicating portfolio theory called a lattice is useful in this case. [Ref. 22.4]. Lattices assume that: a) the evolution of the asset (or stock) is stationary, i.e., the evolution is the same over time; b) there are only states that can result from the evolution of the first state; and c) all new states are multiples of previous states.

A binomial lattice assumes that there are only two possible future states (it is a special case of a binomial decision tree). In binomial lattices, at every time step, the value is multiplied by an up, or down factor as shown in Figure 22.3. A binomial lattice assumes that the up and down factors are the same for every step, so that the lattice "recombines" as shown in Figure 22.3.

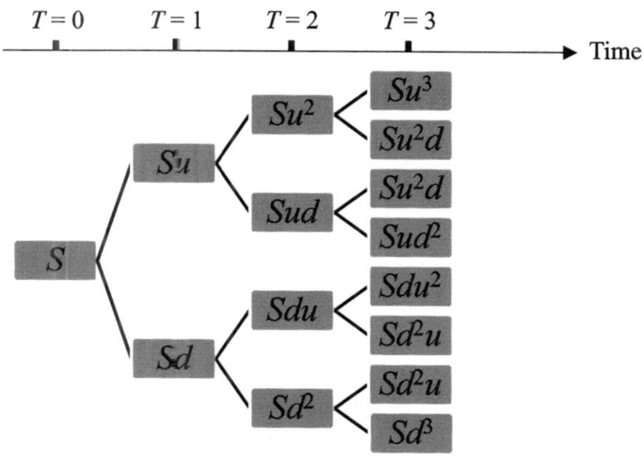

Fig. 22.3. Multi time step binomial lattice.

To derive a binomial lattice, start with a generalization of Equation (22.7a),

$$C = Sm + b_{rb} \tag{22.10}$$

At $T = 1$, Equation (22.7b) becomes,

$$C_u = S_u m + (1 + R_f)^1 b_{rb} \tag{22.11a}$$

$$C_d = S_d m + (1 + R_f)^1 b_{rb} \tag{22.11b}$$

Solving for m and b_{rb} we get,

$$m = \frac{C_u - C_d}{S_u - S_d} \tag{22.12}$$

$$b_{rb} = \frac{C_u - S_u m}{1 + R_f} \tag{22.13}$$

Substituting Equations (22.12) and (22.13) into Equation (22.10) we obtain,

$$C = \frac{pC_u + (1 - p)C_d}{(1 + R_f)} \tag{22.14}$$

where

$$p = \frac{1 + R_f - d}{u - d} \tag{22.15}$$

and $u = S_u/S$, $d = S_d/S$, where S is the initial investment. p in Equation (22.15) is the risk-neutral probability (of an upside result). Equations (22.14) and (22.15) correspond to 1 time step and assume discrete compounding.

Other more complex lattices have been derived, i.e., trinomial lattices. A trinomial lattice implies that there are three future states possible (instead of two).

Single Time Period Binomial Lattice Example

Let's use a binomial lattice to solve the same problem we considered in Section 22.3.1. In this case we set the variables as:

S = 1100 (initial investment).
S_u = 1800 (upside value)
S_d = 675 (downside value).
X = 320 (strike price).
C_u = $Max[(0.25)S_u\text{-}X,0]$ = 130.
C_d = $Max[(0.25)S_d\text{-}X,0]$ = 0.
R_f = 0.04 (riskless rate).
u = 1800/1100 = 1.636.
d = 675/1100 = 0.6136.

Using Equations (22.15) and (22.14) we get, p = 0.4169 and C = \$52.1, where \$52.1 is exactly the same as the result in Equation (22.9).

Now consider a different example. A project is worth \$13 million today. Suppose the value of the project will be worth \$17 million one year from today if there is high demand and \$9 million if there is low demand. Suppose you can buy an option today that allows you to sell the project 1 year from today for \$11 million, this is a "put" option. If the riskless rate is 4% per year, what is the value of this option? In this case we set the variables as:

S = 13 million.
S_u = 17 million (upside value).
S_d = 9 million (downside value).
R_f = 0.04 (riskless rate).
u = 17/13 = 1.308.
d = 9/13 = 0.6923.
C_u = $Max[11m\text{-}17m,0]$ = 0.
C_d = $Max[11m\text{-}9m,0]$ = 2 million.

Using Equations (22.15) and (22.14) we get, p = 0.5647 and C = \$0.8371 million. The key in this example is understanding C_u and C_d. The option allows one to sell the project for \$11 million leading to a \$6 million loss if exercised in the up state and a \$2 million gain if exercised in the down state. In this case the option would not be exercised in the up state, but would be exercised in the down state.

Multiple Time Period Binomial Lattices

Multiple time step can be analyzed using lattices. The solution to multiple time steps in a binomial lattice is obtained by recursion from the last time step to $T = 0$ by comparing the expected cash flow to the exercise price of the option. The analysis follows the steps below:

(1) Compute values at every node using S, u and d (as in Figure 22.4)
(2) Calculate the option value (C) at every node starting at the end date (the right side for Figure 22.4) and working to the start date (left side of Figure 22.4)
    - For the right most nodes (corresponding to the exercise date), $C = Max(\text{Node value} - X, 0)$
    - For other nodes, use Equation (22.14) to calculate C
(3) The final option value is C at the $T = 0$ node (for a European option).

As an example, consider a two time step lattice (Figure 22.4) with,

$S = 20.$
$X = 22.$
$R_f = 0.04.$
$u = 1.284.$
$d = 0.8607.$

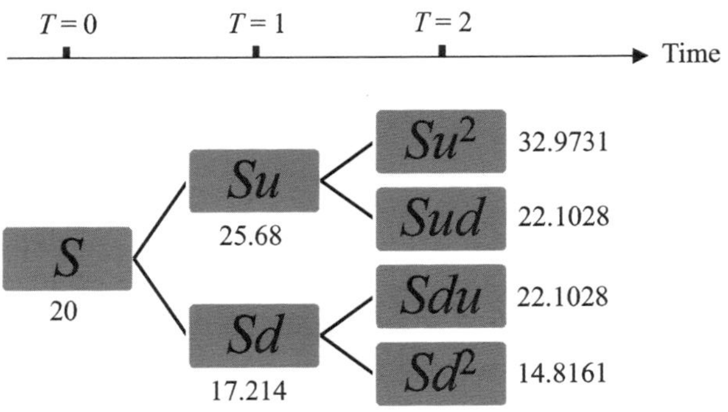

Fig. 22.4. Two time step lattice example with node values computed.

Using these parameters and assuming discrete compounding, we calculate $p = 0.4236$ from Equation (22.15). Now work backwards (start on the right at $T = 2$):

$$C_{u^2} = Max[32.9731 - 22, 0] = 10.9731$$

$$C_{ud} = C_{du} = Max[22.1028 - 22, 0] = 0.1028$$

$$C_{d^2} = Max[14.816 - 22, 0] = 0$$

Now back to $T = 1$ using (22.14):

$$C_u = [pC_{u^2} + (1-p)C_{ud}]/(1+R_f) = 4.5262$$

$$C_d = [pC_{ud} + (1-p)C_{d^2}]/(1+R_f) = 0.0419$$

Finally back to $T = 0$ using Equation (22.14):

$$C = [pC_u + (1-p)C_d]/(1+R_f) = 1.8867$$

The final option value is \$1.8867.

The example just worked is a European option that can only be exercised at $T = 2$ (stopping at $T = 1$ is not allowed). What if it was an American option? An "American" option can be exercised at either $T = 2$ or $T = 1$. Exercising the American option at $T = 2$ is the same as the European option. If you exercised at $T = 1$ (forget the $T = 2$ part of the lattice):

$$C_u = Max[25.680 - 22, 0] = 3.68$$

$$C_d = Max[17.214 - 22, 0] = 0$$

Back to $T = 0$:

$$C = [pC_u + (1-p)C_d]/(1+R_f) = 1.4988$$

Since $C = 1.8867$ (the option value when the option is exercised at $T = 2$) is greater than $C = 1.4988$, the value of the American option is the same as the value of the European option for this example.

22.3.3 Risk-Neutral Probabilities and Riskless Rates

Notice that the lattice analysis in Section 22.3.2 did not involve the objective probabilities of the upside or downside actually occurring or the risk-adjusted discount rate.

Risk-neutral probabilities are from the world of make-believe. We "make believe" that all investors are completely risk neutral, and then we ask, "In this make-believe world, what probabilities would lead to the same asset prices as we observe in the real world?" p is not equal to the objective probability of the upside because in lattice analysis we adjust the p to be consistent with, and calculated from the riskless rate ($R_f = 0.04$).[11]

To understand the connection between the riskless rate and the risk-adjusted discount rate, the risk-neutral probability and the objective probability, set Equation (22.14) equal to Equation (22.5),

$$C = \frac{pC_u + (1-p)C_d}{(1+R_f)} = \frac{qC_u + (1-q)C_d}{(1+r)} \tag{22.16}$$

where q is the objective probability of the upside and r is the risk-adjusted discount rate. Solving Equation (22.16) for r we get,

$$r = \frac{[qC_u + (1-q)C_d](1+R_f)}{[pC_u + (1-p)C_d]} - 1 \tag{22.17}$$

For the first single time period example in Section 22.3.2 (where C = 52.1) with q = 0.5, Equation (22.17) gives r = 0.247. This implies that when we solved the same problem in Equation (22.5) we used the wrong risk-adjusted discount rate (we used 13% and should have used 24.7%). In general, DTA is wrong because it assumes a constant discount rate throughout the decision tree, whereas, in reality the discount rate (risk) varies based on where you are in the tree.

Real options does a risk-neutral valuation that does not depend on the risk-adjusted discount rate. However, this does not mean that cost of money isn't included in the analysis. Real options does the same thing as DTA, but with a path dependent cost of money — more accurately, it

---

[11] The risk-adjusted discount rate and the objective probability go together, and the p goes with the R_f (they are a package deal) — there is a reason why this section is titled "risk neutral probabilities and riskless rates."

applies a risk-adjustment (which we call cost of money) to the source of uncertainty in the cash flow (i.e., it is path dependent). Note, real options and decision tree analysis will give the same answer if you allow for path-dependent discount rates [Ref. 22.5].

Options are priced so that the expected return on the stock (or project) and the option are both equal to the riskless interest rate (this is no-arbitrage). Under this assumption, the option value could be calculated by taking the expected pay-off at expiration and discounting at the riskless rate.

If investors are risk-averse, they will demand a risk premium on a project. This risk premium is determined using the rate of return on a financial asset that is designed to have a systematic risk that is identical to the project's risk. Under the assumption of complete markets (i.e., a market in which the all possible bets on the future state can be created using existing assets), a financial asset with this level of systematic risk exists and so a portfolio or bundle of these financial assets can be created to replicate the risk in the project. In this way, if the project or cash flows were traded, a replicating portfolio can be constructed from the traded project and a riskless bond.

22.4 Black-Scholes

Fischer Black and Myron Scholes at MIT developed a partial differential equation that predicts the price of an option as a function of time, [Ref. 22.6]. The solution to the differential equation is known as the Black-Scholes formula, which provides an estimate of the price of European options. Today the Black-Scholes formula is widely used by the options market.[12]

The Black-Scholes equation (for the price of an option over time) is,

$$\frac{\partial C}{\partial t}+\frac{1}{2}\sigma^2 S^2\frac{\partial^2 C}{\partial S^2}+R_f S\frac{\partial C}{\partial S}-R_f C=0 \qquad (22.18)$$

[12] Robert Merton coined the term "Black-Scholes options pricing model". Merton and Scholes received the 1997 Nobel Prize in Economics for their work (Fischer Black died in 1995).

where

C = price of a derivative (i.e., an option).
S = current stock price.
t = Time.
R_f = riskless rate.
σ = standard deviation of returns on the underlying security (volatility).

The Black-Scholes equation can be transformed into the heat equation,

$$\frac{\partial g}{\partial t} - \alpha\left(\frac{\partial^2 g}{\partial x^2} + \frac{\partial^2 g}{\partial y^2} + \frac{\partial^2 g}{\partial z^2}\right) = 0 \tag{22.19}$$

where t is time and g is a function of x, y, z, and t. Black-Scholes' solution to Equation (22.18) for a <u>call</u> option is,[13]

$$C = SN(d_1) - Xe^{-R_f T} N(d_2) \tag{22.20}$$

where C is the call option price and X is the option strike price.[14] $N(d_1)$ and $N(d_2)$ are cumulative standard normal distribution functions, and d_1 and d_2 are given by,

$$d_1 = \frac{\ln\left(\frac{S}{X}\right) + \left(R_f + \frac{\sigma^2}{2}\right)T}{\sigma\sqrt{T}} \tag{22.21}$$

$$d_2 = d_1 - \sigma\sqrt{T} \tag{22.22}$$

where T is the time to expiration of the option.

---

[13] Black-Scholes also works for European "put" options, where P, the put option price is given by,

$$P = -SN(-d_1) + Xe^{-R_f T} N(-d_2)$$

Black-Scholes formulations for other variations of European options also exist.

[14] Note, the example of a 25% increase in return for $X = 320$ used in Sections 22.1, 22.3.1 and 22.3.2 is <u>not</u> a simple call option, it is an expansion option.

The fundamental assumptions made by Black and Scholes are:

1. The stock price evolves according to geometric Brownian motion.[15] Note, u and d do not appear in the Black-Scholes solution because they are modeled using Brownian motion.
2. Constant "riskless" interest rate
3. No dividends on the stock during the life of the option
4. European-style option.

As an example of Black-Scholes, consider ABC stock that currently trades for $30 per share. A call option on ABC stock has a strike price of $25 and expires in three months (European option). The current riskless rate is 5% (per year), and ABC stock has a standard deviation of 0.45. What should the price of the call option be?

Using Equations (22.21) and (22.22),

$$d_1 = \frac{\ln\left(\frac{30}{25}\right) + \left(0.05 + \frac{0.45^2}{2}\right)0.25}{0.45\sqrt{0.25}} = 0.978 \tag{22.23}$$

$$d_2 = 0.978 - 0.45\sqrt{0.25} = 0.753 \tag{22.24}$$

$N(d_1) = 0.836$, $N(d_2) = 0.774$.[16] Using Equation (22.20) the value of the option is $C = 30(0.836) - 25e^{-0.05(0.25)}(0.774) = \$5.97$.

---

[15] Brownian motion means making rapid movements about the origin. Brownian motion is a random walk occurring in continuous time, with movements that are continuous rather than discrete. For example, a *random walk* can be generated by moving one steps each time period with the direction of the step determined flipping a coin. To generate *Brownian motion*, we would flip the coins infinitely fast and take infinitesimally small steps at each point.

If a stochastic process, S_t, follows a Geometric Brownian Motion then,

$$dS_t = \mu S_t dt + \sigma S_t dW_t$$

Where the first term is the trend (μ is the drift) and the second term is the uncertainty (σ is the volatility). W_t is a Wiener process (a continuous-time stochastic process) and $dW_t = I\sqrt{dt}$ where I is the inverse of a normal cumulative distribution.

[16] Calculated using NORMSDIST(d_x) in Excel.

Some real options fit nicely into the Black-Scholes framework, however, the majority of real options have complexities that are not captured by simple option pricing models. In practice, real options are analyzed using simulation methods.

22.4.1 *Correlating Black-Scholes to Binomial Lattice*

Comparing Equation (22.20) to Equation (22.10), $N(d_1)$ is m (the fraction of the base project in the portfolio). $N(d_2)$ is the probability that the option will be "in the money" at $t = T$. $Xe^{-R_f T}$ is the strike price at $t = T$ discounted back to $t = 0$ using the riskless rate. Replicating portfolio theory (generalized into a lattice) is doing exactly the same thing as Black-Scholes, the difference is that Black-Scholes starts with a stochastic differential equation while replicating portfolio theory is an algebraic solution.

The limit of a binomial tree where the number of periods (n) goes to infinity is the Black-Scholes solution. The relationship between the up and down multipliers in the binomial lattice and the volatility in the Black-Scholes model are given by, [Ref. 22.4],

$$u = e^{\sigma\sqrt{T/n_t}}$$
$$d = \frac{1}{u} \tag{22.25}$$

where T is the time to expiration of the option and n_t is the number of periods in the tree. Using Equation (22.25) and letting $1 + R_f \cong e^{R_f T/n_t}$, Equations (22.14) and (22.20) can be correlated. See [Ref. 22.5] for a complete discussion of the mapping from the binomial lattice to Black-Scholes.

22.5 Simulation-Based Real Options Example: Maintenance Options

In this section we describe a particular type of real option that is specifically applicable to the sustainment of systems. This section also demonstrates how a simulation-based real options solution works.[17]

Maintenance options are created when *in situ* health management (either CBM or PHM, see Section 17.3) is added to systems. In this case the health management approach generates a remaining useful life (RUL) estimate that can be used to take preventative action prior to the failure of a system. The real option is defined by,

- Buying the option = paying to add PHM to the system
- Exercising the option = performing predictive maintenance prior to system failure after an RUL indication
- Exercise price = predictive maintenance cost
- Letting the option expire = do nothing and run the system to failure then perform corrective maintenance.

The value from exercising the option is the sum of the predictive maintenance revenue loss and maintenance cost avoidance.

The predictive maintenance revenue loss is the difference between the cumulative revenue that could be earned by waiting until the end of the RUL to do maintenance versus performing the predictive maintenance earlier than the end of the RUL. Restated, this is the portion of the system's RUL that is thrown away when predictive maintenance is done prior to the end of the RUL.

Maintenance cost avoidance includes: avoided corrective maintenance cost (parts, service, labor, etc.), avoided downtime revenue lost, avoided under-delivery penalty due to corrective maintenance (if any), and avoided collateral damage to the system.

Figure 22.5 graphically shows the construction of maintenance value. The cumulative revenue lost due to predictive maintenance is largest on day 0 (the day the RUL is forecasted). This is because the most remaining

---

[17] In this case, the paths are not binomial and do not represent a lattice.

life in the system is disposed of if mainenance is performed the day that the RUL is predicted. As time advances, less RUL is thown away (and less revenue is lost) until the RUL is reached at which point the revenue lost is zero. The cost avoided is assumed to be constant until the RUL is reached at which point it drops to zero.

When the cumulative revenue lost and the cost avoided are summed, the predictive maintenance value is obtained. If there were no uncertainties, the optimum point in time to perform maintenance would be at the peak value point (at the RUL). Unfortunately, everything is uncertain.

The primary uncertainty is in the RUL prediction. The RUL is uncertain due to inexact prediction capabilities, and uncertainties in the environmental stresses that drive the rate at which the RUL is used up. A "path" represents one possible way that the future could occur starting at the RUL indication (Day 0).[18] The cumulative revenue paths have variations due to uncertainties in the system's ability to operate or uncertainties in how compensation is received for the system's outcome.[19] The cost avoidance path represents how the RUL is used up and varies due to uncertainties in the predicted RUL. Each path is a single member of a population of paths representing a statistically significant set of possible ways the future of the system could play out.

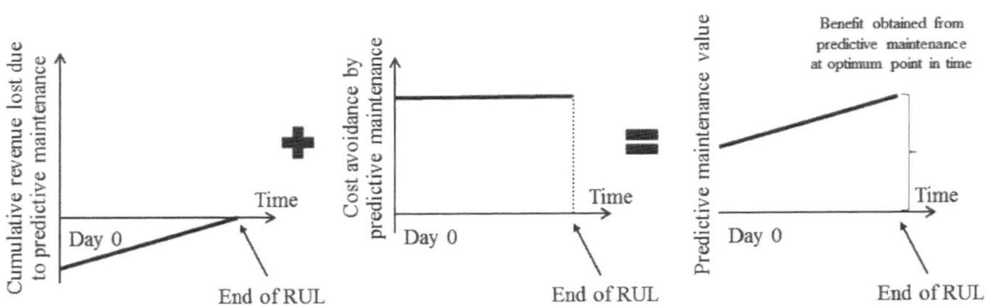

Fig. 22.5. Predictive maintenance value construction (uncertainties ignored).

[18] Note, each path is a branch that DTA would have to explicitly model.
[19] For example, if the system is a wind turbine, revenue path uncertainties could be due to uncertain wind over time.

Due to the uncertainties described above, there are many paths that the system can follow after an RUL indication. Real options lets us evaluate the set of possible paths to determine the optimum action to take, Figure 22.6.

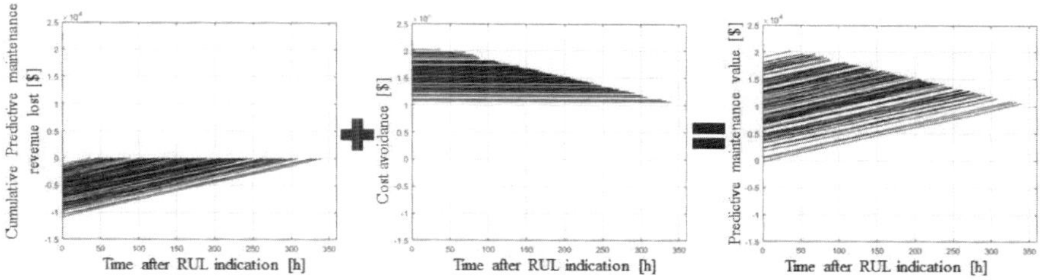

Fig. 22.6. Example of real paths after an RUL indication.

Consider the case where predictive maintenance can only be performed on specific dates. On each date, the decision-maker has the flexibility to determine whether to implement the predictive maintenance (exercise the option) or not (let the system run to failure, i.e., let the option expire). This makes the option a sequence of "European" options that can only be exercised at specific points in time in the future. The left side of Figure 22.7 shows two example paths (diagonal lines) and the predictive maintenance cost (the cost of performing the predictive maintenance). Real options analysis is performed for the option valuation where the predictive maintenance option value is given by,

$$C = Max(C_{PMV} - C_M, 0) \tag{22.26}$$

where C_{PMV} is the value of the path (right most graph in Figure 22.6 and the diagonal lines in Figure 22.7), and C_M is the predictive maintenance cost. The values of C calculated for the two example paths shown on the left side of Figure 22.7 are shown on the right side of Figure 22.7. Note that there are only values of C plotted at the maintenance opportunities (not in between the maintenance opportunities). Equation (22.26) only produces a value if the path is above the predictive maintenance cost, i.e., the path is "in the money".

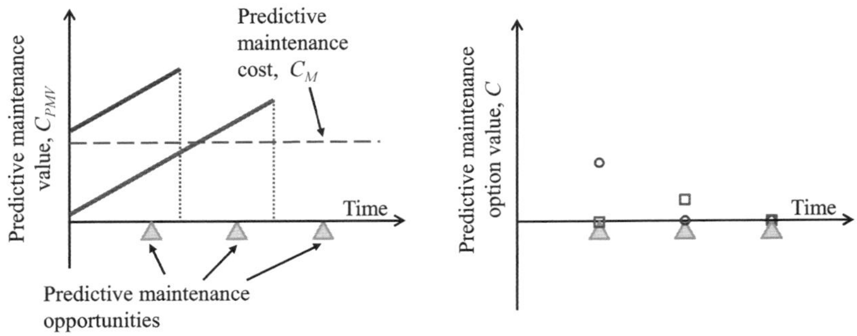

Fig. 22.7. Real options analysis valuation approach. Right graph: circles correspond to the upper path and the squares correspond to the lower path in the left graph.

Each separate maintenance opportunity is treated as a European option. The results at each separate maintenance opportunity are averaged to get the expected predictive maintenance option value of a European option expiring on that data. Using this process, the predicted maintenance option value is determined for all maintenance opportunity dates. The optimum predictive maintenance date is determined as the one with the maximum expected option value. Figure 22.8 shows an example for a wind turbine.

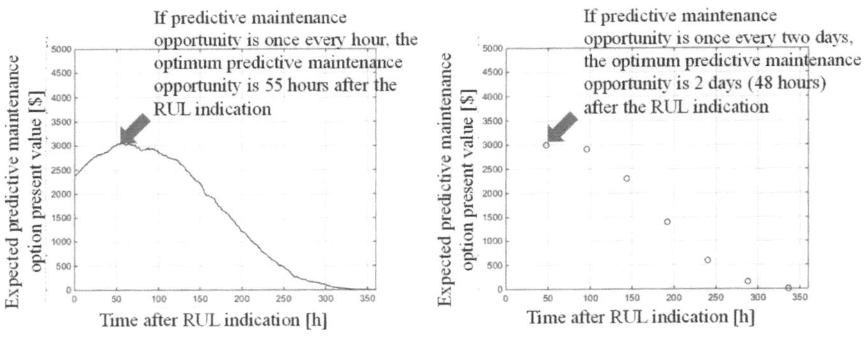

Fig. 22.8. Optimum maintenance time after an RUL indication for a wind turbine.

In this example, the real options approach is not trying to avoid corrective maintenance, but rather to maximize the predictive maintenance option value. In this example, at the optimum maintenance date the predictive maintenance will be implemented on only 65.3% of the paths. 32.0% of the paths chose not to implement predictive maintenance and in 2.7% of the paths the turbine failed prior to the predictive maintenance.

22.6 Closing Comments

Real options provides a way to account for management flexibility and accommodate path specific risk. This is appealing for many applications, however, if there is inherently no flexibility after a project starts then real options analysis yields the same solution as discrete-event simulation. Similarly in maintenance problems like the example in Section 22.5, if the penalties for corrective maintenance become very large, the real options solution will become the same as a discrete-event simulation solution because every path will opt to avoid all risk of having to perform corrective maintenance, effectively removing all flexibility from the problem.

DCF is a special case of real options analysis.[20] DCF is a real options analysis with no flexibility in project decision making.

$$NPV = \sum_{t=1}^{N} \frac{\text{Expected Cash Flow}_t}{(1+r)^t} - \text{Investment at time 0} \quad (22.27)$$

where r is the discount rate per time period. Equation (22.27) is exactly what we did in Section 22.1 for $t = 1$ time period. Uncertainties can be folded into this calculation using DES (Appendix C).

Because DCF does not accommodate any future flexibility and has to make a decision for the future based on only today's data, it is defined by,

$$Max(\text{at } t = 0)\left[\text{E}[V_T] - X, 0\right] \quad (22.28)$$

where V_T is the value at time T, X is the strike price and E[ ] is the expectation value. Translated, Equation (22.28) means that the outcome (or value) of all mutually exclusive management solutions are evaluated and the best one is chosen. It is a "maximum of expectations".

Alternatively, real options analysis is defined by,

$$E\left[Max_{\text{at } t=0}\left(V_T - X, 0\right)\right] \quad (22.29)$$

Real options analysis is an "expectation of maximums". In real options analysis, the option is exercised at time T only if $V_T > X$, i.e., "in the money". For DCF, the option is exercised if $E[V_T] > X$ at $t = 0$. If there is not uncertainty then the two rules are the same.

---

[20] The discussion accompanying Equations (22.27)-(22.29) follows [Ref. 22.5].

This chapter has only introduced call, put and expansion options. There are a large variety of options that are used in real world applications including options to defer and abandon. There are also compound options whose value depends on other options. There are switching options that allow the mode of operation to be changed. In the financial world there is a whole host of exotic options — see a texts on financial derivatives for a complete treatment, e.g., [Ref, 22.7].

References

22.1 Myers, S. C. (1977). Determinants of corporate borrowing, *Journal of Financial Economics*, 5(2), pp. 147-175.

22.2 Samis, M., Laughton, D. and Poulin, R. (2003). Risk discounting: The fundamental difference between the real option and discounted cash flow project valuation methods, *Kuiseb Minerals Consulting Working Paper No. 2003-1.*

22.3 Investopedia, http://www.investopedia.com/terms/a/arbitrage.asp Accessed on April 20, 2016.

22.4 Cox, J., Ross, J. and Rubinstein, M. (1979). Option pricing: A simplified approach, *Journal of Financial Economics*, 7(3), pp. 229-264.

22.5 Copeland, T. and Antikarov, V. (2003). *Real Options: A Practitioner's Guide*, TEXERE.

22.6 Black, F. and Scholes, M. (1973). The pricing of options and corporate liabilities, *Journal of Political Economy*, 81(3), pp. 637-654.

22.7 Sundaram, R. K., and Das, S. R. (2011). *Derivatives: Principles and Practice*, McGraw Hill.

Bibliography

In addition to the sources referenced in this chapter, there are many good sources of information on real options analysis, including:

Brandao, L. E., Dyer, J. S., and Hahn, W. J. (2005). Using binomial decision trees to solve real-option valuation problems, *Decision Analysis*, 2(2), pp. 69-88.

Kodukula, P. and Papudesu, C. (2006). *Project Valuation Using Real Options*, J. Ross Publishing, Inc.

Problems

22.1 Suppose you invest \$100 today ($T = 0$) and obtain \$180 one year from today. The risk-adjusted discount rate is 14%/year. What is the NPV from this investment?

22.2 Suppose you invest \$100 today ($T = 0$) and one of two outcomes are possible one year from today: either you get \$180 back or \$94 back. The objective probability of getting \$180 is known to be 65%. The risk-adjusted discount rate is 14%/year for both paths. What is the NPV from this investment?

22.3 What if $S_d = S_u$ in the replicating portfolio case in Section 22.3.1? What is the value of V_0 in this case?

22.4 What is V_0 in the example case in Section 22.3.1 if $X = 0$ (the strike price)? Does this make sense?

22.5 Rederive Equations (22.14) and (22.15) assuming continuous compounding.

22.6 What are the relative magnitude restrictions between u, d, and $1+R_f$ implied by the binomial lattice formulation?

22.7 What if $X = 0$ in the Black-Scholes solution, is this valid?

22.8 Assume that the current price of a stock is \$80 and that 1 year from now the stock will be worth either \$90 or \$75. The exercise price of a call option for this stock is \$74. Assuming a riskless interest rate of 6% per year (and discrete compounding), what is the call option price?

a) Work the problem using a binomial lattice.
b) Work the problem using replicating portfolio theory.
c) Work the problem using Black-Scholes, assume that $u = e^{\sigma\sqrt{dt}}$ with an incremental time step of $dt = 1$ year.

Hint: The solutions to parts a) and b) should be exactly the same. The Black-Scholes solution will be a bit larger.

22.9 A company is considering making an investment in new processing equipment. The value of the future cash flow one year in the future that results from this investment is either \$12,000 if the market goes up or \$7000 if the market goes down. The capital investment at time 0 is \$10,000.

a) Determine the present value of the equipment investment at time 0 using decision tree analysis (DTA). Assume that the objective probability that the market goes up is 0.7. The risk-adjusted discount rate is 20% per year.
b) What is the net present value (NPV) of the equipment investment at time 0 (from part a)?
c) Assume that the company can purchase an option for this investment. The option allows the company to abandon the investment after 1 year and sell the equipment for 50% of its original cost (i.e., 0.5 x \$10,000); OR, it can expand, which will result in twice the cash flow value (i.e., 2 x \$12,000, or 2 x \$7000). To expand, the company will have to make an additional capital investment of \$4500. What should the price of this option be (i.e., if the company has to pay

up-front in year 0 for an "option" that allows the flexibility described, what should it pay)? The riskless rate is 2% per year.

d) What is the risk-adjusted discount rate corresponding to part c)? Use the objective probability from part a).

22.10 A company is considering developing of a new product. Based on its experience with similar products, it believed that it can wait for five years (T = 5) before releasing the new product. An analysis using an appropriate risk-adjusted discount rate indicates that the present value of the expected future cash flows for the new product will be S = \$160 million, while the investment to develop and market the new product is X = \$200 million. The annual volatility of the future cash flows is estimated to be σ = 30% and the continuous annual risk-free rate over the option's life is R_f = 5%/year. What is the value of the option to wait?

a) Use a Binomial Lattice to solve this problem, with the other necessary parameters given as below and assuming continuous compounding:

$$u = e^{\sigma\sqrt{dt}}$$

Incremental time step dt = 1 year

$$d = \frac{1}{u}$$

b) Use the Black-Scholes method to solve the problem.

22.11 A project is worth \$13 million today. Suppose the value of the project will be worth \$17 million one year from today if there is high demand and \$9 million if there is low demand. Suppose you can buy an option today that allows you to sell the project 1 year from today for \$11 million. If the riskless rate is 4% per year, use Black-Scholes to determine the value of this option. Hint: This is a "put" option (see footnote 13) and this problem was worked with a binomial lattice in Section 22.3.2. You may assume that $u = e^{\sigma\sqrt{dt}}$ and an incremental time step of 1 year.

22.12 A major semiconductor manufacturer is impressed with a new packaging technology developed at the University of Maryland. To block their competitor from getting the new technology, the semiconductor manufacturer purchased an exclusive license from the University of Maryland for \$30 million. The semiconductor manufacturer figures that they will have to spend an additional \$100 million to implement the new technology, BUT, their decision whether to move forward or not depends on what their competitor does in the next year. On the upside, if implemented, the new technology could increase the value of the semiconductor manufacturer's project by 50% (u = 1.5). If the riskless rate is 5%/year and assuming that d=1/u, what does the semiconductor manufacturer think the present value of the project must be?

Appendix A

Notation

The notation (symbols) used in each chapter are summarized in this Appendix. Every attempt has been made to make the notation consistent from chapter-to-chapter; however, there are common symbols that have slightly different meanings in different chapters.

Chapter 1 – Introduction

b = labor burden rate.
C_L = labor cost of manufacturing or assembly (per unit).
C_{OH} = overhead cost.
L_R = labor rate.
L_{RB} = burdened labor rate.
N_{pm} = total number of units produced during the lifetime of the product.

Chapter 2 – Process-Flow Analysis

C_C = capital cost of a process step associated with one product instance.
C_e = purchase price of the capital equipment or facility.
C_L = labor cost of a process step associated with one product instance.
C_m = unit cost of the material per count, volume, area, or length.
C_M = material cost of a process step associated with one product instance.
C_{manuf} = total manufacturing cost associated with one product instance.
C_{OH} = overhead (indirect) cost allocated to each product instance.
C_t = cost of the tooling object or activity.
C_T = tooling cost of a process step associated with one product instance.
C_W = waste disposition cost per product instance.
D_L = depreciation life in years.
D_W = wafer diameter.
E = edge scrap (unusable wafer edge).
F = flat edge length (wafer).
K = minimum spacing between die (kerf).
L = die or board length.
L_R = labor rate.

N_e = number of wafers or panels concurrently processed by the step.
N_p = number of product instances that can be treated simultaneously by the activity (capacity).
N_u = number-up (number of die or boards per wafer or panel).
N_t = number of tooling objects or activities necessary to make the quantity Q of products.
P_L = panel length.
P_W = panel width.
Q = quantity of products that will be made.
Q_t = number of objects that can be made for one tooling cost.
S = die dimension.
T = length of time taken by the step (calendar time).
T_{op} = operational time per year of the equipment or facilities.
U_L = number of people associated with the activity (operator utilization).
U_M = quantity of the material consumed as indicated by its count, volume, area, or length.
W = die or board width.
$\lceil \ \rceil$ = ceiling function.
$\lfloor \ \rfloor$ = floor function.

Chapter 3 – Yield

A = area.
α = clustering parameter.
C_i = cost of the ith process step.
C_{in} = cost of a unit entering a process step.
C_{out} = cost of a unit exiting a process step.
C_p, C_{pk} = process capability metrics.
C_{Step} = cost of a process step.
C_Y = yielded cost.
$C_{Y_{Step}}$ = yielded cost of a process step.
D = defect density.
D_i = defect density of the ith process step.
D_0 = fixed defect density value.
$\delta(\ )$ = Dirac delta function.
$\mathrm{erf}(\ )$ = error function.
$f(\ )$ = probability distribution, PDF.
F = flat edge length (wafer).
HSL = high specification limit.
L = die length.
LSL = low specification limit.
λ = average number of fatal defects per item.
μ = the mean of the process.
m = number of process steps.
n, N = count.

p = individual event probability.
$\Pr(\,)$ = probability.
q = individual event probability.
R = wafer radius.
σ = the standard deviation of the process.
W = die width.
Y = yield.
Y_i = yield of the ith process step.
Y_{in} = yield of units entering a process step.
Y_{out} = yield of units exiting a process step.
Y_{Step} = yield of a process step.
$\lfloor \; \rfloor$ = floor function.

Chapter 4 – Equipment/Facilities Cost of Ownership

b = burden on the labor rate.
C_C = capital cost of a process step associated with one product instance.
C_{cap} = capital cost contribution to COO.
$C_{change\text{-}overs}$ = change overs contribution to COO.
C_D = cost of repairing one defect.
C_{fixed} = fixed cost: purchase, installation, etc.
$C_{lp\text{-}co}$ = lost production due to change overs contribution to COO.
$C_{lp\text{-}maint}$ = lost production due to maintenance contribution to COO.
$C_{lp\text{-}s}$ = lost production due to scrap contribution to COO.
$C_{ownership}$ = cost of ownership (COO).
$C_{production\ penalty}$ = production penalty contribution to COO.
$C_{repairable\ defects}$ = repairable defects contribution to COO.
$C_{sched\ maint}$ = scheduled maintenance contribution to COO.
C_{scrape} = scrap contribution to COO.
$C_{unsched\ maint}$ = unscheduled maintenance contribution to COO.
$C_{variable}$ = variable cost: labor, material, utilities, overhead, etc.
$C_{yield\ loss}$ = cost due to yield loss: money invested into scrapped parts and production lost by producing defective parts.
D_L = depreciation life.
D_{nr} = rate at which non-repairable defects are produced.
D_r = rate at which repairable defects are produced.
I = investment in the product up to the scrap point, i.e., how much has been spent on one product instance.
L_R = labor rate for maintenance activities.
$MTBF$ = mean time between failure.
$MTTR$ = mean time to repair (per unscheduled maintenance instance).
N_{co} = number of change-overs during production hours.
N_{off} = number of scheduled shutdowns for maintenance during off-production hours.
N_{on} = number of unscheduled shutdowns for maintenance during production hours = Production time/$MTBF$, where $MTBF$ is the mean time between failure for the machine, facility and/or process.

N_p = number of product instances that can be treated simultaneously by the activity (capacity).
P = purchase price of the machine, facilities, and/or process and is assumed to include installation and any extra facilities needed to make it operational.
R = residual value of the machine, facilities, and/or process at the end of the depreciation life.
T_{co} = time to perform a change-over (per change-over instance).
T_{cool} = time for the process (and/or the specific tool) to cool down before maintenance can begin.
T_i = the effective time interval between the completion of product instances by the process that the machine, facility or subprocess is associated with.
TPT = throughput.
T_R = time to perform scheduled maintenance activity (per scheduled maintenance instance).
T_{start} = time for the process (and/or the specific tool) to warm up after the maintenance is completed.
U = utilization: ratio of production time to total available time.
V = value of the product (profit that can be made on one instance of the product).
Y = composite yield.
$\lceil \; \rceil$ = ceiling function.

Chapter 5 – Activity-Based Costing (ABC)

A_R = activity rate.
b = burden rate.
C_A = activity cost.
C_L = labor cost of a process step associated with one product instance.
C_M = material cost of a process step associated with one product instance.
C_{OH} = overhead cost.
CC_R = capacity cost rate.
L_R = labor rate for maintenance activities.
N_A = number of times an activity is performed.
N_{tp} = total number of instances of the product manufactured
T = length of time taken by the step (calendar time).
U_L = number of people associated with the activity (operator utilization).

Chapter 6 – Parametric Cost Modeling

A_{die} = die area.
C_{die} = die cost.
$C_{disposal}$ = cost to dispose of drummed hazardous waste.
C_{test} = cost of performing testing on one unit (one product instance).

C_w = cost of processing one wafer.
D = defect density.
D_W = the diameter of the wafer.
D_r = number of drums.
E = edge scrap allowance (unusable wafer edge).
f_c = fault coverage.
K = scribe street (minimum distance between adjacent die).
M_l = number of miles between the location that generated the waste and the hazardous waste disposal facility.
N_G = gate count.
N_u = number up (number of die per wafer).
OEW = operating empty weight (of an aircraft) in millions of kilograms.
R^2 = coefficient of determination.

Chapter 7 – Test Economics

A = die area.
A_{DFT} = die area when DFT is included.
A_{noDFT} = die area when DFT is not included.
b_t = base cost of a test system with zero pins (scales with capability, performance and features).
B_{waf_die} = die tiling fraction, i.e., accounts for wafer edge scrap, scribe streets between die and the fact that rectangular die cannot be perfectly fit into a circular wafer.
C = conversion matrix.
C_c = the portion of the test cost incurred to apply the fault coverage.
C_{design} = cost of designing a die.
C_{DFT} = die cost when DFT is included.
C_{equip} = the cost of purchasing the tester, facilities needed by the tester, and maintenance of the tester minus the residual value of the tester at the end of its depreciation life.
C_{fab} = yielded cost of fabricating a die.
C_{ij} = element of the conversion matrix that relates fault type i to defect type j.
C_{in} = cost of a unit entering a test step.
C_{noDFT} = die cost when DFT is not included.
C_{out} = cost of a unit exiting a test step.
$C_{out\ per\ die}$ = cost of individual die after wafer probing.
C_p = the portion of the test cost incurred to create the false positives.
C_{probe} = probe card cost.
C_{saw} = cost of sawing a wafer (per wafer).
C_{sort} = cost of sorting die (per wafer).
C_{step} = process step cost (per wafer).
C_{test} = cost of performing testing on one unit (one product instance).
C_{tester} = the portion of the tester cost that should be allocated to each die that is tested.
d = defect spectrum (vector of defect types).

d_{cover_j} = fraction of all devices under test with detected defects of defect type j.
d_j = number of defects of defect type j in the device under test.
dpm_j = number of defects of defect type j per million elements (ppm).
D = defect density (defects per area).
D_L = depreciation life of the tester in years.
E = escape fraction, fraction of product that enters the test step that is defective, but is passed by the test step.
f = fault spectrum.
$f(\,)$ = probability density function.
f_c = fault coverage.
f_{c_i} = fault coverage for fault type i.
f_{cover_i} = fraction of all devices under test with detected faults of fault type i.
f_i = fraction of devices under test faulty due to fault type i.
f_{ij} = the fraction of devices under test faulty due to fault type i that are related to defect type j.
f_p = false positives fraction, the probability of testing a good unit as bad.
$f_{p\text{-}coverage}$ = false positive coverage.
M = number of units that are passed by a test step.
n_e = number of elements in the device under test.
n_o = the average number of defects per part.
N = number of units that enter a test step.
N_B = the number of bad (defective) parts entering a test step.
N_D = the quantity of die to be fabricated.
N_G = the number of good (non-defective) parts entering a test step.
N_{in} = number of parts that come into the test affected by the false positives.
N_{inb} = number of units that enter a test step that are bad (defective).
N_{ing} = number of units that enter a test step that are good (not defective).
N_{out} = number of parts exiting a test step (after false positives are created).
N_{outb} = number of units that pass a test step that are bad (defective).
N_{outg} = number of units that pass a test step that are good (not defective).
N_P = the number of parts passed by a test step.
N_S = the number of parts scrapped by a test step.
N_u = number of die on a wafer (number up).
p = probability of a single fault occurring.
P = pass fraction (fraction of the product that enters a test step that is passed by the test step).
P_{bad} = probability of accepting a die with one or more faults.
$\Pr(\,)$ = probability.
Q = quantity of products that will be made.
Q_{wafer} = fabricated wafer cost.
R_{wafer} = radius of the wafer.
S = scrap fraction (fraction of the product that enters a test step that is scrapped by the test step).
T_{die} = effective time to load, unload, and test one die.
T_f = average fail time.
T_h = handling time (loading the tester).
T_{op} = effective operational time of the tester per year.
T_p = average pass time.

T_t = dead time (between samples).
TPT_t = throughput rate (parts/time).
Y = yield.
Y_{bg} = the probability (or yield) of a bad part being tested as good.
Y_{BP} = bonepile yield, yield (fraction of good parts) in the set of parts scrapped by the test activity.
Y_{DFT} = yield of a die that has DFT (Design for Test).
Y_{in} = yield of units entering a test step.
Y_{noDFT} = yield of a die that does not have DFT (Design for Test).
Y_{out} = yield of units exiting a test step.
Y_{saw} = die yield from sawing a wafer (per wafer).
Y_{sort} = die yield from sorting die (per wafer).
$\binom{k}{x}$ = binomial coefficient.

Chapter 8 – Diagnosis and Rework

C_{device} = the cost of a device when it enters the board assembly process.
C_{diag} = cost of performing diagnosis on one unit (one product instance).
$C_{diag/rew}$ = cost of performing diagnosis and rework on one unit (one product instance).
C_{in} = cost of a unit entering a test step.
C_{out} = cost of a unit exiting a test/diagnosis/rework process.
C_{rew} = cost of performing rework on one unit (one product instance).
$C_{rework\ fixed}$ = the fixed cost per unit instance to perform a replacement.
C_{test} = cost of performing testing on one unit (one product instance).
C_Y = yielded cost.
d_i = number of tests on the branch from the root to the ith leaf node.
D_{avg} = average diagnostic length (i.e., the depth) of a diagnosis tree.
f_c = fault coverage.
f_d = fraction of units that are diagnosible.
f_{dr} = fraction of units that are diagnosible and reworkable.
f_p = false positives fraction, the probability of testing a good unit as bad.
f_r = fraction of units that are reworkable.
N_d = number of units diagnosed.
N_{device} = total number of devices on the board.
N_f = number of distinguishable fault sets.
N_{gout} = number of no fault found units.
N_{in} = number of parts entering a test step.
N_{out} = number of units passed by a test/diagnosis/rework process.
N_r = number of units to be reworked.
N_{rout} = number of units reworked.
N_s = number of units scrapped.
p_i = probability of occurrence of the fault (or fault set) represented by the ith leaf node.
P = pass fraction.
S = scrap fraction.

S_{total} = total scrap from a test/diagnosis/rework process.
T_{device} = time to rework a single device in a unit.
T_{diag} = diagnosis time per unit.
T_{rew} = rework time per unit.
T_{test} = test time per unit.
$T_{total\ diag}$ = total time spent in diagnosis per unit.
$T_{total\ rew}$ = total time spent in rework per unit.
$T_{total\ test}$ = total time spent in test (on the tester) per unit.
$Y_{aftertest}$ = yield of processes that occur exiting the test.
$Y_{beforetest}$ = yield of processes that occur entering the test.
Y_{device} = the yield of a device when it enters the board assembly process.
Y_{in} = yield of units entering a test step.
Y_{out} = yield of units exiting a test/diagnosis/rework process.
Y_{rew} = yield of the rework process.
$Y_{rework\ process}$ = yield of a single device replacement action.

Chapter 9 – Uncertainty Modeling – Monte Carlo Analysis

A = area of a board.
α = minimum of a triangular distribution.
β = mode of a triangular distribution.
C_{in} = cost of board entering a test step.
C_{out} = cost of board exiting a test step.
C_{test} = cost of performing test on one board.
D = Pearson's cumulative test statistic.
D_0 = defect density.
E_j = expected frequencies (for the jth bin).
$f(\,)$ = probability density function, PDF.
f_c = fault coverage.
$F(\,)$ = cumulative distribution function, CDF.
γ = maximum of a triangular distribution.
h = probability corresponding to the mode of a triangular distribution.
LCL = lower confidence limit.
μ = mean as the sample.
n = number of samples.
n_I = number of intervals.
O_j = number of observations in the jth bin.
P_m = scaled and shifted uniform random number.
σ = standard deviation.
U, U_m = uniform random number between 0 and 1 inclusive.
UCL = upper confidence limit.
$\chi^2_{a,v}$ = chi-square distribution.
z = the z-score (standard normal statistic, which is the distance from the sample mean to the population mean in units of standard error), two-sided.
$\lceil\ \rceil$ = ceiling function.

$\lfloor\ \rfloor$ = floor function.

Chapter 10 – Learning Curves

A = die area.
A_{ji} = critical areas for each defect type.
α = cluster factor.
β = learning constant.
B, C = general coefficients in parametric models.
D = defect density.
D_i = defect density for defect type i.
F = first unit.
H = time or cost of the first unit.
k = "midpoint" unit, $F < k < L$.
L = last unit.
$L_e(Y)$ = learning effects (Gruber's learning curve for yield).
λ_j = average number of faults for circuit type j.
P = productivity.
r_l = learning rate.
$r(t)$ = error term.
R^2 = coefficient of determination.
s = learning index (slope) of the learning curve.
t = the time that a product has been in production.
$T_{F,L}$ = time or cost of manufacturing units F through L inclusive.
T_i = total time for i units.
$\overline{T}_i$ = average time for i units.
τ = time constant.
θ = α set of parameters unique to the specific yield model.
U_i = time or cost of the ith unit.
V = volume (in yield space).
$V_E(t)$ = mean individual volume.
V_L = total volume inside V that has been mastered or "learned".
Y = yield.
Y_0 = asymptotic yield.
Y_t = the instantaneous (average) yield during time period t.
Y_c = yield of products produced by a process.

Part II – Life-Cycle Cost Modeling

f = inflation rate (per time period).
n_t = number of time periods.
r = discount rate (per time period).
$r^{nominal}$ = nominal discount rate (per time period).

r^{real} = real discount rate (per time period).

V_n = future value.
$V_n^{nominal}$ = nominal future value.

V_n^{real} = real future value.

Chapter 11 – Reliability

β = shape parameter (Weibull distribution).
E[ ] = expectation value.
$f(t)$ = PDF, fraction of products failing at time t.
$f(t,T)$ = conditional PDF, fraction of products failing at time $t+T$ given that the product survived to time T.
$F(t)$ = CDF, cumulative failures to time t, unreliability at time t.
γ = location parameter (Weibull distribution).
$h(t)$ = hazard rate at time t.
λ = failure rate.
$MTBF$ = mean time before failure.
$MTTF$ = mean time to failure.
$N_s(t)$ = the number of the N_0 product instances that survived to t without failing.
$N_f(t)$ = the number of the N_0 product instances that failed by t.
η = scale parameter (Weibull distribution).
N_0 = total number of tested product instances.
Pr( ) = probability.
$R(t)$ = reliability at time t.
$R(t,T)$ = conditional reliability at time $t+T$ given that the product survived up to time T.
t, τ = time.
T = failure time.

Chapter 12 – Sparing

C_h = holding (or carrying) cost per period per spare (cost of storage, insurance, taxes, etc.).
C_p = cost per order (setup, processing, delivery, receiving, etc.).
C_{Total}, C_{Totalj} = total cost of spares for one spared item (in the jth period of time).
d_r = demand rate.
D_j = number of spares needed (demanded) in period j for one spared item.
$f(t)$ = PDF, fraction of products failing at time t.
k = number of spares.
λ = failure rate (more generally the replacement or removal rate).
m = number of items in a kit.
n_t = number of time periods.
$MTBF$ = mean time before failure.
$MTBUR$ = mean time between unit removals.

n = number of unduplicated (in series, non-redundant) units in service.
P = purchase price of the spare.
PL = probability that k is enough spares <u>or</u> the probability that a spare will be available when needed.
PL_{item} = protection level for an item.
PL_{kit} = protection level for a kit.
Pr( ) = probability.
Q = quantity per order.
r = discount rate.
$R(t)$ = reliability at time t.
t, τ = time.
u_r = usage rate.
z = the number of standard deviations from the mean of a standard normal distribution (the standard normal deviate from 1-α, where α = 1-desired confidence level), single-sided.
$\lceil \; \rceil$ = ceiling function.

Chapter 13 – Warranty Cost Analysis

α = quantity of products sold.
β = shape parameter (Weibull distribution).
C_{cw} = average cost of servicing one warranty claim (manufacturer's cost).
C_{dw} = cost of resolving a denied warranty claim.
C_{fw} = fixed cost of providing warranty coverage.
C_{pw} = effective warranty cost per product instance.
C_{rw} = total cost of warranty coverage.
$D(T_W)$ = expected number of denied warranty claims per product.
E[ ] = expectation value.
$f(t)$ = PDF, fraction of products failing at time t.
$F(t)$ = CDF, cumulative failures to time t, unreliability at time t.
$G(t)$ = cumulative distribution of usage rates.
γ_l = usage rate.
Γ( ) = gamma function.
λ = failure rate.
$m(t)$ = renewal density function.
$M(t)$ = renewal function.
$MTBF$ = mean time before failure.
μ = mean.
σ = standard deviation.
$N(t)$ = number of failures in (0,t].
n_t = number of time periods.
η = scale parameter (Weibull distribution).
Pr( ) = probability.
r = discount rate.
$R(t)$ = reliability at time t.
Rb = pro-rated customer rebate at time t.
s = variable in Laplace domain.

S_n = total time to the nth renewal.
t = time.
T_i = failure time.
T_W = warranty period.
θ = product price (including warranty).
θ' = product price without warranty included.
u = usage rate.
U = usage limit (2-D warranty).
V_n = present value of an investment.
W = age limit (2-D warranty).
$\hat{X}(s)$ = Laplace transform of $X(s)$.
$\mathcal{L}[\,]$ = Laplace transform.

Chapter 14 – Burn-in Cost Analysis

A_F = acceleration factor associated with the burn-in.
C_1 = fixed and non-recurring cost per unit.
C_B = recurring burn-in cost per unit (energy costs, etc.).
C_{BD} = fixed cost of burn-in development.
C_{BI} = cost of performing burn-in (all units).
$C_{BI/unit}$ = cost of performing burn-in (per unit).
C_{BNR} = non-recurring burn-in cost — includes the cost of qualifying, calibrating and maintaining the burn-in equipment and facilities, and training people.
C_{Bt} = recurring burn-in cost per unit per time.
C_{CS} = customer satisfaction value (allocated per unit).
C_{cw} = average cost of servicing one warranty claim on the unit.
C_{fw} = fixed cost of providing warranty coverage.
C_{LR} = cost associated with life removed by the burn-in from non-failed units.
C_{manuf} = manufacturing cost per unit.
$C_{manuf+burn\text{-}in}$ = manufacturing and burn-in cost per unit.
C_O = opportunity cost associated with the unit (profit that could have been made by selling the unit that failed during burn-in) — this assumes that all manufactured units can be sold.
C_{OBF} = operational cost of the burn-in facility per hour (varied in the results that follow).
C_P = unit cost.
$C_{TB}(t)$ = cost of burning-in one unit for the equivalent of t.
E[ ] = expectation value.
$f(t)$ = PDF, fraction of products failing at time t.
$F(t)$ = CDF, cumulative failures to time t, unreliability at time t.
λ = failure rate.
$M(t)$ = renewal function, mean number of renewal events (warranty claims) that occur in the interval $(0,t]$.
n_u = number of units being burned-in.

ROI = return on investment.
t = time.
t_{bd} = equivalent burn-in time.
t_s = time under stress (burn-in test time).
T_W = warranty period.
V_B = value (per unit) of performing a burn-in.

Chapter 15 – Availability

a = number of units under repair.
$A, A(t)$ = availability (generic).
$\overline{A}, \overline{A(t)}$ = average availability.
$A(\infty)$ = steady-state availability.
ADT = administrative delay time.
A_a = achieved availability.
A_E = energy-based availability.
A_i = inherent availability.
A_m = materiel availability.
A_o = operational availability.
A_s = supply availability.
ADT = administrative delay time.
E_{real} = actual energy generated.
$E_{theoretical}$ = theoretical maximum energy that could be generated.
E[ ] = expectation value.
EBO = expected backorders.
erf( ) = error function.
$f(t), \hat{f}(s)$ = PDF in the time and Laplace domains.
F_t = failures that need to be repaired per unit per unit time.
$g(t), \hat{g}(s)$ = repair time distribution in the time and Laplace domains.
k = number of spares.
l = number of unique repairable items in a system.
LDT = logistics delay time.
λ = failure rate.
$m(t)$ = renewal density function.
$\hat{m}(s)$ = Laplace transform of the renewal density function.
m_b = number of backorders.
$M(t)$ = renewal function.
$\overline{M}$ = mean active maintenance time.
$M_a(t)$ = maintainability.
$\overline{M}ct$ = mean corrective maintenance time (same as *MTTR*).
MDT = mean maintenance downtime.
$\overline{M}pt$ = mean preventative maintenance time.

MSD = mean supply delay.
$MTBF$ = mean time before failure.
$MTBM$ = mean time between maintenance.
$MTPM$ = mean time to perform preventative maintenance.
$MTTR$ = mean time to repair.
μ = repair rate, mean of $\ln(t)$, location parameter in the lognormal distribution.
μ_r = mean repair time (mean time to repair one unit).
n = number of identical systems in the fleet.
N = number of fielded units, number of available systems.
Φ = standard normal CDF.
p_{ij} = probability that the state is j at T given that it was i at time T-1.
Pr( ) = probability.
$R(t)$ = reliability at time t.
s = variable in Laplace domain.
σ = standard deviation of $\ln(t)$, scale parameter in the lognormal distribution.
t, τ = time.
T = time (actual repair time).
$w(t), \hat{w}(s)$ = time-to-failure distribution in the time and Laplace domains.
Z_i = number of instances of item i in each system.

Chapter 16 – The Cost Ramifications of Obsolescence

C_{DR} = design refresh cost.
C_{DR_0} = design refresh cost in year 0.
C_h = holding (or carrying) cost per part per year.
C_{LTB} = cost of a last time buy.
C_O = overstock cost.
C_{Pi} = cost of the action defined in profile p in period i.
C_{Total} = total cost for managing obsolescence.
C_U = understock cost.
D = demand.
E[ ] = expectation value.
$f(\,)$ = PDF.
$F(\,)$ = CDF.
i = years until refresh.
L = total loss.
n = total number of profiles in the application.
N_P = number of instances of the profile p in the application.
OR_{Pi} = OR (obsolescence risk) for profile p in period i.
P_0 = price of the obsolete part in the year of the last time buy.
Pr( ) = probability.
Q = quantity ordered.
Q_i = number of parts needed in year i.
q_{ij} = quantity for the ith discrete demand in the jth year.

Q_{opt} = value of Q that minimizes the total loss.
r = discount rate.
Y = number of years the part needs to be supported for.
Y_R = year of the design refresh.
$\lceil \; \rceil$ = ceiling function.

Chapter 17 – Return on Investment

C_{INF} = PHM management infrastructure costs.
C_{NRE} = PHM management non-recurring costs.
C_{PHM} = life cycle cost of the system when managed using a PHM approach.
C_{REC} = PHM management recurring costs (cost of putting PHM hardware into each instance of the system).
C_u = life cycle cost of the system when managed using unscheduled maintenance.
D_L = depreciation life in years.
D_s = average die shrink (% area decrease).
I = investment.
I_{PHM} = investment in PHM when managing the system using a PHM approach.
I_u = investment in PHM when managing the system using unscheduled maintenance.
M_s = average market share increase (%) — one time increase.
N_c = number of chips effected.
P = average profit per chip.
r = discount rate.
R = return.
ROI = Return on Investment.
S = average original sales volume per chip per year.
V = volume.
V_f = final value of an investment.
V_i = initial value of an investment.

Chapter 18 – The Cost of Service

b = labor burden rate
C_A = total accommodation costs
$C_{average}$ = the average service cost per failure for machines
C_{BO} = total bonus for providing a good service
C_j = the total cost for a population of machines in their j^{th} year of service
C_L = total labor costs
C_P = total telephone service costs
C_S = total subsidies for travelling
C_{SP} = total costs for spare parts
C_{TP} = total transportation costs
C_{TR} = total training costs

I_s = the year the machines is sold (and enters service)
i = years in service
j = year of service
λ_j = the failure rate for machines in their j^{th} year of service
N_{f_j} = the total number of failures for machines in their j^{th} year of service
N_i = the total number of machines in service for at least i years
N_{ij} = the number of failures of machines in service for at least i years in their j^{th} year of service
N_s = the total number of machines sold in year I_s
T = the service contract length (in years)

Chapter 19 – Software Development and Support Costs

ACT = annual change traffic.
CW = complexity weight.
DSI = delivered source instructions.
E = effort adjustment factor.
F_i = components of TCF.
FP = function point count.
$KDSI$ = thousands of delivered source instructions.
PM = effort in person months.
$SLOC$ = source lines of code.
TCF = technical complexity-weighting factor.
$TDEV$ = software development time.
UFC = unadjusted function point count.

Chapter 20 – Total Cost of Ownership Examples

C_{a_i} = assembly cost of one instance of the part in year i
C_{ap_i} = purchase order generation cost
C_{as_i} = annual cost of supporting the part within the organization
$C_{assembly_i}$ = total assembly cost (for all products) in year i
C_{design_i} = non-recurring design-in costs associated with the part
$C_{field\ use_i}$ = total field failure cost in year i
C_{ia_i} = initial part approval and adoption cost
C_{in_i} = incoming cost/part
$C_{ink/toner}$ = total cost of ink or toner
C_{nonPSL_i} = setup and support for all non-PSL (Preferred Supplier List) part suppliers

C_{or_i} = obsolescence case resolution costs
C_{out_i} = output cost/part
C_{pa_i} = product-specific approval and adoption
C_{paper} = total cost of paper
$C_{printer}$ = total cost of printers
C_{proc_i} = cost of processing the warranty returns in year i
C_{ps_i} = all costs associated with production support and part management activities that occur every year that the part is in a manufacturing (assembly) process for one or more products
C_{repair} = cost of repair per product instance
$C_{replace}$ = cost of replacing the product per product instance
$C_{support_i}$ = total support cost in year i
C_{TCO} = total cost of ownership
E_i = quantity of energy produced in year i
f = fraction of failures requiring replacement (as opposed to repair) of the product
f_p = false positives fraction, the probability of testing a good unit as bad
F_i = fuel expenditures in year i
FIT = failures in time
i = year
I_i = investment expenditure in year i
$I_{ink/toner}$ = cost of an inkjet cartridge set or toner cartridge set
$L_{printer}$ = lifetime of the printer measured in the number of printed pages
$LCOE$ = levelized cost of energy
M_i = operations and maintenance expenditures in year i
n = number of years over which the LCOE applies
N_{f_i} = number of failures under warranty in year i
N_i = total number of products assembled in year i
N_j = the number of part sites assembled in a particular year j
N_{pages} = total number of pages printed
$N_{printers}$ = number of printers needed
N_{refill} = number of ink refills needed
$N_{withprinter}$ = number of pages that can be printed with ink/toner cartridge set that comes with the original printer purchase
P_i = purchase price of one instance of the part in year i
$P_{printer}$ = purchase price of the printer
r = after tax discount rate on money
$TLCC$ = total life-cycle cost
$Y_{aftertest}$ = yield of processes that occur exiting the test
$Y_{beforetest}$ = yield of processes that occur entering the test
Y_{out} = yield of units exiting a test step
Y_{rew} = yield of the rework process
YTO = years to obsolescence

Z = number of pages that can be printed with one ink/toner cartridge set

$\lceil \ \rceil$ = ceiling function

Chapter 21 – Cost, Benefit and Risk Tradeoffs

A = annual value.
A_m = annual maintenance cost.
BCR = benefit-cost ratio.
C_{fail} = cost per failure.
$C_{Risk\ Total}$ = total money spent on risk mitigation activities.
E_{fail} = expected number of failures per product service life.
$f(\ )$ = probability density function, PDF.
$fp\ rate$ = false positive rate.
F = future value.
FN = number of false negatives.
FP = number of false positives.
F_I = increased fare collection due to increased number of trips.
m = number of severity levels.
n = number of boards.
n_t = time periods.
N = number of years.
Ng = value per day of removing single-tracking delays for non-rush hour trips after improvement.
Nu = value per day of removing single-tracking delays for non-rush hour trips that would be taken anyway.
P = present value.
$PCFC$ = projected cost of failure consequence.
P_i = increase in probability of death.
r = discount rate.
Rg = value per day of removing single-tracking delays for rush hour trips after improvement.
ROI = return on investment.
Ru = value per day of removing single-tracking delays for rush hour trips that would be taken anyway.
T_A = test accuracy.
TN = number of true negatives.
TP = number of true positives.
$tp\ rate$ = true positive rate.
VSL = value of a statistical life.
W_p = wage premium.
y = year.
Y = yield.

Chapter 22 – Real Options Analysis

b_{rb} = dollar holdings of a riskless bond in the portfolio.
C = call option value (price) at $T = 0$.
C_d = downside value of an option.
C_M = predictive maintenance cost.
C_{PMV} = value of the path.
C_u = upside value of an option.
C_{u^2}, C_{d^2}, C_{ud} = binomial lattice option values at $T = 2$.
d = downside multiplier.
dt = time step.
d_1, d_2 = factors appearing in the Black-Scholes solution.
E[ ] = expectation value.
I = inverse of a normal cumulative distribution.
m = fraction of the base project in the portfolio.
μ = drift.
n_t = number of time periods.
$N()$ = cumulative standard normal distribution function.
NPV = net present value.
p = risk-neutral probability (of an upside result).
P = put option price.
$PV_{investment}$ = present value of the payoff from an investment or project.
PV_{option} = present value of the option.
q = objective probability.
r = risk-adjusted discount rate.
R_f = interest rate paid by the riskless bond (riskless rate), also called the risk-free rate.
RUL = remaining useful life.
S = value of the investment or project at $T = 0$.
S_d = downside value of an investment or project.
S_u = upside value of an investment or project.
S_0 = investment or project value at $T = 0$.
S_1 = investment or project value at $T = 0$.
σ = standard deviation of returns on the underlying security (volatility).
t = time.
T = time (time to expiration of the option).
u = upside multiplier.
V_0 = portfolio value at $T = 0$.
V_1 = portfolio value at $T = 1$.
V_T = portfolio value at T.
$WACC$ = weighted average cost of capital.
W_t = Wiener process.
X = strike price.

Appendix B – Weighted Average Cost of Capital (WACC)

β = sensitivity (also called volatility).
D = debt.
E = equity.
r = discount rate.
R_e = cost of equity.
R_f = risk-free (or riskless) interest rate, the interest rate of U.S. Treasury bills or the long-term bond rate is frequently used as a proxy for the risk-free rate.
R_m = market return.
R_p = equity market risk premium (EMRP).
T_e = effective marginal corporate tax rate.
V = the company's total value (equity + debt).
$WACC$ = weighted average cost of capital.

Appendix C – Discrete-Event Simulation (DES)

$Cost_i$ = individual event costs.
$f(t)$ = PDF, fraction of products failing at time t.
$F(t)$ = CDF, cumulative failures to time t, unreliability at time t.
λ = failure rate.
$MTBF$ = mean time before failure.
r = discount rate.
$R(t)$ = reliability at time t.
t = time.
t_c = cumulative failure time at the ith event.

Appendix B

Weighted Average Cost of Capital (WACC)

The inclusion of cost of money within cash flow analyses in engineering economics and life-cycle costing is a very important (and in many cases dominate) contributing factor in understanding the respective costs. Cost of money reflects the fact that the use of money to support a product (e.g., to fund design, manufacturing, and sustainment) is not free, i.e., the money has to come from some source and it is likely that that source will require some form of compensation over time.

In general there are three sources of funding available to a company to fund its operations: retained earnings, borrowed money (debt financing) , and selling equity (e.g., stocks).

If the money to support a project is obtained via a loan (debt financing), then the cost of that money is the interest paid to the loan provider. If all of the money is obtained via a loan then the interest rate on the loan is set when the money a company uses is obtained and the interest rate can simply be used to modify future cash flows as in Equation (II.1), however, rarely is the case this simple. Usually companies are funded by, and fund projects via, a combination of debt and equity capital.

Most engineering economics texts refer to the rate paid for money as simply the "interest rate" and many engineers more generally call it the "discount rate" . Both of these terms infer the source of the money — interest rate infers debt financing, while the discount rate is defined as the interest rate charged on loans made by the Federal Reserve Bank's discount window to commercial banks and other depository institutions. A more general term is the "weighted average cost of capital" or WACC, which captures and combine the cost of all the sources of money that a company uses.

This appendix describes the general calculation and use of the WACC. It also describes how the WACC can change over time and issues with using the WACC in long (calendar) time calculations.

B.1 The Weighted Average Cost of Capital (WACC)

While many methods can be used to determine the rate for the cost of money, it should be pointed out that in many cases, these methods are more art than science.

A common strategy is to calculate a weighted average cost of capital (WACC). The WACC represents a weighted blending of the cost of equity and the after-tax cost of debt.

B.1.1 *Cost of Equity*

Equity is a stock or any other security representing an ownership interest in a company. Companies, whether public or private, raise money by selling equity. Unlike debt, for which the company pays a set interest rate, equity does not have a predefined price. However, this doesn't mean that equity has no cost to a company. Equity holders (e.g., shareholders) expect a return on their investment in a company.

The equity holders' required rate of return represents a cost to the company, because if the company cannot provide the expected return, the equity holders may sell their equity, which will cause the stock price to drop.

The effective cost of equity is the company's cost of maintaining a share price that meets the expectations of the investors. A common method for calculating the cost of equity uses the capital asset pricing model (CAPM),[1]

$$R_e = R_f + \beta(R_m - R_f) \tag{B.1}$$

[1] Developed by William Sharpe from Stanford University who shared the 1990 Nobel Prize in Economics for the development of CAPM [Ref. B.1]. Other models exist including: APM, multi-factor and proxy models.

where

R_e	=	cost of equity.
R_f	=	risk-free interest rate, the interest rate of U.S. Treasury bills or the long-term bond rate is frequently used as a proxy for the risk-free rate (R_f is referred to as the "riskless" rate in Chapter 22).
R_m	=	market return.
β	=	sensitivity (also called volatility).
$(R_m - R_f)$	=	R_p, Equity Market Risk Premium (EMRP).

In Equation (B.1), the sensitivity (β) models the correlation of the company's share price with the market. $\beta = 1$ indicates that the company is correlated to the market ($\beta = 0$ indicates a riskless investment); $\beta > 1$ means that the share price exaggerates the market's movements; and $\beta < 1$ means that the share price is more stable than the market. A $\beta < 0$ indicates a negative correlation with the broader market. The EMRP is the return that investors expect above the risk-free interest rate. The EMRP is the compensation that investors require for taking extra risk (above the risk-free rate) by investing in the company's stock, i.e., EMRP is the difference between the risk-free rate and the market rate. There are several services (e.g., Barra and Ibbotson) that provide EMRP and β for public companies.[2]

Adjustments are commonly made to the cost of equity calculated in Equation (B.1) to account for various company-specific risk factors including: the company's size, lawsuits that may be pending against the company (or lawsuits that the company has pending against others), the company's dependence on key employees, and customer base concentration. The magnitude of these adjustments are often based on investor judgment and will vary significantly from company to company.

---

[2] If you are interested in finding EMRP or β for a non-public company, you should search for a public company with a similar business and use their EMRP or β.

B.1.2 *Cost of Debt*

Debt is an amount of money borrowed by one party from another. Corporations use debt as a method for making large purchases that they could not afford under normal circumstances. A debt arrangement gives the borrowing party permission to borrow money under the condition that it is to be paid back at a later date, usually with interest. Compared to the cost of equity, the cost of debt is more straightforward to calculate. The cost of debt (R_d) is the market rate the company is paying on its debt.

Because companies benefit from tax-deductible interest payments on debt, the net cost of debt is the interest paid less the taxes paid — this is the "tax shield" that arises from the interest expense. As a result, the after-tax cost of debt is R_d (1 - corporate tax rate).

B.1.3 *Calculating the WACC*

Combining the cost of debt and equity together based on the proportion of each, we obtain the overall cost of money to the company. WACC, the weighted average of the cost of capital is given by,[3]

$$\text{WACC} = R_e\,(E/V) + R_d\,(1 - T_e)\,(D/V) \tag{B.2}$$

where

V = the company's total value (equity + debt).
D/V = the proportion of debt (leverage ratio).
E/V = the proportion of equity.
T_e = effective marginal corporate tax rate.[4]

Figure B.1 shows the variation of WACC with the ratio of D to E. Note, in Figure B.1 that the costs of equity and debt vary with the company's debt to equity mix, for example, the cost of debt for a company increases as more of the company is financed via debt (because lenders infer more risk and therefore charge a higher interest rate). Also note that as the cost of debt increases, the cost of equity also increases –

---

[3] In the Part II introduction, Chapters 1,12, 13, 16, 17, 20, 21, and 22, and Appendix C of this book, WACC is referred to as the "discount rate" and represented with the symbol r.

[4] The *effective tax rate* is the actual taxes paid divided by earnings before taxes.

why? The costs of debt and equity track each other because equity holders are always taking more risk than debt holders and therefore require a premium return above that of debt holders. It is also important to point out that there is an implicit assumption in Figure B.1 that the company's value does not change with the *D/E* ratio.

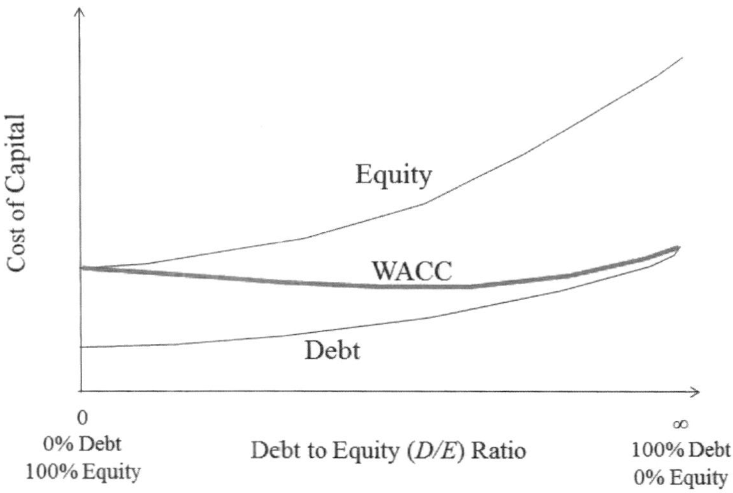

Fig. B.1. Variation in WACC with *D/E* ratio.

In the calculation of the WACC one can subdivide the cost of equity into different types of equity, e.g., common and preferred stock. Also, sometimes the rate of return on retained earnings is also included as a separate term in Equation (B.1).

Be careful: Equation (B.2) appears easier to calculate than it actually is. No two people will calculate the same value of WACC for a company due to their unique judgments about the circumstances of the company and the valuation methods that they use.

As a simple example of computing the WACC, consider a semiconductor manufacturer that has a capital structure that consists of 40% debt and 60% equity, with a tax rate of 30%. The borrowing rate (R_d) on the company's debt is 5%. The risk-free rate (R_f) is 2%, the β is 1.3 and the risk premium (R_p) is 8%. Using these parameters the following can be computed:

$$R_e = R_f + \beta(R_m - R_f) = 0.02+1.3(0.08) = 0.124$$

$$D/V = 0.4/(0.6{+}0.4) = 0.4$$

$$E/V = 0.6/(0.6{+}0.4) = 0.6$$

$$\text{WACC} = R_e\,(E/V) + R_d\,(1 - \text{corporate tax rate})\,(D/V)$$

$$= 0.124(0.6){+}0.05(1{-}0.3)(0.4) = 0.0884$$

The WACC comes to 8.84% (this is a "beta-adjusted discount rate" or "risk-adjusted discount rate").

Actual values of WACC for companies vary widely. It is not uncommon for WACCs to range from 3-4% up to 20% or more. Various web sites provide WACC estimates for publicly traded companies.[5]

All the discussion in this section assumes there is no time dependence in the WACC, i.e., this is all valid at an instant in time and may have no validity at other times. The irony of the WACC calculation is that WACC is used to model the time value of money as in Equation (II.1), but the WACC that is calculated is only valid at one instant in time.

B.2 Forecasting Future WACC

One of the biggest problems with WACC is that while it may accurately reflect what a company believes its cost of money is at the current time, the dynamics of the broader economy and the company's capital structure change with time. Therefore the WACC is not constant over time. Specifically the WACC is dynamic because: 1) a company's debt to equity ratio changes over time;[6] 2) the cost of equity (R_e) may change with time; 3) the cost of debt (R_d) may change over time; and 4) the tax rate (T_e) will be a function of profitability and tax breaks allowed for certain industries in certain locations during certain periods of time. Computing the WACC for a future time is difficult, but really important.

---

[5] Does the US Government have a WACC? Yes, it's the rate on 3, 5, 7, 10, and longer-term treasury securities (T-Bills).

[6] Depending on the form that the debt takes the *D/E* ratio may or may not remain constant. For example, the *D/E* ratio remains unchanged for debt in the form of a bond for which only the interest (coupon) payments are made, which is replaced by an equivalent bond at its maturity date. In the case of a loan whose balance reduces as payments are made, the *D/E* ratio drops over time.

Assuming that today's WACC will remain constant into the future may be a source of significant errors in life-cycle cost modeling. For example, at a macro-level, world economics dictate whether interest rates on debt rise or fall and high profile corporate disasters increase the perceived risk of equity investments.

Many other factors affect the WACC associated with specific companies in specific business sectors. For example, for companies that operate wind farms (a relatively new and growing business sector),

- Increasing experience amongst operators of wind farms will reduce the risk premium that investors can demand thus lowering R_e over time.
- In 2014-2015 interest rates for debt are climbing due to the recovery of the global economy, increasing R_d.
- The equity and debt ratios will change over time as well. For example, as risk decreases, companies are able to take on a larger share of debt (D/V increases and E/V decreases), which companies tend to do because usually $R_d < R_e$.
- The corporate tax rate will change because the company becomes profitable and the expiration of tax breaks granted by local and national governments.

The trends over time in R_d can be modeled with a yield curve.[7] R_e has to be modeled using a capital asset pricing model (CAPM), in which β is the primary parameter that trends over time.

In reality all the parameters used to determine WACC are probability distributions. Therefore, the resulting WACC is a probability distribution. Monte Carlo analysis can be used to determine the appropriate probability distribution for the WACC in each year of an analysis. In addition, the WACC is a non-stationary process.[8] In the case of WACC, not only does the distribution's mean shift over time (driven

---

[7] Found by calculating a forward interest rate, which is an interest rate that is applicable to a future financial transaction.

[8] Stationary processes are stochastic processes whose joint probability distributions do not change when shifted in time or space (time is the relevant parameter for us).

by the trends in the parameters), but its variance also becomes larger as time progresses. Note, if non-stationary methods are used to estimate future WACC, the coupling (non-independence) of parameters must be respected.

B.3 Comments

What engineers often call "discount rate" would be referred to as "weighted average cost of capital (WACC)" by business analytics people. The WACC is not the inflation rate! In actuality, "WACC is neither a cost nor a required return, it is a weighted average of a cost and required return" [Ref. B.2].

The net present value (NPV) is the difference between the present value of future net cash inflows (the benefits) and the present value of implementation costs (the investment costs). However, in many instances both investments and costs are discounted using the same WACC, which may be incorrect.[9]

B.3.1 *Trade-off Theory*

The cost of debt is lower than the cost of equity. Does this mean that a company (or projects) should be financed only with debt? What is the fallacy here? In reality, using cheap debt increases the cost of equity (because its financial risk increases).

Company management seeks to find a debt/equity ratio (*D/E*) that balances the risk of bankruptcy (i.e., large *D/E*) with the risk of using too little of the least expensive form of financing, which is debt (i.e., small *D/E*).[10] According to the trade-off theory [Ref. B.4], there is a best way to finance a company, i.e., an optimal *D/E* ratio that minimizes a company's cost of capital — Fig. B.1 shows this concept graphically.

---

[9] It is more correct to discount the benefits at the WACC, and discount the investment at a reinvestment rate that is similar to the risk-free rate [Ref. B.3].

[10] The after-tax cost of debt will always be lower than the cost of financing with equity.

B.3.2 *Social Opportunity Cost of Capital (SOC)*

The concept of the Social Opportunity Cost of Capital (SOC) is sometimes invoked to specify the return governments require when making investments on behalf of the community, [Ref. B.5]. The social opportunity cost of capital is the discount rate that reduces the net present value of the best alternative private use of the funds to zero [Ref. B.6]. It is the rate at which society is willing to forgo present consumption for the sake of future consumption. With this discount rate, the discounted value of future consumption goods equals the value of forgone present consumption goods. The SOC is the consumption-based opportunity cost of capital.

References

B.1 Sharpe, W. F. (1964). Capital asset prices – A theory of market equilibrium under conditions of risk. *Journal of Finance*, XIX(3): pp. 425–442.

B.2 Fernandez, P. (2011). WACC: Definition, misconceptions and errors, *IESE Business School, University of Navarra, Working Paper WP-914.*

B.3 Mun, J. (2006). Real options analysis versus traditional DCF valuation in layman's terms. http://www.realoptionsvaluation.com/attachments/whitepaperlaymansterm.pdf

B.4 Kraus, A. and Litzenberger, R. H. (1973). A state-preference model of optimal financial leverage, *Journal of Finance*, 28(4): pp. 911-922.

B.5 Harberger, A. C. (1969). The discount rate in public investment evaluation. Proceedings of the Committee on the Economics of Water Resources Development, Western Agricultural Economics Research Council, Report No. 17, Denver Colorado, pp. 1-24.

B.6 Young, L. (2002). Determining the Discount Rate for Government Projects, *New Zealand Treasury Working Paper 02/21.*

Problems

B.1 Why does paying more taxes reduce the WACC? Explain this. Companies want to decrease their WACC, so why is moving the company to a state with a higher tax rate not a good approach for reducing the WACC?

B.2 Why do equity holders require a greater return than debt holders?

B.3 If a company borrows money at a 6.5%/year rate (after taxes), pays 9% for equity, and raises its capital in equal proportions of debt and equity, what is its WACC?

B.4 A company currently has the following capital structure:

Source of Funding	Amount of Funding	Expected Rate of Return
Retained Earnings	$100M	11%
Loans	$35M	3.4%
Bonds	$150M	8.75%
Preferred Stock	$60M	7%
Common Stock	$110M	10.5%

Note, the interest paid on the bonds is not tax deductible. Assuming a corporate tax rate of 25%:

a) What is the current WACC of the company?

b) If the company expects the total capital to remain the same, but the debt to equity ratio to increase 10% (via an increase in the debt and an across the board decrease in equity financing) and the cost of debt to increase 50% in the next 2 years, what will the WACC be after these changes?

B.5 A semiconductor fabrication company is installing a new process that requires $2,000,000 in new equipment.

a) The company has two financing options: 1) 40% equity funds at 9% per year and a loan for the rest at 10% per year. 2) 25% equity funds at 9% per year and the rest of the money borrowed at 10.5% per year. Which alternative results in a smaller WACC? Assume a corporate tax rate of 5% per year.

b) Yesterday the finance committee in the company decided that the WACC for all new projects must not exceed the 5 year historical average WACC in the company of 10% per year. With this restriction what is the maximum loan interest rate that can be incurred for the two options in part a)?

B.6 A contract manufacturer of printed circuit boards plans to raise $5 million in debt capital by issuing five thousand bonds (each has a face value of $1000). The bonds pay 8%/year, paid annually (coupon interest rate) and have a 10 year life. Assume the effective tax rate of the company is 23% and the bonds are sold at a 2% discount. Calculate the cost of this debt equity before and after taxes. Assume discrete compounding.

B.7 If the tax rate (T_e) is zero, under what conditions is WACC independent of D/V?

Appendix C

Discrete-Event Simulation (DES)

Life-cycle cost modeling generally involves modeling systems (more specifically system costs) that evolve over time. For complex systems with high electronics content the time dependent costs usually involve the operation and support of the system. Depending on the type of system, operation may involve the purchase of fuel, the training of people, the cost of various consumable materials, etc. Maintenance costs that occur over time are combinations of labor, equipment, testing, and spare parts. f the life cycle of a system is relatively short (i.e., less than a couple of years), then direct calculation methods work well, however, when the modeled life cycle extends over significant periods of time and the cost of money is non-zero, the calculation of life-cycle cost changes from a multiplication problem into a summation problem and the dates of cost events become important, e.g., the cost of individual maintenance events differ based on when they occur due to the cost of money. Discrete-event simulation is commonly used to model life-cycle costs that are accumulated over time when time spans are long and the cost of money is non-zero.

When we simulate a system that evolves over time, the system either changes continuously or discontinuously. An example of a continuous system is the weather — temperature, humidity, wind speed, etc., all change in a continuous way. Other types of systems change in a discontinuous (or discrete) manner, for example, an inventory system that decreases or increases at specific points in time when parts are demanded or replenished. In this case a graph of the quantity of parts in the inventory as a function of time would look like a series of step functions separated by periods of time where there was no change in the inventory.

Discrete-event simulation (DES) is the process of codifying the behavior of a complex system as an ordered sequence of well-defined events. In the context of cost modeling, an event represents a particular change in the system's state at a specific point in time, and the change in state generally has cost consequences. Discrete-event simulation utilizes a mathematical/logical model of a physical system that portrays state changes at precise points in simulated time called events [Ref. C.1].[1] Discrete means that successive changes are separated by finite amounts of time, and by definition, nothing relevant to the model changes between events.

Time may be modeled in a variety of ways within the DES. Alternate treatments of time include: time divided into equal increments, i.e., a time step; unequal increments; or cyclical (periodic), e.g., as in a traffic light or bus schedule. In DES the system "clock" jumps from one event to the next, periods between events are ignored. A timeline is defined as a sequence of events and the times that they occur.

At each event, various properties of the system can be calculated and accumulated. Accumulated parameters of interest as the simulation proceeds along the timeline could be: time (system "clock"), cost, system up or down time, inventory levels, throughput, defects, resources consumed (material, energy, etc.), waste generated, etc. Using the accumulated parameters, one could generate various important results as a function of time: total cost, resources consumed, availability, return on investment, etc. Everything in the discrete-event simulation is uncertain and can be represented by probability distributions. This means that we model the timeline (and accumulate relevant parameters) many times (through many possible time histories) in order to build a statistical model of what will happen.

[1] It is difficult to pinpoint the exact origin of discrete-event simulation, however Conway, Johnson and Maxwell's 1959 paper [Ref. C.2] discusses many of the key points of a discrete-event simulation, including managing the event list (they call it an element-clock) and methods for locating the next event. It is evident that many of the concepts of discrete-event simulation were being practiced in industry in the late 1950s.

C.1 Events

An event represents something that happens to the system at an instant in time that may change the state of the system where, by definition, nothing relevant to the model changes between events. Relevant types of events in the life cycle of a complex electronic system include:

- Scheduled maintenance (preventative)
- Unscheduled maintenance (failures)
- Spares purchases
- Upgrades (or other scheduled system changes)
- Annual charges (e.g., inventory holding).

Events have various properties that include: costs and durations (even though events occur at an instant in time, they can have a finite duration). The event costs include the same costs that are articulated in Chapter 2, namely: labor, materials (e.g., spare parts), capital (equipment, inventory), and tooling, plus business interrupt. These are summed to get the total event cost. As described in previous chapters, possible modifiers to these costs include: learning curves, volume pricing, inflation/deflation, and cost of money.

An important note here is that each event is dependent on the previous events that have occurred on the timeline. The dependency may simply be timing (see the examples in Section C.2), or it may be more complex — the previous events may change the state of the system in such a way as to influence the type of event that occurs next.

Events may have start and end times if the events are not instantaneous (see Problem C.3).

C.2 DES Examples

This section presents several DES examples beginning with a very simple (trivial) example followed by more complex examples that can be used to analyze the life cycle of a system.

C.2.1 *A Trivial DES Example*

Assume that we have some type of system whose failure rate is constant. The reliability of the system is given by Equation (11.16) as,

$$R(t) = e^{-\lambda t} \tag{C.1}$$

where t is time and λ is the failure rate. As shown in Equation (11.17) the mean time between failures for this system is $1/\lambda$ (known as the MTBF). Suppose, for simplicity, failures of this system are resolved instantaneous at a maintenance cost of \$1000/failure. If we wish to support the system for 20 years, how much will it cost? Assuming that the discount rate is zero, this is a trivial calculation:

$$\textit{Total Cost} = 1000(20\lambda) \tag{C.2}$$

The term in parentheses is the total number of failures in 20 years. If $\lambda=2$ failures per year, the *Total Cost* is \$40,000.

This example is very easy and we certainly do not need any sort of fancy DES to solve it, but what if the discount rate (r) was 8%/year compounded discretely? Now the solution becomes a sum, because each maintenance event costs a different amount of money,

$$\textit{Total Cost} = \sum_{i=1}^{20\lambda} \frac{1000}{(1+r)^{i/2}} \tag{C.3}$$

where $i/2$ is the event date in years.[2] The *Total Cost* is now \$20,021.47 in year 0 dollars.

Even though the two cases described so far are pretty easy and we don't need DES to solve them, let's use DES to illustrate the process. To create a DES for these simple cases, we start at time 0 with a cumulative cost of 0, advance the simulator to the first failure event, cost that event and add it to the cumulative cost, and then repeat the process until we reach 20 years. Table C.1 shows the discrete-event simulation events and costs.

---

[2] The $i/2$ assumes that $\lambda = 2$ and the failures are uniformly distributed throughout the year.

Table C.1. Simple example described in terms of events.

Event Number	Event Date (years)	Event Cost ($r = 0$)	Cumulative Cost ($r = 0$)	Event Cost ($r = 8\%$)	Cumulative Cost ($r = 8\%$)
0	0	0	0	0	0
1	0.5	$1000	$1000	$962.25	$962.25
2	1	$1000	$2000	$925.93	$1888.18
3	1.5	$1000	$3000	$890.97	$2779.15
....					
40	20	$1000	$40,000	$214.55	$20,021.47

Obviously there are several implicit assumptions about exactly when the events take place and other things (we will leave these to a homework problem). At this point Table C.1 is just a rather arduous way of performing the calculations in Equations (C.2) and (C.3). However, Table C.1 is a DES. In this case each failure has a specific date on which a maintenance cost is charged and added to the cumulative maintenance cost. Nothing (that costs money) is assumed to happen to the system between events.

C.2.2 *A Not So Trivial DES Example*

Suppose that the actual event dates in the example presented in the previous sub-section are not known, rather the time-to-failures are represented by a failure distribution. For our simple case, the corresponding failure distribution is given by Equation (11.14),

$$f(t) = \lambda e^{-\lambda t} \tag{C.4}$$

Now, instead of assuming that the failures of the system take place at exactly MTBF intervals (the MTBF is just the expectation value of the time to failure), they take place at intervals determined by sampling (using Monte Carlo) the $F(t)$ distribution. Now the total cost is given by the sum in Equation (C.3), but the event dates come from sampling; so there is no simple analytical sum to use for the solution.

Let's solve this problem using DES. First we need to generate the failure times. For this we use the CDF of the exponential distribution from Equation (11.15),

$$F(t) = 1 - e^{-\lambda t} \tag{C.5}$$

Rearranging Equation (C.5) to solve for t we get,

$$t = \frac{-\ln[1 - F(t)]}{\lambda} \tag{C.6}$$

To sample this, we choose a random number between 0 and 1 (inclusive) that we assign to $F(t)$, then solve Equation (C.6) for t, which is the failure time sampled from the exponential distribution. Note, t is not the next event date, it is the time measured from the previous event, so the ts need to be accumulated to produce the event dates.[3]

Using the event dates, we can now calculate the individual event costs using,

$$Cost_i = \frac{1000}{(1+r)^{t_c}} \tag{C.7}$$

where t_c is the cumulative failure time at the ith event. Table C.2 shows an example of the first three events and two final events in the process.

Table C.2. Time-to-failure distribution sampling example.

Event Number	Random Number ($F(t)$)	Time-to-Failure Sample (years) (t)	Event Date (years) (t_c)	Event Cost ($Cost_i$)	Cumulative Cost
0	-	-	0	0	0
1	0.194981	0.108445	0.108445	\$991.69	\$991.69
2	0.430298	0.281321	0.389765	\$970.45	\$1,962.14
3	0.978275	1.914642	2.304407	\$837.49	\$2,799.62
...					
41	0.197316	0.109897	18.85356	\$234.34	\$20,826.08
42	0.971349	1.776292	20.62985	\$204.40	\$21,030.48

In this case there is no set number of events that need to be generated to reach the 20 year support life considered in this problem; i.e., you may need more or less than 40 events to get there, so the simulation needs a stopping criteria: stop when the Event Date > 20 years and do not cost the final event. In this example, the total cost is \$20,826.08.

The example described in this section samples reliability distributions to generate a sequence of events. The sequence of failure events generated

[3] You may not need to manually sample the distribution as we have done in Equations (C.5) and (C.6). Excel, for example, has commands that will return a sample from an exponential distribution for you.

represents one possible future scenario ("path") for the system. Embedding this process within a broader Monte Carlo analysis would allow the generation of many future paths for the system.

C.3 Discussion

Other approaches exist for modeling the dynamics of systems, e.g., Markov chains (see Section 15.4). Discrete-event simulation is a scenario-based simulation method that simulates each item of the system separately through different event-paths/sample paths. In general, simulation-based methods consider components with differing attributes that move from one event to another in time while including modeling parameters of each part, such as age, maintenance history, and usage profile. Many analyses use simulation for optimization of stochastic problems. Simulation-based approaches are especially useful and common when the model grows in size or the integration of multiple disciplines is required. Monte Carlo sampling is usually used for sampling from probability distributions of each parameter, as long as one can estimate reasonable distributions.

Discrete-event simulators represent a straightforward method of solving many real-world problems — they are effectively an emulation of the real world. The arguments against using DES are that they can become cumbersome and lead to long simulations for large systems — because they are "brute force" emulations of the real world, they need not oversimplify a problem in order to obtain a solution. DES can be used to find practical optimums to problems, but cannot be used to obtain provable optima. DES is simply a discounted cash flow (DCF) analysis and will yield the same result.

In the example cases in Section C.2, only one type of event generating action was present (a system failure for which a maintenance action was necessary). For real systems, multiple types of events may occur concurrently on the same timeline. If one is only interested in the final cost at the end of a defined period of time, then it may be possible to simulate separate independent timelines and simply add the final results together. However, if one wishes to see the cost as a function of time, or if the different types of events are not independent (i.e., if the next event of type A depends on the occurrence and/or timing of an event of a type B), then the timeline has to be modeled sequentially for all events. An example of this would be multiple system instances drawing spares from a common inventory. In this case, separate DESs for each system instance cannot be generated and then added because the timing of spare replenishment (which represents an event that costs money) depends on the demands from all of the system instances. In this case, both system instances have to be simulated concurrently.

Discrete-event simulators also suffer from the constraint that they only operate in one direction, i.e., forward in time. Because of this, there are many outputs of discrete-event simulators that are straightforward to generate (e.g., cost and availability) that become very difficult to use as inputs to a design process. For example, availability requirements can be satisfied by running discrete-event simulators in the forward direction (forward in time) for many permutations of the system parameters and then selecting the inputs that generate the required availability output. Such "brute force" search-based approaches are computationally impractical for real problems (particularly for real-time problems), and are unable to deal with general uncertainties. There have been attempts to perform reverse simulation (run discrete-event simulators backwards in time) but this has only been demonstrated on extremely simple problems with limited applicability to the real world systems.

References

C.1 Nance, R. E. (1993). A History of Discrete Event Simulation Programming Languages, TR 93-21, Virginia Polytechnic Institute and State University, Department of Computer Science.

C.2 Conway, R. W., Johnson, B. M. and Maxwell, W. L. (1959). Some problems in digital systems simulation, *Management Science*, 6(1), pp. 92-110.

Bibliography

In addition to the sources referenced in this chapter, there are many books and other good sources of information on discrete-event simulation, including:

Banks, J., Carson II, J. S., Nelson, B. L., and Nicol, D. M., (2009). Discrete-Event System Simulation, 5th Edition, Prentice Hall.

Leemis, L. M., and Park, S. K., (2006). Discrete-Event Simulation: A First Course, Person Prentice Hall.

Problems

C.1 In the simple example in Section C.2.1, several implicit assumptions were made about when failures occur and how they have to be fixed. Identify and discuss these assumptions.

C.2 Rework the example in Section C.2.2 assuming that the time to failure is given by a Weibull distribution with the following parameters: location parameter = 500 hours, shape parameter = 4, and the scale parameter = 10,000 hours.

C.3 Rework the example in Section C.2.2 (with the constant failure rate), assuming that the time to resolve the failures (which was previously assumed to be instantaneous) is given by a triangular distribution with a lower bound of 30 days, an upper bound of 60 days and a mode of 45 days. Is the cumulative cost larger or smaller than the cumulative cost when the failures are resolved instantaneously?

C.4 Calculate the final (after 20 years) time-based availability of the system in Problem C.3.

C.5 What if an infrastructure charge of $150/month is incurred in the example in Section C.2.2 (with the constant failure rate)? What is the total cost after 20 years? Hint: the infrastructure charge represents an event that is independent of the maintenance events.

C.6 Starting with the example in Section C.2.2 (with the constant failure rate), assume that each maintenance event requires one spare. For simplicity, assume that the spare costs $1000 and the spare is the only maintenance cost – this is effectively identical to the solution in Section C.2.2. Now assume that the spares are kept in an inventory and that the inventory initially has 5 spares in it (purchased for $1000 each at time 0). Whenever the inventory drops below 3 spares, 5 more replenishment spares are ordered (for $1000 each). Assume that the replenishment spares arrive instantaneously. What is the total cost after 20 years?

C.7 Suppose that the time-to-failure distribution used in the simulation in Section C.2.2 was for a particular part in a system and that the part becomes obsolete (non-procurable) at the instant the simulation begins. If you had to make a lifetime buy of parts to support this system through 20 years, how many would you buy?

Index